MW00791612

The MAMMALS of TRANS-PECOS TEXAS

Second Edition, Revised

INTEGRATIVE NATURAL HISTORY SERIES

Sponsored by the Museum of Natural History Collections
Sam Houston State University
William I. Lutterschmidt and Brian R. Chapman, General Editors

The
Mammals of
Trans-Pecos
Texas

Including Big Bend and Guadalupe Mountains National Parks

Second Edition, Revised

Franklin D. Yancey II,
David J. Schmidly,
Richard W. Manning,
and Stephen Kasper

Illustrations by Chester O. Martin

Foreword by William I. Lutterschmidt
and Brian R. Chapman

TEXAS A&M UNIVERSITY PRESS | COLLEGE STATION

(∞) This paper meets the requirements
of ANSI/NISO Z39.48–1992 (Permanence of Paper).
Binding materials have been chosen for durability.
Manufactured in China through Martin Book Management

Library of Congress Cataloging-in-Publication Data

Names: Yancey, Franklin D., 1960– author. | Schmidly, David J., 1943– author. | Manning,
 Richard W., 1949– author. | Kasper, Stephen, 1956– author. | Martin, Chester O.,
 1943– illustrator.

Title: The mammals of Trans-Pecos Texas: including Big Bend and Guadalupe Mountains
 National Parks.

Other titles: Integrative natural history series.

Description: Second edition, revised / Franklin D. Yancey II, David J. Schmidly, Richard W.
 Manning, and Stephen Kasper ; illustrations by Chester O. Martin. | College Station:
 Texas A&M University Press, [2023] | Series: Integrative natural history series | Includes
 bibliographical references and index.

Identifiers: LCCN 2023001826 | ISBN 9781648430244 | ISBN 9781648430251 (ebook)

Subjects: LCSH: Mammals—Trans-Pecos (Tex. and N.M.) | Mammals—Texas—Big Bend
 National Park. | Mammals—Texas—Guadalupe Mountains National Park. | BISAC:
 NATURE / Animals / Mammals | NATURE / Regional

Classification: LCC QL719.T4 Y36 2023 | DDC 599.09764/9—dc23/eng/20230117

LC record available at https://lccn.loc.gov/2023001826

In loving memory of my mom, Donna B. Yancey,
who triggered my interest in science and natural history at an early age and
who continues to inspire me throughout my life.   —FDY

⁓

In loving memory of my brother, Steven,
who left us far too soon.   —DJS

⁓

In loving memory of my parents, James and Rosemarie,
for all their patience and support.   —RWM

⁓

In loving memory of my late wife, Dawn Marie Kasper,
for her love and care of the family while I studied and
measured specimens late into the night.   —SK

⁓⁓⁓⁓⁓⁓⁓⁓⁓⁓⁓⁓⁓⁓

To the memory of our good friend and mentor Dr. Clyde Jones,
Horn Professor of Biology at Texas Tech University. For more than 25 years,
Clyde and his wife, Mary Ann, traveled all over the Trans-Pecos
collecting and documenting natural history information about mammals in
the region. It was our pleasure to accompany them on many of these trips and
enjoy their companionship and the immense knowledge they had of
the region. Clyde's publications and those of his students, including
Frank Yancey and Rick Manning, constitute a major portion of the
information base about the natural history of Trans-Pecos mammals.

⁓

To the memory of Dr. James Scudday,
who spent his lifetime and career in the Trans-Pecos as a
member of the Biology Department at Sul Ross State University,
who taught students to ensure a lasting legacy, and who provided
leadership to preserve the natural history legacy of the region.

CONTENTS

A gallery of photos follows page 58.

FOREWORD

We are honored by the opportunity to include the revision of *The Mammals of Trans-Pecos Texas* within the Integrative Natural History Series. This book, coauthored by noted mammalogists Franklin D. Yancey II, David J. Schmidly, Richard W. Manning, and Stephen Kasper, exemplifies how biologists integrate art with science to develop a wider appreciation for nature. Natural historians have long appreciated the art of illustration and how such talent and a trained eye for detail are used in describing morphology and highlighting important characteristics. The collaboration between David J. Schmidly and wildlife artist Chester O. Martin "brought mammals to life" in their initial 1977 publication. In this edition, Martin once again employs his exceptional talents as a natural historian and artist to accurately depict the mammalian species that inhabit the Trans-Pecos region.

The first edition of this book, authored by David Schmidly, provided clearly written descriptions of the mammalian species found in the Trans-Pecos region and summarized the important details about their habits and habitats. The book was widely appreciated by biologists, residents of the region, and interested visitors who desired to learn more about the Chihuahuan Desert fauna west of the Pecos River. During the 45-plus years since the original volume appeared, the distributional patterns, taxonomy, and scientific names of many species have changed and new species from the region have been described. Many would agree that a revision of this important book was long overdue. This revised edition includes information about 17 additional species and is written and illustrated in the same clear style that will appeal to general readers and scientists alike.

As editors of the Integrative Natural History Series, we are proud to recognize this book as a significant contribution to the art and science of mammalogy and the natural history of the Trans-Pecos region. Such a work brings together an appreciation and understanding for the diversity of fauna, life histories, and ecologies within such a fascinating landscape. We embrace the integrative nature of this book and welcome the opportunity to promote such a meritorious collaboration within the series.

—William I. Lutterschmidt and Brian R. Chapman
General Editors

PREFACE TO THE SECOND EDITION

The term *Trans-Pecos* is somewhat loosely applied to the triangular portion of Texas west of the Pecos River, bounded by the Rio Grande on the south and west, and on the north by the 32nd parallel, which forms the boundary with the state of New Mexico. This area includes the main representation of the Chihuahuan Desert in the United States and is distinct from other parts of the state in its physical and biological conditions. Among the 10 ecoregions in Texas, the Trans-Pecos stands out as the most iconic, the most biologically diverse, and the most pristine. Traversed from north to south by an eastern arm of the Rockies, it contains all of Texas' true mountains. The climate provides a botanical assemblage that is unusual when compared to that of the more densely populated areas in Texas. Pine-oak forests, reminiscent of those in the western United States, occur on the tops of the highest peaks, but most of the surface is covered by desert or grassland vegetation.

Because of the country's harsh terrain, scrubby vegetation, scarcity of water, and severe climate, many people consider the Trans-Pecos to represent the last frontier of Texas. Centuries ago, when the Spaniards first encountered the barren and desolate landscape, they called it *el despoblado*, the uninhabited land. Even today, it remains the least populated region of the state, with only 3 percent of Texans (850,000) living west of the Pecos River. More than 90 percent of these residents live in El Paso County, leaving only about 80,000 residents distributed across the remainder of the region. The overall population density of around 27 persons per mile^2 (10 persons per km^2) is less than a third of the population density of the state as a whole.

The subject of this book is the mammals that occur in the Trans-Pecos. By *mammal* we are referring to the class of vertebrate animals possessing hair, with the females having milk-secreting glands. There are 104 species of native mammals documented from the Trans-Pecos, which represent about 72 percent of the 145 species of native terrestrial mammals known from Texas (see chapter 2). Of these species, 32 (about 31 percent) have all or at least 90 percent of their entire geographic range in the Trans-Pecos. In addition, the region is home to 27 species of bats (79 percent of Texas's total) and 46 species of rodents (about 65 percent of the state's total), making it, without question, the most species-rich ecoregion for small mammals in the state. But conservation concerns have developed in the twenty-first century, with 6 species having become extirpated, although 2 of these have recently been reintroduced back into the region, and another 11 species appearing on various lists of mammals with critical conservation concerns (see chapter 4).

The first edition of the book, authored by Dr. Schmidly and published in 1977, was based on his work in the region in the 1960s and 1970s while conducting extended studies of mammals along the Rio Grande in Big Bend National Park (BBNP) and the Pecos River near Amistad National Recreation Area (ANRA). Since its publication over four decades ago, scientific knowledge of the mammalogy of the region has changed substantially—the scientific names of three orders have changed, two taxonomic families have been added, and 18 additional species (8 new native species, 1 reintroduced previously extirpated native species, 6 nonnative species, and 3 feral domestic species) have been recorded in the region. In addition to the taxonomic changes and faunal additions, the literature about Trans-Pecos mammals has grown dramatically. The 1977 book included 102 published references; this new edition has almost 400, a nearly fourfold increase in just over 45 years. This growth in knowledge has been especially prevalent for charismatic (mountain lion, bighorn sheep, and elk) and large game (pronghorn, mule deer, and white-tailed deer) mammals. Naturalists and mammalogists working in the Trans-Pecos have made use of new, powerful technologies (e.g., radiotelemetry, DNA methods, camera traps, and computer modeling) to better understand the natural history of mammals in the region.

In this second edition, Dr. Schmidly is joined by three other mammalogists (Dr. Frank Yancey, Dr. Rick Manning, and Stephen Kasper) who began their work in the Trans-Pecos in the 1980s and continue to work there today. Their work has involved major natural history surveys of BBNP and Big Bend Ranch State Park (BBRSP), as well as the Chinati and Davis Mountains. Collectively, the authors have more than a century of firsthand, direct field experience studying mammals in the Trans-Pecos.

Chester Martin has provided the illustrations for the second edition as he did for the first one. While many of his illustrations have been preserved from the original version of the book, he has prepared several new ones for this edition. Chester, in addition to a career as a mammalogist, has received many accolades for his wildlife illustration work. In 2018, he became the third person to receive the prestigious Jay N. "Ding" Darling Memorial Award for Wildlife Stewardship Through Art given by The Wildlife Society.

We have also included color plates of some of the most iconic mammals that occur in the Trans-Pecos region. These photographs were taken by Frank Yancey and Rick Manning during their various field trips and studies of mammals in the Big Bend area, by Mark Lockwood (former regional TPWD director for the Trans-Pecos region), who spent most of his professional career in the region, and by interested individuals who were fortunate to record a few excellent camera shots.

Our purpose in writing this book has been to provide a guide to the identification and distribution of the mammals of the Trans-Pecos, and to summarize important facts about their lives. In designing the book, we have attempted to consider the interests of both professional and nonprofessional biologists. For the benefit of the nonprofessional, illustrations as well as a brief description of each species are included. The accounts pertaining to habits and geographic distribution of the species have been written for people with little or no technical knowledge of biology. Unlike in the first edition, literature references are cited by author and date, rather than by number, to improve the utility of the species accounts for professional mammalogists, and a complete list of citations is provided at the end of the book. To the extent possible, the species keys have been prepared to facilitate field identification.

Simplicity has been our basic goal in organizing this book. We have followed Schmidly and Bradley (2016) in the use of scientific and common names unless otherwise noted. Accounts for each species are arranged so that they contain in sequence (1) a brief description of the mammal, with special emphasis given to distinguishing features, accompanied by an illustration; (2) a description of the geographic distribution of the species in the Trans-Pecos, with reference to a map; (3) a discussion of some of the basic life history of the mammal, including habitat preferences, reproduction, food habits, and predators; and (4) where necessary an explanation of the taxonomic intricacies of some of the species as well as population trends and conservation status. Unless otherwise indicated, taxonomic assignments, including subspecies, are taken from Schmidly and Bradley (2016). The life histories include observations recorded by other researchers working in the Trans-Pecos and reported in the literature as well as our personal experiences based on fieldwork in the area.

Additional general natural history information has been adopted from Schmidly and Bradley's 2016 account of Texas mammals; the *Smithsonian Book of North American Mammals*, compiled and edited by Wilson and Ruff

(1999); and the Mammalian Species series, published by the American Society of Mammalogists. The latter provides detailed references and information for individual mammal species, including 100 species from the Trans-Pecos.

The projected geographic range for most species is shaded on the distribution maps that accompany the species accounts. We have not placed black dots on the distribution maps to indicate the localities represented by scientific specimens, as was done in the 1977 edition. However, for a few species with limited records in the region, and for which projecting a geographic range would amount to pure speculation, we have used dots to recognize places where they have been found. For several species, a list of counties from which the species has been documented (either by specimens or in the scientific literature) is presented in addition to a distribution map. Counties that lack records may, in some situations, be partially or completely shaded if the species is projected to occur there. In other words, maps have been designed to depict what we perceive as the overall range of the species in the Trans-Pecos. If suitable habitat and other conducive biogeographic factors exist between or beyond counties of record, shading may be extended to include counties from which there are no documented records.

A list of specimens collected and examined, which was included in the first edition, is not included in this edition. Those readers interested in specific locality and specimen records can use Schmidly (1977b) for early historical records, and they can access VertNet for the most recent records. VertNet is a publicly accessible database of vertebrate biodiversity data from natural history collections around the world. Users can capture data associated with vertebrate specimens either by major taxonomic group (birds, mammals, etc.), by species, or by geographic region. The data portal can be accessed online at vertnet.org. For specific locality and specimen records of bats, readers can access batsof texas.com, which provides similar data but for bats only.

Specific locality and specimen records are also available from published surveys of mammals in the following regions of the Trans-Pecos: Guadalupe Mountains National Park (GMNP) (Genoways et al. 1979), the Apache Mountains (Dalquest and Stangl 1986; Stangl et al. 1994), the Beach Mountains (Stangl et al. 1993), the Delaware Mountains (Stangl et al. 1994), the Davis Mountains (Stangl et al. 1994; DeBaca 2008), BBRSP (Yancey 1997; Jones and Lockwood 2008; Yancey and Manning 2018), BBNP (Yancey et al. 2006), and Chinati Mountains State Natural Area (CMSNA) (C. Jones et al. 2011; Yancey and Manning 2018; Yancey et al. 2019).

In this edition, we use primarily the metric system, which is the official standard of measurement for scientific communication. For those who are less familiar with the metric system, a table of metric-to-imperial conversion factors is provided in appendix 1. A list of the scientific and vernacular names of all plants referred to in the text is provided in appendix 2. The glossary has been updated and expanded to include definitions of new terms in the book. A list of abbreviations and acronyms used throughout the text is provided following this preface.

Finally, two chapters are devoted to the most significant conservation issues in the Trans-Pecos, particularly those pertaining to mammals. Chapter 4 outlines the major landscape changes that have occurred in the twentieth and early twenty-first centuries, and chapter 7 examines some of the current conservation pressures and the best strategies for addressing those challenges.

Our hope is that readers of this treatise, including visitors to the two national parks and numerous state parks and wildlife management areas in the region, will gain a better appreciation of the mammals that inhabit the last frontier of Texas.

ACKNOWLEDGMENTS

We are grateful for the assistance and support numerous individuals have provided to our research in the Trans-Pecos over the years. We are especially indebted to personnel from two entities, Texas Tech University and the Texas Parks and Wildlife Department (TPWD). While professor of biological sciences and curator of mammals at the Natural Science Research Laboratory (NSRL) at the Museum of Texas Tech, Clyde Jones, to whom this book is in part dedicated, designed and directed several mammal studies in the Trans-Pecos. He and his wife, Mary Ann, spent countless hours in the field with colleagues and graduate students studying mammalian natural history and demonstrating the art of specimen preparation. It is from these research projects that much of the information in this book was gleaned.

We also appreciate the support provided by Robert Bradley, professor of biological sciences and director of the NSRL. He generously furnished field equipment and supplies, made graduate student assistance and his laboratory available, and was always accessible for consultation on mammal identifications and habits. In addition, Heath Garner, collections manager at the NSRL, helped us by cataloging our specimens, checking on museum holdings, and furnishing copies of field notes archived at the NSRL. Lisa Bradley of the NSRL kindly assisted with locating many of the published sources of information about Trans-Pecos mammals.

For three decades, David Riskind, former director of the Natural Resources Program at the TPWD, supported our work in the Trans-Pecos. During this time, he was responsible for funding (through the TPWD) several research projects in the region including studies at Big Bend Ranch, Davis Mountains, and Balmorhea State Parks, and Chinati Mountains State Natural Area. He also granted us state lands access, lodging, and collecting permits. Mark Lockwood, former State Parks Region 1 (Trans-Pecos) director for the TPWD, worked closely with us in multiple ways. He provided collecting permits, arranged for lodging, and escorted us into otherwise restricted areas. Moreover, he was always eager to join us in the field and assist with small mammal trapping, bat netting, and camera trap setup. Mark provided some spectacular photographs of Trans-Pecos mammals and their habitats to enhance this book. Shawn Gray, Region 1 (Trans-Pecos) Mule Deer and Pronghorn Program director for the TPWD, provided the photograph of the juniper-roughland grasslands near Marathon. Nicolas Havlik, natural resources coordinator for the TPWD, assisted with the generation of distribution maps. Other TPWD personnel who supported us over the years include Luis Armendariz, David Alloway, Tony Gallego, and David Dotter.

Bonnie R. McKinney of El Carmen Land & Conservation Co. LLC, kindly provided information about the distribution of mammals immediately south of the Rio Grande in the Sierra del Carmen and surrounding area in northern Coahuila, Mexico. Randy Simpson of San Marcos, Texas, provided the photographs taken at the Pecos River Bridge in Val Verde County and at Santa Elena Canyon in Big Bend National Park in Brewster County. We also thank Maryann Eastman for the use of her exceptional kit fox photograph.

We are extremely fortunate that award-winning artist Chester Martin agreed to contribute drawings to this project. His works provide accurate visual images that will assist users of this book tremendously in identifying unfamiliar species of mammals.

ABBREVIATIONS

ANRA	Amistad National Recreation Area
BBNP	Big Bend National Park
BBRSP	Big Bend Ranch State Park
BGWMA	Black Gap Wildlife Management Area
CMSNA	Chinati Mountains State Natural Area
DSHS	Department of State Health Services
EMWMA	Elephant Mountain Wildlife Management Area
ESA	Endangered Species Act
GMNP	Guadalupe Mountains National Park
IUCN	International Union for Conservation of Nature
NGO	nongovernmental organization
NPS	National Park Service
TNC	The Nature Conservancy
TPWD	Texas Parks and Wildlife Department
USDA	United States Department of Agriculture
USDOI	United States Department of the Interior
USFWS	United States Fish and Wildlife Service
USGS	United States Geological Survey
YBP	years before present

The **MAMMALS** of
TRANS-PECOS TEXAS

1

DESCRIPTION OF THE AREA

A rougher, more rocky, more mountainous, and rugged country, can scarcely be imagined.

—LT. EDWARD L. HARTZ,
participant in the Big Bend camel expedition of the 1850s

The Trans-Pecos lies in the extreme western part of Texas and comprises that portion of the state west of the Pecos River (map 1). Bounded on the north by New Mexico and on the south and west by the Republic of Mexico, the Trans-Pecos contains more than one-ninth of the state of Texas and is nearly as large as the state of Maine. It is an irregularly shaped area encompassing nine entire counties (Brewster, Culberson, El Paso, Hudspeth, Jeff Davis, Pecos,

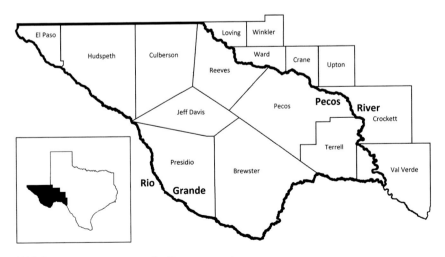

MAP 1. The Trans-Pecos and adjacent counties

Presidio, Reeves, and Terrell) and part of another (Val Verde). The total area covers approximately 82,880 km², or about 8.3 million ha.

Most of the Trans-Pecos is typified by the Texas portion of the Basin and Range physiographic province. Toward the northeast, along the western margins of the Pecos River valley, the region grades into the Great Plains. Farther south the Edwards Plateau forms the eastern boundary. A western extension of the plateau, known as the Stockton Plateau, ends at the eastern boundary of the Basin and Range. The Rio Grande, one of the longest rivers in North America, runs along the southern part of the Trans-Pecos and forms the boundary between Texas and Mexico. In Presidio and Brewster Counties the region is called the "Big Bend" because here the river turns in a big bend where it forms the gorge of the Rio Grande, part of which has been designated a National Wild and Scenic River.

With an elevation range from less than 600 m above sea level in the eastern Big Bend to as high as 2,667 m in the northern Guadalupe Mountains, this is a region of dramatic topographic relief characterized by rugged, wooded mountains, plateau grasslands, volcanic outcrops, limestone canyons, desert valleys, flat-topped mesas, undulating dune fields, and seemingly interminable salt flats. Plant communities are varied, with over 2,000 known plant species. These diverse habitats also support a large variety of wildlife species. The region supports, or has in the past supported, almost 500 species of birds, some 170 species of reptiles and amphibians, and 104 native species of mammals. Because of its diverse habitats, the Trans-Pecos is regarded by many as one of the most species-rich deserts in the world (Chapman and Bolen 2018).

The vast majority of the Trans-Pecos consists of privately owned ranchland, although the region contains the most public lands in the state, including more than 567,000 ha of rugged wilderness. While ranching has been the primary industry in the region, ecotourism has grown rapidly and is now a mainstay of the economy. Iconic parks and wilderness areas, such as those listed below, define the region and draw thousands of visitors to the area in the growing business of ecotourism. Protected natural areas make up about 6.8 percent of the region, a number that is considerably higher than the 2.8 percent overall average of protected places in the state (Schmidly et al. 2022).

NATURAL AREAS OF THE TRANS-PECOS	HECTARES (HA)
Big Bend National Park	324,267
Big Bend Ranch State Park	125,857
Black Gap Wildlife Management Area	41,683
Guadalupe Mountains National Park	34,951

Chinati Mountains State Natural Area	15,433
Franklin Mountains State Park	9,812
Elephant Mountain Wildlife Management Area	9,367
Sierra Diablo Wildlife Management Area	4,447
Davis Mountains State Park	1,096
Balmorhea State Park	304
Hueco Tanks State Park	299
	Total: 567,516

Dramatic mountain vistas with a special, rugged grandeur as well as unique vegetation, varied wildlife, and the hostile beauty of sun-washed deserts await visitors to these places. The breathtaking and unspoiled landscapes attract hundreds of thousands of visitors each year to the region. Big Bend National Park (BBNP), established in 1944 in the great bend of the Rio Grande in Brewster County, has spectacular mountain and desert scenery plus a variety of unusual geological structures. Guadalupe Mountains National Park (GMNP), established in 1972, is located in Hudspeth and Culberson Counties on the Texas–New Mexico border. It is noted for its tremendous earth faults, high jagged peaks, and unusual flora and fauna, which are more characteristic of the southern Rocky Mountains of New Mexico than of any other place in Texas.

Private lands make up more than 70 percent of the Trans-Pecos. These lands are owned and operated by about 2,000 individual operators, and 600 of these individuals own properties larger than 800 ha. The average ownership size in the Trans-Pecos is about 3,035 ha. Current trends show that many large ranches in the region are being fragmented into smaller ownerships and switching to other uses.

Current Climate

The current climate of the Trans-Pecos is arid to semiarid and is characterized by scant precipitation; cool, dry winters; hot, dry summers; and a large proportion of bright, sunny days (table 1). Rainfall is the most variable climatic factor, and it, more than the length of the growing season, determines the amount of plant growth. Rainfall, as a rule, decreases from east to west, although this trend is interrupted by local mountainous areas, where the precipitation is uniformly greater than elsewhere. In the eastern part of the Trans-Pecos the annual rainfall is about 30.5 cm, but in the western part it averages less than 25.4 cm. The highest annual rainfall, about 48.3 cm, occurs in the Davis Mountains in Jeff Davis County. Fluctuations in precipitation are great from year to year. Droughts

TABLE 1. Average temperature, average precipitation, and elevation for various locations in the Trans-Pecos

Location	Average high temperature (July) in °F (°C)	Average low temperature (January) in °F (°C)	Average annual high temperature in °F (°C)	Average annual low temperature in °F (°C)	Average annual precipitation in inches (centimeters)	Elevation in feet (meters)
Alpine	89 (31.6)	30 (-1.1)	76 (24.4)	47 (8.3)	17.0 (43.2)	4,481 (1,366)
Balmorhea	94 (34.4)	30 (-1.1)	79 (26.1)	48 (8.9)	13.5 (34.4)	3,205 (977)
El Paso	95 (35.0)	33 (0.6)	77 (25.0)	52 (11.1)	9.7 (24.7)	3,762 (1,147)
Fort Davis	87 (30.6)	29 (-1.7)	75 (23.9)	45 (7.2)	17.5 (44.4)	5,000 (1,524)
Fort Stockton	94 (34.4)	33 (0.6)	79 (26.1)	52 (11.1)	15.2 (38.5)	3,050 (930)
Presidio	101 (38.3)	35 (1.7)	87 (30.6)	55 (12.8)	10.7 (27.2)	2,594 (791)
Sanderson	92 (33.3)	31 (-0.6)	78 (25.6)	51 (10.6)	14.7 (37.4)	2,777 (846)
Sierra Blanca	92 (33.3)	25 (-3.9)	76 (24.4)	43 (6.1)	11.9 (30.2)	4,512 (1,375)
Van Horn	92 (33.3)	28 (-2.2)	77 (25.0)	47 (8.3)	11.6 (29.4)	4,010 (1,222)
Big Bend National Park	92 (33.3)	38 (3.0)	79 (26.0)	53 (12.0)	13.8 (35.0)	7,825 (2,385)
Guadalupe Mountains National Park	86 (30.0)	32 (0)	70 (21.0)	48 (9.0)	18.3 (46.0)	8,751 (2,667)

of short duration occur frequently, while prolonged ones are expected every 10 to 15 years. Most thunderstorms occur during the "monsoon" season—the summer months from June to September—when warm, moist air often accompanies remnants of tropical storms. Rare deluges from intensive summer cloudbursts, called flash floods, are known locally as "gully washers." For the remainder of the year, scattered light showers and occasional light snowfalls, combined with high evapotranspiration rates, provide little surface or soil moisture. Evaporation is enhanced by strong summer winds whose intensity increases as they blow through narrow canyons (Chapman and Bolen 2018).

Temperature is less variable than rainfall, averaging about 17.8°C annually over the entire region. The climate is subject to considerable seasonal fluctuation. The summers are marked by comparatively high temperatures, modified, especially at night, by the high elevations of the region. The winters are relatively short and cool, with occasional cold periods during "northers," which bring cold north winds and frequently rain or snow. Some years are marked with exceptional snowfall occurring mostly in January. The average frost-free season is from March 29 to October 26, but the growing season conforms more closely to the summer rainy season than to the frost-free season.

Past Climate

While the Trans-Pecos region today conjures up severe desert images, its rugged terrain was once dominated by piñon pine, oak, and juniper woodlands. The relatively lush landscape was marked by small streams, springs, and seeps. Grasslands dominated the lower elevations, invading valleys and canyon floors, and supporting resident and migratory herds of bison, pronghorn, camel, mammoth, and horse. The mountain slopes and hillsides were home to deer, bear, mountain sheep, and the occasional ground sloth. A host of smaller, more secretive animals such as chipmunk, water shrew, vole, and marmot also found conditions to their liking and shared the place as home.

The entire region has been influenced profoundly by climatic changes over the past 18,000 years, during late Pleistocene and Holocene times, which produced landscapes quite different from those seen today. Fred Stangl, Walter Dalquest, and Robert Hollander have produced an excellent account of the evolution of the plant and animal communities of the Trans-Pecos during this interval, and much of the account that follows has been taken from their 1994 publication (Stangl et al. 1994).

What happened, in a nutshell, is that about 16,000 to 18,000 years before present (YBP), during the Wisconsin glaciation, a massive ice sheet dominated the North American landscape, with the largest area covering the Rocky Mountains and extending as far south as the modern boundary of northern Nebraska. During this time, Texas became cooler and the climate was more equable, with less seasonal differentiation. The effects on the latitudinal and elevational distribution of plant associations were profound. Woodlands generally occurred at lower elevations than today, and grasslands were more extensive. The higher elevations supported extensive mixed coniferous forests. All but perhaps the lowest hills were cloaked in piñon-juniper woodland, which reached the lower basins and continued east of the Pecos River. Available water was plentiful and allowed for the presence of permanent streams.

The Holocene represented a transition between Pleistocene and modern times. It began with a brief warming period (10,000–8,000 YBP), followed by a reversal to cooler conditions (8,000–4,000 YBP), and finally a continuation of the warmer climate, culminating in today's current conditions (4,500 YBP–present). Semiarid woodlands of piñon, juniper, and oak dominated the landscape with conifers at the higher elevations. The lowlands supported well-developed grasslands. The middle Holocene featured pronounced seasonal temperature extremes (warm summers and cold winters) coupled with an increasingly dry cycle. In response, the woodlands began to retreat into the higher, wetter elevations of the mountains. In the face of the increasing aridity, grasslands began to dominate the region. Approximately 7,000 YBP, a warming and drying trend changed the Trans-Pecos from a mesic landscape dominated by grasslands and woodlands to a semiarid and arid landscape dominated by grasses and shrubs. It was during this time that creosote bush first arrived in the region, and the grasslands began to be replaced by desert scrub vegetation adapted to more xeric conditions and the scarcity of topsoils (Warnock and Loomis 2002). By the late Holocene, the climate was essentially modern as the trend toward greater warmth and aridity continued. As a result, the Chihuahuan Desert of today occupies a greater area than at any time in the past.

Physiography

The greater part of the Trans-Pecos differs from the rest of the state because it contains all of Texas's true mountains. The area includes three outstanding physical features: plateaus, mountainous ridges, and lowland basins. The major physiographic features of the area are shown in map 2.

The mountainous topography is the result of tectonic uplift and volcanic eruptions combined with erosion by the Rio Grande and its tributaries. River

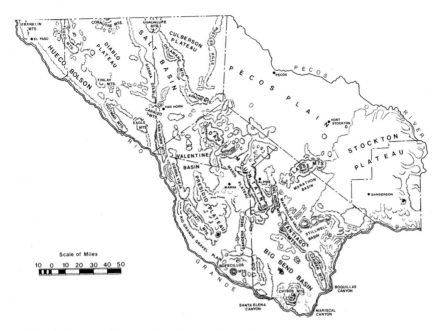

MAP 2. The major physiographic features of the Trans-Pecos

erosion, however, is minimal because the region is arid. Thus, the landforms take on a sharp and angular appearance. Fault rocks provide the topographic relief, while physical weathering forms the cliffs and talus slopes at the mountain bases.

The Trans-Pecos contains two physiographic subdivisions, the Stockton Plateau and the Mountain and Basin sector. The Stockton Plateau is the western extension of the Edwards Plateau of central Texas across the Pecos River, reaching nearly to Fort Stockton and bounded on the west by the Pecos Plain, the Glass Mountains, the Marathon Basin, and the Stillwell Basin. It consists of an imperfectly dissected plateau (except for a narrow fringe a few miles wide along the Rio Grande) ranging in elevation between 610 and 762 m above sea level. The geological formation that makes up most of the plateau consists of limestone deposits, in contrast to the igneous formations of the Davis Mountains farther west.

The Mountain and Basin physiography characteristic of the southwestern United States reaches its eastern limits in the Trans-Pecos and occupies most of the region. It is made up of irregular areas of mountain ridges and isolated groups of roughlands between which lie flat, desertlike basins. The mountains of the ridge belts are formed principally by two kinds of rocks. sedimentary limestones, and igneous, mainly basaltic formations. The basin floors are areas of sediment accumulation instead of tectonic construction.

Three principal mountain areas cross the Trans-Pecos from northwest to southeast. The easternmost ridge, known as the Front Range, consists of, from north to south, the Guadalupe, Delaware, Apache, Davis, Del Norte, Santiago, and Carmen Mountains, all but the Davis Mountains being made of thick strata of hard limestone. The Davis Mountains are the most extensive mountains of the Front Range, and they are basically rugged masses of primarily igneous rocks. They lie mainly in Jeff Davis County but extend southward into adjoining counties. The Guadalupe Mountains, which extend into Texas from New Mexico, are the highest mountains in the area, with Guadalupe Peak rising to 2,667 m above sea level.

The second mountainous ridge, known as the Central Range, is located west of and parallel to the Front Range, from which it is separated by a vast trough occupied by several basins. The Central Range consists of the Diablo Plateau, the Hueco Mountains, and the Sierra Diablo, succeeded southward by the Quitman, Eagle, and Van Horn Mountains, the Sierra Vieja, and the Chinati, Bofecillos, and Chisos Mountains. These mountains are made up largely of limestone rocks, although the Sierra Vieja and the Bofecillos, Chinati, and Eagle Mountains, and portions of the Quitmans, are composed of igneous rocks.

The third and most westerly of the mountain ridges is formed by the Franklin Mountains. These consist of various kinds of rock, but limestones predominate. Only the south end of these mountains lies within the Trans-Pecos.

Flat plains or basins surround the mountain ranges. The easternmost basin belt consists of the Pecos Plain, which constitutes its largest area; the Marathon Basin, separated from the Pecos Plain by the Glass Mountains; the Maravillas Valley lowland; and the Stillwell Basin. A series of limestone hills, the Rustler Hills, bound the Pecos Plain on the west and separate it from the slightly higher areas of the Culberson Plateau, a high, rolling plain of soft gypsum. Lying between the Front and Central Ranges is the second basin belt, which consists of the Salt Basin at its northern end, and the Valentine Basin, Presidio Plateau, Marfa Plateau, and the less distinctly defined Big Bend Basin to the south. The Salt Basin has no drainage outlet; large salt lakes are interspersed throughout much of the area. To the south the Valentine Basin is free of salt or alkali. This basin reaches almost to the Davis Mountains, but to the southeast it increases in elevation and forms the well-defined Marfa Plateau. The more broken areas of adjacent gravelly hills make up the Presidio Plateau. The flat trough between the Front and Central Ranges extends into the Big Bend, where it is characterized by diverse forms of relief, including eroded areas of badlands and beds of volcanic ash and eroded lava plains. The Chisos Mountains rise out of the Big Bend Basin. The western basin area, lying between the Central and Western Ranges, consists chiefly of Hueco Bolson, a large, flat sand plain. A fourth basin belt is represented by a short stretch of the Rio Grande Valley lying west of the Franklin Mountains and north of El Paso.

Several narrow valleys have been cut by streams through the region of mountains and basins. Most of them are gorges where the Rio Grande, the Pecos River, and other streams have cut through rough highlands and mountains. Several short, intermittent streams originate within the mountain sections and carry drainage water to these rivers. Surface water is seldom permanent in the Trans-Pecos, and even the heavy runoff from thunderstorms or snowmelt quickly evaporates. Nonetheless, somewhat more permanent sources of water exist as springs, seeps, ciénagas, and tinajas—all surprisingly more abundant in some areas than might be imagined (Chapman and Bolen 2018). These sources not only provide drinking water during the hottest and driest seasons of the year but also form secluded microhabitats that sustain moisture-dependent communities.

Soils

The soils of the region have been formed mostly by mountain outwash materials. Textures and profile characteristics of the soils are varied, but most of the soils are generally alkaline. Caliche soils, rich in calcium carbonate precipitates, are widespread in the arid and semiarid areas.

The soils can be divided into highland and lowland groups. The lowland group, which includes the soils of the basins and plains, is subdivided into three series: Reeves (very light brown), Reagan (light brown), and Verhalen (chocolate or reddish brown). For the most part, these three are deep surface soils and subsoils resting on beds of hard or soft caliche (calcium carbonate) and rounded gravel (Carter et al. 1928). In some places the soils are shallow and contain large quantities of gravel and caliche fragments. The texture of the surface soils ranges from fine sand to clay, though silty clay loam predominates.

The Reeves soils are dominated by desert scrub vegetation and occupy large areas in the Hueco Bolson, the lower parts of the Diablo Plateau in northern Hudspeth County, the Culberson Plateau, the Salt Basin, the Pecos Plain, the Valentine Basin, and the Stockton Plateau. The Reagan soils are slightly darker than the Reeves soils, with more organic matter and higher fertility; they are deeper and more developed and support grass vegetation instead of desert scrub. Reagan soils occur principally on flats near mountainous areas such as the Diablo Plateau and the basins of the Davis Mountains. Verhalen soils consist of chocolate-brown or reddish-brown clays. Large tracts of this soil are often covered with tobosagrass, giving rise to the name "tobosa flats." Verhalen soils are found in the Pecos Plain north of the Davis Mountains, large areas in the Marathon Basin, the western part of the Hueco Bolson, and the Diablo Plateau. The soils of the alluvial valleys, also included in the lowland group, are represented by the Toyah soils in the valleys of the Davis Mountains; the Rio Grande, Gila, and Anthony soils along the Rio Grande; and undifferentiated alluvial soils along the Pecos River.

The highland group, which includes the soils of the rough highlands and rolling hills, is divided into two series: Ector and Brewster. A large portion of the Trans-Pecos is covered by these shallow and undeveloped soils. The limestone mountain and hill soils are classified in the Ector series; they are light brown, calcareous, and friable. The Ector soils support primarily grassland and montane vegetation interspersed with extensive growths of xeric shrubs. The soils of the Brewster series, found on igneous mountains and mesas, are brown or red, noncalcareous, and friable. Although sparse, the Brewster soils are sometimes rich in humus and often covered by leaf litter. Forest communities at higher elevations, as well as grasslands with moderate growths of xeric shrubs, are found on these soils.

The upland soils of the southern desertic basins are mostly well-drained, light reddish-brown to brown clay loams, clays, and sands (some have large amounts of lime, and some have a large amount of gypsum and other salts). Many areas have shallow soils and rock outcrops, and sizable areas have deep sands. Bottomland soils are deep, well-drained, dark grayish-brown to reddish-brown silt loams, loams, clay loams, and clays. On the Stockton Plateau, upland soils are mostly shallow, very stony or gravelly, dark alkaline clays and clay loams underlain by limestone. Lighter-colored soils are on the steep slopes, and deep, less stony soils are in the valleys. Bottomland soils are mostly deep, dark gray or brown, alkaline loams and clays.

Vegetation

With Trans-Pecos elevations ranging from 305 to 2,667 m and extreme variations in topography, soils, precipitation, and vegetation, ecologists have taken great latitude in describing the region's ecological subunits. A recent classification of plant associations in the region, for example, recognized no fewer than 117 vegetation cover types (Chapman and Bolen 2018). Our treatment focuses on seven general vegetation groups that seem to best represent the

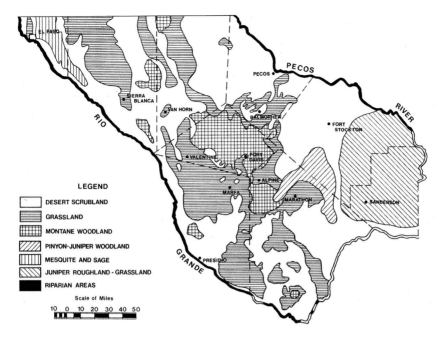

MAP 3. The major vegetative regions of the Trans-Pecos

biophysiographic associations: (1) desert scrubland; (2) areas of mesquite and sage; (3) grassland; (4) juniper roughland-grassland; (5) montane woodland; (6) piñon-juniper woodland; and (7) riparian areas or bottomland (map 3).

Desert scrub covers about half the Trans-Pecos, most of it as an invasion of shrubby plants over former grassland. The kinds of brush species vary considerably, and three different ecological associations may be recognized. The creosote bush association, consisting of pure stands of creosote bush with scarcely a weed or blade of grass covering the ground, occupies a considerable area in the Trans-Pecos, occurring on the poorest soils and where the annual precipitation is only 20 to 25 cm (fig. 1). A creosote bush–ocotillo association is common in many lowland desert regions of the Trans-Pecos (fig. 2). A mixed scrub association, dominated by blackbush, occurs where the soils are best developed and where the land has not been extensively overgrazed. Common shrubs in this association include catclaw, skunkbush, lotebush, allthorn, creosote bush, javelina bush, mesquite, and various yuccas (fig. 3). Among the dominant grasses are blue grama, black grama, tridens, and sideoats grama. A sotol-lechuguilla association is confined to slopes above the mixed scrub association; it occurs throughout much of the Trans-Pecos region on limestone hills and mountain slopes where the soils are undeveloped accumulations of stony debris (fig. 4)

FIG. 1. Creosote bush association in Big Bend Ranch State Park. Photograph by Franklin D. Yancey II

FIG. 2. Creosote bush–ocotillo association in Big Bend Ranch State Park. Photograph by Franklin D. Yancey II

FIG. 3. Mixed desert scrub habitat in Chinati Mountains State Natural Area. Photograph by Franklin D. Yancey II

FIG. 4. Sotol-lechuguilla association in the Christmas Mountains. Photograph by Franklin D. Yancey II

The dominant plants are sotol and lechuguilla; important subdominants include sacahuista, cholla, ocotillo, and Thompson's yucca. Hairy grama, several kinds of three-awn grasses, slim tridens, and fluffgrass form a scanty cover of grass in the spaces between the dominants.

The deep, sandy soils of the Hueco Bolson in El Paso and Hudspeth Counties, although constituting a true desert, support a different type of vegetation from that of the creosote bush–blackbush desert that covers most of the Trans-Pecos. The most abundant plant in the bolson is honey mesquite, but saltbush, sage, and ephedra are common in its north-central region (fig. 5). Sage and honey mesquite are found most often around sand dunes, whereas ephedra is found exclusively between the dunes. Soaptree yucca is frequently found in this area, as are several grama grasses. The extensive central portion of the bolson is characterized florally primarily by mesquite. Rarely, ephedra and soaptree yucca can be found. The southern portion of the bolson is similar to the northern area except that sage is much more abundant; grass, however, is less evident.

Desert grasslands occur on the better soils of the Trans-Pecos, particularly where the annual rainfall is about 25 to 43 cm (fig. 6). Rainfall of 15 to 20 cm a year is usually inadequate for grassland, producing instead desert scrub vegetation. In the mountains where the rainfall is 43 cm or more, the moisture is sufficient to produce woodlands. Since the introduction of the livestock industry, much grassland has been invaded by desert scrub. Perhaps 30 to 40 percent of former grassland in the Trans-Pecos is now covered by desert scrub. Different types of grasslands are scattered throughout the roughlands of the region, and they are usually found in association with two or more species of woody plants. The dominant grasses, depending on the depth and character of the soil, include several species of grama grass, three-awn grass, tobosagrass, needlegrass, and bluestem. In a few places where runoff from mountain slopes flows far onto the dry lands of the basins there are large tobosa flats. Grasslands of relatively thick cover occur chiefly in the Davis, Chisos, Bofecillos, Chinati, and Eagle Mountains and the Sierra Vieja as well as in some other smaller roughland areas of igneous rocks. Associated with grasslands are many shrubs and succulents, of which the more common are lechuguilla, sotol, mesquite, catclaw, soaptree yucca, Spanish dagger, buckthorn, sacahuista, cholla, prickly pear, javelina bush, woolly senecio, and ocotillo. Grasslands with large amounts of associated shrubs are common in the Carmen, Santiago, Delaware, Glass, Hueco, and Franklin Mountains and in the Sierra Diablo.

In some places, especially near the beds of salt lakes, there is a considerable growth of saltbush, alkali sacaton, and other salt-tolerant plants. This type

FIG. 5. Saltbush-mesquite habitat in Guadalupe Mountains National Park. Photograph by Mark W. Lockwood

FIG. 6. Desert grassland habitat in Presidio County. Photograph by Mark W. Lockwood

FIG. 7. Gypsum dunes habitat in Guadalupe Mountains National Park. Photograph by Mark W. Lockwood

of vegetation grows entirely on areas of "gypland" in the Salt Basin and on the Culberson Plateau, where much of the land is bare, soft gypsum with only a slight layer of soil (fig. 7).

The middle and upper slopes of the higher mountains and narrow mountain valleys in the Trans-Pecos support woodland vegetation in which three different associations (piñon-oak, piñon-juniper, and oak-juniper) may be recognized. This type of vegetation varies considerably, ranging from dense woodlands on the north slopes of mountains to open woodland savannas with only a few scattered trees on the south and west slopes (fig. 8).

Pine-oak forests occur at the highest elevations of the Guadalupe Mountains, at the top of Mount Livermore in the Davis Mountains, and in some of the highest reaches of the Chisos Mountains (fig. 9). This association is characterized by ponderosa pine, limber pine, Douglas fir, chinquapin oak, buckthorn, quaking aspen, and mountain snowberry. Douglas fir, an important dominant of this association, occurs in the Guadalupe and Chisos Mountains but not in the Davis Mountains.

The piñon-juniper association occurs at elevations ranging from about 1,676 to 2,286 m (fig. 10). Grasses form a conspicuous ground cover beneath the trees and on open slopes. The dominant plants include piñon pine, alligator juniper, gray oak, Emory oak, mountain grass, and various species of

FIG. 8. Piñon-oak woodland habitat in the Davis Mountains. Photograph by Mark W. Lockwood

FIG. 9. Ponderosa pine–oak habitat in the Davis Mountains. Photograph by Mark W. Lockwood

FIG. 10.
Piñon-juniper
habitat in Big
Bend National
Park. Photograph
by Franklin D.
Yancey II

FIG. 11.
Juniper
roughland-
grassland habitat
in the Glass
Mountains,
near Marathon,
Brewster County.
Photograph by
Shawn S. Gray

muhly and hairy grama. The oak-juniper association is found at lower elevations (1,372 to 1,676 m) than the piñon-juniper. Piñon pine is absent from the
oak-juniper association, and alligator juniper is replaced by one-seeded juniper.
Gray oak and Emory oak are the dominant oaks. In some moist canyons Chisos
red oak is locally dominant. The higher mesas of the Trans-Pecos may be completely covered by oak-juniper vegetation. The landscape is open and parklike,

FIG. 12.
Riparian habitat
along the Rio
Grande in Big
Bend Ranch State
Park. Photograph
by Franklin D.
Yancey II

FIG. 13. Riparian habitat along the Rio Grande in Big Bend Ranch State Park.
Photograph by Franklin D. Yancey II

consisting essentially of grassland with varying densities of juniper and oak, the trees being scattered on the lower slopes and moderately dense on higher hills of rough, stony soil. Oaks are abundant in the larger draws and canyons.

The Stockton Plateau combines, to some extent, vegetative characteristics of both deserts and grasslands. However, because it does contain some unique features, it has been placed in a different category of vegetation, the juniper-roughland-grassland (fig. 11). The rolling limestone tablelands and mesa tops are covered by low red-berry or Ashe junipers, with Wheeler's sotol, chino grama, and ocotillo dominating on the sloping ridges that flank the shallow valley bottomlands. Shin oak and persimmon are common on the rimrocks that surround the margins of the mesa tops. Several grama grasses occur in this area, including black grama, sideoats grama, and red grama. The canyon floors between the mesas are covered with a mesquite–creosote bush association that is only a modification of the widespread creosote bush–blackbush desert formation. Many of the shrubs that share the mesa tops with the junipers are important constituents of the mesquite–creosote bush association. These include lotebush, allthorn, buckthorn, various cacti, and guayacan.

Riparian vegetation in the Trans-Pecos occurs primarily in the alluvial bottomlands of the major streams and rivers and adjacent streams in mountain valleys (figs. 12 and 13). The bottomlands of the Rio Grande are choked with salt cedar, honey and screwbean mesquite, some willows, and occasional cottonwoods. Two large grasses, the giant reed and the common reed, grow on the banks of the Rio Grande. In the Pecos River valley, mesquite, desert willow, and salt cedar prevail. Riparian areas in mountain valleys support some oaks, Mexican walnut, mountain cottonwood, willow, ash, mountain maple, and Texas madrone. In addition, some continuously flowing springs will maintain short stretches of riparian habitat that include dense stands of willows and cottonwoods.

2

SYNOPSIS OF TRANS-PECOS MAMMALS

Diversity

There are 104 kinds of native mammals that presently live in the Trans-Pecos or have inhabited the area within the past 125 years. If you add in 10 introduced species, 5 domesticated species, and 3 feral species plus the mule (technically the mule is not considered a species), which collectively constitute a total of 19 mammals, then the total number of kinds of mammals presently or recently recorded from the region increases to 123. Another 10 species occur so close to the region that they could potentially be found there, bringing the total of possible kinds of mammals to 133. From a taxonomic perspective, these mammals represent 9 orders, 29 families, and 84 genera. Three orders contain almost all of the region's mammal species—Chiroptera (bats) with 27 species, Carnivora (carnivores) with 21 species, and Rodentia (rodents) with 46 species. Collectively, these three orders include 86 percent of the native species and 76 percent of the total mammals known from the region.

A checklist is presented to equip both residents and visitors to the region with an easy-to-use resource to quickly learn about the mammal species known to occupy the area. Its purpose is to provide a skeleton classification of a group, in this case mammals, listed by taxa for quick reference. In our checklist, the taxa have been organized hierarchically according to our current view of classification. The checklist includes the scientific name, common name, and status of each species according to a list of categories defined below. It is arranged into three sections, one for native mammals (both extant and extirpated), a second for introduced species, and a third for feral and domestic mammals. Status is defined using the following categories, with the total number of species given in brackets:

Common. A common species is abundant wherever it lives in the region. Most common species are widely distributed over the area. [45 native + 8 nonnative]

Uncommon. An uncommon species may or may not be widely distributed, but it does not occur in large numbers and is not well represented in museum collections. Uncommon species are not necessarily rare or endangered. [24 native + 7 nonnative]

Rare. A rare species is present in such small numbers throughout the region that it is seldom seen or collected. Although not presently threatened with extinction, a rare species may become endangered if conditions in its environment worsen. [11 native + 4 nonnative]

Endangered or Threatened. The USFWS publishes a list of endangered and threatened species that includes mammals listed in those categories in the *Federal Register*, and the TPWD has a list of protected nongame wildlife. These are the official lists governed by federal and state laws, statutes, and regulations. Species on these lists are those whose prospects of survival and reproduction are in immediate jeopardy. [see table 3 in chapter 4]

Migrant. A migrant species is not a permanent resident of the region and occurs there only periodically. [2 native]

Peripheral. A peripheral species is one whose range only barely enters (usually by only a few kilometers) the boundary of the region, but it exists in viable, breeding populations. [5 native]

Possible. A possible species has not yet been recorded from the region but is known from closely adjacent areas; additional collecting may very well document its occurrence in the region. Species in this category are indicated by the superscript 1 (east of the Pecos River) or 2 (south of the Rio Grande). [10 native: 7 east of the Pecos River and 3 south of the Rio Grande]

Extirpated. An extirpated species was once present but no longer occurs in the region and has not been reestablished. The native populations of two species (bighorn and elk), categorized as "extirpated/reestablished," became extirpated and were subsequently reestablished by introduced populations. [8 extirpated, 2 of which are reestablished]

Extralimital. An extralimital species is known based on only one or a few records that probably resulted from unusual wanderings of animals into the region from other areas. There is no indication these species have viable, reproducing populations in the Trans-Pecos. [2 native]

Enigmatic. An enigmatic species is one whose occurrence is sporadic and difficult to explain with any degree of biological certainty. [3 native]

Feral. A feral mammal was domesticated by humans for agricultural purposes and subsequently became established in the wild, where it now lives without animal husbandry assistance. [4 nonnative]

Domestic. A domestic species has been domesticated by humans to live and breed in a tame condition and depends on humankind for survival. [5 nonnative]

Of the 98 extant, native species in the checklist, 45 (45.9 percent) are considered common today, 24 (24.5 percent) are uncommon, and 11 (11.2 percent) are considered rare. Assignment of species to these categories is based on the extent of specimen records housed in museum collections, the published opinions and information from naturalists who have collected mammals in the Trans-Pecos, or the experience of the authors based on their field studies of mammals in the region. Ten species are listed as of possible occurrence, and there are two categories of these—those species that have been recorded in counties immediately east of the Pecos River (indicated by the superscript 1) and those known immediately south of the Rio Grande in northern Coahuila, Mexico (indicated by the superscript 2).

NATIVE SPECIES

ORDER DIDELPHIMORPHIA—Opossums and Allies
    Family Didelphidae—Opossums

Didelphis virginiana, Virginia Opossum	Rare

ORDER CINGULATA—Armadillos and Allies
    Family Dasypodidae—Armadillos

Dasypus novemcinctus, Nine-banded Armadillo	Rare

ORDER LAGOMORPHA—Pikas, Hares, and Rabbits
    Family Leporidae—Hares and Rabbits

Sylvilagus audubonii, Desert Cottontail	Common
Sylvilagus floridanus, Eastern Cottontail	Common
Sylvilagus holzneri, Holzner's Mountain Cottontail	Rare
Lepus californicus, Black-tailed Jackrabbit	Common

ORDER EULIPOTYPHLA—Shrews, Moles, and Relatives
    Family Soricidae—Shrews

Cryptotis parva, Least Shrew	Possible[1]
Notiosorex crawfordi, Crawford's Desert Shrew	Uncommon

Family Talpidae—Moles

Scalopus aquaticus, Eastern Mole   Enigmatic

ORDER CHIROPTERA—Bats

Family Molossidae—Free-tailed Bats

Tadarida brasiliensis, Brazilian Free-tailed Bat   Common

Nyctinomops femorosaccus, Pocketed Free-tailed Bat   Uncommon

Nyctinomops macrotis, Big Free-tailed Bat   Uncommon

Eumops perotis, Western Bonneted Bat   Rare

Family Mormoopidae—Leaf-chinned Bats

Mormoops megalophylla, Ghost-faced Bat   Common

Family Phyllostomidae—New World Leaf-nosed Bats

Choeronycteris mexicana, Mexican Long-tongued Bat   Rare

Leptonycteris nivalis, Mexican Long-nosed Bat   Endangered

Leptonycteris yerbabuenae, Lesser Long-nosed Bat   Extralimital

Desmodus rotundus, Common Vampire Bat   Possible[2]

Diphylla ecaudata, Hairy-legged Vampire Bat   Possible[1]

Macrotus californicus, California Leaf-nosed Bat   Possible[2]

Family Vespertilionidae—Vesper Bats

Myotis auriculus, Southwestern Myotis   Possible[2]

Myotis occultus, Southwestern Little Brown Myotis   Extralimital

Myotis yumanensis, Yuma Myotis   Common

Myotis velifer, Cave Myotis   Common

Myotis thysanodes, Fringed Myotis   Common

Myotis volans, Long-legged Myotis   Common

Myotis californicus, California Myotis   Common

Myotis ciliolabrum, Western Small-footed Myotis   Uncommon

Lasionycteris noctivagans, Silver-haired Bat   Migrant

Lasiurus borealis, Eastern Red Bat   Uncommon

Lasiurus frantzii, Western Red Bat   Enigmatic

Lasiurus seminolus, Seminole Bat   Possible[1]

Aeorestes cinereus, Hoary Bat   Migrant

Dasypterus xanthinus, Western Yellow Bat   Rare

Parastrellus hesperus, American Parastrelle (Canyon Bat)   Common

Perimyotis subflavus, American Perimyotis (Tricolored Bat)   Rare

Eptesicus fuscus, Big Brown Bat   Common

Nycticeius humeralis, Evening Bat   Rare

Euderma maculatum, Spotted Bat   Threatened

Corynorhinus townsendii, Townsend's Big-eared Bat   Uncommon

Antrozous pallidus, Pallid Bat   Common

ORDER CARNIVORA—Carnivores
    Family Canidae—Dogs, Foxes, and Wolves

Canis latrans, Coyote	Common
Canis lupus, Gray Wolf	Extirpated
Urocyon cinereoargenteus, Common Gray Fox	Common
Vulpes macrotis, Kit Fox	Uncommon

    Family Felidae—Cats

Leopardus pardalis, Ocelot	Extirpated
Lynx rufus, Bobcat	Common
Panthera onca, Jaguar	Extirpated
Puma concolor, Mountain Lion	Common

    Family Mephitidae—Skunks

Spilogale leucoparia, Desert Spotted Skunk	Uncommon
Mephitis mephitis, Striped Skunk	Common
Mephitis macroura, Hooded Skunk	Rare
Conepatus leuconotus, Hog-nosed Skunk	Common

    Family Mustelidae—Weasels, Otters, and Badgers

Mustela frenata, Long-tailed Weasel	Rare
Mustela nigripes, Black-footed Ferret	Extirpated
Taxidea taxus, American Badger	Common

    Family Procyonidae—Raccoons, Ringtails, and Coatis

Bassariscus astutus, Ringtail	Common
Nasua narica, White-nosed Coati	Threatened
Procyon lotor, Northern Raccoon	Common

    Family Ursidae—Bears

Ursus americanus, American Black Bear	Threatened
Ursus arctos, Grizzly or Brown Bear	Extirpated

ORDER ARTIODACTYLA—Even-Toed Ungulates
    Family Antilocapridae—Pronghorn

Antilocapra americana, Pronghorn	Uncommon

    Family Bovidae—Cattle, Antelope, Sheep, Goats, and African Exotics

Bos bison, American Bison	Extirpated
Ovis canadensis, Bighorn Sheep	Extirpated/Reestablished

    Family Cervidae—Deer and Allies

Cervus canadensis, Elk or Wapiti	Extirpated/Reestablished
Odocoileus hemionus, Mule Deer	Common
Odocoileus virginianus, White-tailed Deer	Uncommon

    Family Tayassuidae—Peccaries

Pecari tajacu, Collared Peccary	Common

Order Rodentia—Rodents
  Family Sciuridae—Squirrels and Allies
    Tamias canipes, Gray-footed Chipmunk   Peripheral
    Ammospermophilus interpres, Texas Antelope Squirrel   Common
    Ictidomys parvidens, Rio Grande Ground Squirrel   Common
    Xerospermophilus spilosoma, Spotted Ground Squirrel   Common
    Otospermophilus variegatus, Rock Squirrel   Common
    Cynomys ludovicianus, Black-tailed Prairie Dog   Uncommon
    Sciurus niger, Eastern Fox Squirrel   Uncommon
  Family Geomyidae—Pocket Gophers
    Thomomys baileyi, Bailey's Pocket Gopher   Common
    Geomys arenarius, Desert Pocket Gopher   Peripheral
    Geomys knoxjonesi, Jones's Pocket Gopher   Possible[1]
    Geomys personatus, Texas Pocket Gopher   Possible[1]
    Cratogeomys castanops, Yellow-faced Pocket Gopher   Common
  Family Heteromyidae—Pocket Mice and Kangaroo Rats
    Perognathus flavus, Silky Pocket Mouse   Common
    Perognathus merriami, Merriam's Pocket Mouse   Common
    Perognathus flavescens, Plains Pocket Mouse   Peripheral
    Chaetodipus hispidus, Hispid Pocket Mouse   Uncommon
    Chaetodipus eremicus, Chihuahuan Desert Pocket Mouse   Common
    Chaetodipus intermedius, Rock Pocket Mouse   Common
    Chaetodipus collis, Highland Coarse-haired Pocket Mouse   Common
    Dipodomys spectabilis, Banner-tailed Kangaroo Rat   Uncommon
    Dipodomys ordii, Ord's Kangaroo Rat   Common
    Dipodomys merriami, Merriam's Kangaroo Rat   Common
  Family Castoridae—Beavers
    Castor canadensis, American Beaver   Uncommon
  Family Cricetidae—New World Mice, Rats, and Voles
    Baiomys taylori, Northern Pygmy Mouse   Possible[1]
    Reithrodontomys fulvescens, Fulvous Harvest Mouse   Uncommon
    Reithrodontomys montanus, Plains Harvest Mouse   Uncommon
    Reithrodontomys megalotis, Western Harvest Mouse   Common
    Peromyscus attwateri, Texas Deermouse   Possible[1]
    Peromyscus eremicus, Cactus Deermouse   Common
    Peromyscus laceianus, Lacey's White-ankled Deermouse   Common
    Peromyscus boylii, Brush Deermouse   Common
    Peromyscus nasutus, Northern Rock Deermouse   Uncommon
    Peromyscus labecula, Elliot's Deermouse   Uncommon

Peromyscus leucopus, White-footed Deermouse	Common
Peromyscus truei, Piñon Deermouse	Peripheral
Onychomys arenicola, Mearns's (Chihuahuan) Grasshopper Mouse	Uncommon
Onychomys leucogaster, Northern Grasshopper Mouse	Uncommon
Sigmodon hispidus, Hispid Cotton Rat	Common
Sigmodon ochrognathus, Yellow-nosed Cotton Rat	Uncommon
Sigmodon fulviventer, Tawny-bellied Cotton Rat	Threatened, Enigmatic
Neotoma micropus, Southern Plains Woodrat	Common
Neotoma leucodon, White-toothed Woodrat	Common
Neotoma mexicana, Mexican Woodrat	Uncommon
Microtus mogollonensis, Mogollon Vole	Peripheral
Ondatra zibethicus, Common Muskrat	Rare
Family Erethizontidae—New World Porcupine	
Erethizon dorsatum, North American Porcupine	Uncommon

---

[1] Species of possible occurrence. Known from counties immediately east of the Pecos River.
[2] Species of possible occurrence. Known from south of the Rio Grande.

INTRODUCED MAMMALS

ORDER CARNIVORA—Carnivores	
Family Canidae—Dogs, Foxes, and Wolves	
Vulpes vulpes, Red Fox	Uncommon
ORDER ARTIODACTYLA—Even-Toed Ungulates	
Family Bovidae—Cattle, Sheep, Goats, and African Exotics	
Ammotragus lervia, Barbary Sheep or Aoudad	Common
Antilope cervicapra, Blackbuck	Rare
Oryx dammah, Scimitar-horned Oryx	Rare
Taurotragus oryx, Common Eland	Rare
Family Cervidae—Deer and Allies	
Cervus elaphus, Red Deer	Uncommon
ORDER RODENTIA—Rodents	
Family Muridae—Old World Mice and Rats	
Mus musculus, House Mouse	Common
Rattus norvegicus, Norway or Brown Rat	Uncommon
Rattus rattus, Black Rat	Uncommon
Family Echimyidae—Nutria	
Myocastor coypus, Nutria	Rare

DOMESTIC (D) AND FERAL (F) MAMMALS

ORDER CARNIVORA—Carnivores
    Family Canidae—Dogs, Foxes, and Wolves
        Canis familiaris, Domestic Dog   D–Common
    Family Felidae—Cats
        Felis catus, Domestic Cat   D–Common
ORDER ARTIODACTYLA—Even-Toed Ungulates
    Family Bovidae—Cattle, Sheep, and Goats
        Bos taurus, Domestic Cattle   D–Common
        Capra hircus, Domestic Goat   D–Common
        Ovis aries, Domestic Sheep   D–Common
    Family Suidae—Pigs
        Sus scrofa, Feral Hog   F–Common
ORDER PERISSODACTYLA—Odd-Toed Ungulates
    Family Equidae—Feral Horses and Asses
        Equus asinus, Feral Ass or Burro   F–Uncommon
        Equus caballus, Feral Horse   F–Uncommon
        Equus caballus × *Equus asinus*, Mule   F–Uncommon

Geographic and Faunal Affinities

Climatic oscillations throughout the Pleistocene and Holocene, as described in chapter 1, have exerted an influence on mammalian distributions throughout the Trans-Pecos. A climatic trend toward increasing aridity, which began more than 10,000 YBP, gradually transformed the region from woodlands and grasslands to the desert terrain of today, and many of the resident animal species of long ago vanished from the scene, dwindled in numbers, or followed the receding vegetation into the higher mountains of modern New Mexico. Cut off from this retreat, others were forced into the higher elevations of the resident mountain ranges in the Trans-Pecos, where they remain to this day among the isolated woodlands. This regional reduction or loss of species was simultaneously countered by the proliferation of some resident species and the invasion of others, primarily desert and grassland invaders from the surrounding geographic areas (Dalquest and Stangl 1986; Stangl et al. 1994).

For many montane mammals, these oscillating climatic changes had a direct effect on their patterns of distribution. The disjunct distributional patterns in the Trans-Pecos of the brush deermouse (*Peromyscus boylii*), the rock deermouse (*Peromyscus nasutus*), the yellow-nosed cotton rat (*Sigmodon ochrognathus*), the Mexican woodrat (*Neotoma mexicana*), and Holzner's mountain cottontail (*Sylvilagus holzneri*) may be attributed to climatic changes as the area gradually switched from woodlands and grasslands to the desert terrain of today. Each of these species enjoyed a more widespread and continuous distribution across the intervening lowlands during the more mesic pluvial stages, when woodland habitats were displaced to lower elevations. Today they are restricted mostly to pockets of favorable woodland habitat at higher elevations in the isolated mountains of the region.

The various climatic changes and factors, as described above, have produced a native mammal fauna of the Trans-Pecos today that can be characterized as a transitional mixture of species derived from at least four different sources: steppe or grassland species from the central prairies, including the Edwards Plateau and Rio Grande Plains; northern montane species from the Rocky Mountains; southern species characteristic of the Mexican tablelands; and southwestern desert and grassland species characteristic of the southwestern United States and northwestern Mexico. There are no endemic mammal species in the Trans-Pecos, and it appears that only one mammal taxon, a subspecies of Holzner's mountain cottontail (*Sylvilagus holzneri robustus*), possibly evolved within the region.

The largest segment of the mammal fauna is made up of species whose affinities are with the southwestern desert and grasslands characteristic of Arizona, New Mexico, and northwestern Mexico. Examples of representative species, most of which live in desert and grassland habitat, include the following:

Notiosorex crawfordi	Crawford's Desert Shrew
Lepus californicus	Black-tailed Jackrabbit
Antrozous pallidus	Pallid Bat
Parastrellus hesperus	American Parastrelle
Antilocapra americana	Pronghorn
Odocoileus hemionus	Mule Deer
Peromyscus eremicus	Cactus Deermouse
Thomomys baileyi	Bailey's Pocket Gopher
Chaetodipus eremicus	Chihuahuan Desert Pocket Mouse
Ammospermophilus interpres	Texas Antelope Squirrel

Another assemblage of species is associated with the arid tablelands of central Mexico south of the Trans-Pecos. Most of the species of this faunal element probably entered the Trans-Pecos from the south or southeast, and many of those that occur in the area today are in the Big Bend region. Among the mammals represented in this category are the following:

Mormoops megalophylla	Ghost-faced Bat
Leptonycteris nivalis	Mexican Long-nosed Bat
Bassariscus astutus	Ringtail
Pecari tajacu	Collared Peccary
Cratogeomys castanops	Yellow-faced Pocket Gopher
Chaetodipus collis	Highland Coarse-haired Pocket Mouse
Otospermophilus variegatus	Rock Squirrel
Xerospermophilus spilosoma	Spotted Ground Squirrel

Species associated with the steppe or grassland habitats of central North America entered the Trans-Pecos from the north, the east (via the Edwards Plateau), or the southeast along the Rio Grande. Representative mammals in this category include the following:

Sylvilagus floridanus	Eastern Cottontail
Nycticeius humeralis	Evening Bat
Mustela frenata	Long-tailed Weasel
Cynomys ludovicianus	Black-tailed Prairie Dog
Perognathus flavescens	Plains Pocket Mouse
Ictidomys parvidens	Rio Grande Ground Squirrel

A fourth component consists of an assemblage of montane species that reached the Trans-Pecos via a northern route along the eastern escarpment of the Rocky Mountains in New Mexico. Representative mammals in this category include the following:

Myotis volans	Long-legged Myotis
Castor canadensis	American Beaver
Microtus mogollonensis	Mogollon Vole
Neotoma mexicana	Mexican Woodrat
Peromyscus boylii	Brush Deermouse
Tamias canipes	Gray-footed Chipmunk

Finally, several widespread mammals with broad geographic ranges occur in the Trans-Pecos, but they are of little value in determining the relationships of its fauna. Some of the representative Trans-Pecos mammals in this category include the following:

Lasionycteris noctivagans	Silver-haired Bat
Canis latrans	Coyote
Puma concolor	Mountain Lion
Mephitis mephitis	Striped Skunk
Lynx rufus	Bobcat

3

MAMMALIAN HABITATS AND FACTORS INFLUENCING DISTRIBUTION

The principal habitats occupied by mammals in the Trans-Pecos are (1) deserts, (2) grasslands, (3) montane woodlands, (4) riparian areas, (5) rock associations, and (6) sand dunes. The first four of these habitats are defined primarily by vegetative factors, whereas the last two consider substrate and edaphic factors.

Plant-Mammal Relationships

The distribution of many mammals is correlated with the variety and abundance of vegetation, which in turn depends largely on physiographic and climatic factors. Vegetation has an important influence on mammalian distribution chiefly because plants furnish shelter and food. Different kinds of mammals vary in their responses to vegetation changes. Carnivores and bats (except for the genus *Myotis*) seem less sensitive to vegetational influences than are other mammals living in the Trans-Pecos. The distribution of rodents, in particular, is heavily influenced by vegetation. Many species depend on a particular floral aggregation in which both the kinds of plants and the growth form of the vegetation are important. A modification in the distribution of the floral aggregation is often followed by a similar displacement in the occurrence of the rodent species.

Unlike the distinct faunal transition between the montane woodland and the desert, the transition in the mammalian fauna between grassland and desert habitats shows little appreciable change. With only a few exceptions, most mammals that occur in the desert also occur in the grassland, and vice versa. Some of the more common mammals that inhabit the grasslands include the following:

Myotis ciliolabrum	Western Small-footed Myotis
Canis latrans	Coyote

Antilocapra americana	Pronghorn
Reithrodontomys montanus	Plains Harvest Mouse
Sigmodon hispidus	Hispid Cotton Rat
Chaetodipus hispidus	Hispid Pocket Mouse
Cynomys ludovicianus	Black-tailed Prairie Dog

Few mammals are actually restricted to the desert, but some that commonly occur in desert areas include the following:

Lepus californicus	Black-tailed Jackrabbit
Notiosorex crawfordi	Crawford's Desert Shrew
Antrozous pallidus	Pallid Bat
Myotis yumanensis	Yuma Myotis
Parastrellus hesperus	American Parastrelle
Canis latrans	Coyote
Vulpes macrotis	Kit Fox
Odocoileus hemionus	Mule Deer
Neotoma micropus	Southern Plains Woodrat
Peromyscus eremicus	Cactus Deermouse
Dipodomys merriami	Merriam's Kangaroo Rat
Perognathus flavus	Silky Pocket Mouse

The most marked faunal change is from the desert to the montane forest. Montane plant communities are scattered throughout the Trans-Pecos on the tops of the higher mountain slopes. For mammals inhabiting these communities, the intervening desert constitutes a formidable barrier. The isolation and scattering of mountainous regions in the desert have resulted in an island-like distribution for the mammals living there. Conversely, the cooler and moister climates of higher elevations serve as a barrier to the dispersal of desert mammals. Desert species are often not completely isolated, however, because at lower elevations in the mountains, erosion and other geological processes have produced passes or corridors (with desert vegetation) that provide avenues for their dispersal. Some mammals that commonly occur in montane habitats in the Trans-Pecos include the following:

Sylvilagus holzneri	Holzner's Mountain Cottontail
Eptesicus fuscus	Big Brown Bat
Lasionycteris noctivagans	Silver-haired Bat
Myotis volans	Long-legged Myotis

Puma concolor	Mountain Lion
Odocoileus virginianus	White-tailed Deer
Microtus mogollonensis	Mogollon Vole
Neotoma mexicana	Mexican Woodrat
Peromyscus boylii	Brush Deermouse
Peromyscus nasutus	Rock Deermouse
Sigmodon ochrognathus	Yellow-nosed Cotton Rat
Tamias canipes	Gray-footed Chipmunk

Another aspect of vegetation important to the distribution of mammals is the spacing and size of the plants in the habitat and the resulting cover they provide. The role played by plant cover in the life of animals is not well known, but all species of harvest mice (*Reithrodontomys*) in the Trans-Pecos are almost always found in places with high and thick vegetation. Similarly, the eastern cottontail (*Sylvilagus floridanus*) is not known to venture far from brush or other protective cover where it occurs. While these factors may not in and of themselves determine the distribution of these mammals, they undoubtedly influence the local distribution and abundance. Dense plant cover, on the other hand, seems not to be favored by most of the hetero-myid rodents (pocket mice and kangaroo rats). For example, the silky pocket mouse (*Perognathus flavus*) lives on sandy desert soils and avoids areas where vegetation is dense; these small mice prefer open places where clumps of plant growth are widely scattered.

The distribution of a few kinds of mammals in the Trans-Pecos coincides with the occurrence of riparian and heavily wooded habitats along permanent streams that drain the area. It seems that these species rarely move into the nearby desert plains or arid foothills. Several species characteristic of the eastern deciduous forests of the United States are rare or uncommon in the Trans-Pecos, where they occur primarily in riparian areas. These include the Virginia opossum (*Didelphis virginiana*) and the eastern fox squirrel (*Sciurus niger*). The northern raccoon (*Procyon lotor*) occurs throughout the Trans-Pecos along the major streams and in other places where some permanent water is present. Small rodents that are common in riparian as well as desert habitats include the Chihuahuan Desert pocket mouse (*Chaetodipus eremicus*), the white-footed deermouse (*Peromyscus leucopus*), the hispid cotton rat (*Sigmodon hispidus*), and the southern plains woodrat (*Neotoma micropus*).

Substrates and Edaphic Influences

Cover, and the resulting shelter it provides, is an important environmental feature for many mammals. Shelters may be of various types. In the desert they may be grouped as shelters in soils, those in and under rocks, and those built in plants. The number and kinds of shelters in soils and those in and under rocks are determined to a large extent by edaphic factors, which have considerable influence on the geographic distribution of mammals.

Stones, boulders, and rocks are abundant in the Trans-Pecos, and they are essential environmental features for several mammals such as the following:

Spilogale leucoparia	Desert Spotted Skunk
Bassariscus astutus	Ringtail
Neotoma mexicana	Mexican Woodrat
Peromyscus nasutus	Rock Deermouse
Chaetodipus intermedius	Rock Pocket Mouse
Chaetodipus collis	Highland Coarse-haired Pocket Mouse
Ammospermophilus interpres	Texas Antelope Squirrel
Otospermophilus variegatus	Rock Squirrel

The presence of rocky cliffs containing cracks and crevices may be a positive distributional factor for certain bats (such as the spotted bat, *Euderma maculatum*; the American parastrelle, *Parastrellus hesperus*; the big free-tailed bat, *Nyctinomops macrotis*; and the western bonneted bat, *Eumops perotis*) that use these places as roosting sites (fig. 14).

FIG. 14.
Rocky cliff habitat in Big Bend National Park. Photograph by Franklin D. Yancey II

Soil type may influence mammalian distribution in a number of ways. As mentioned earlier, many mammals burrow into the soil and construct nests for shelter. For these species, soil type and texture are important in determining local occurrence and abundance. Soils also affect the distribution of plants that furnish homes as well as emergency shelter, food, and moisture to desert-dwelling mammals.

Some species, such as the coyote (*Canis latrans*), seem to be unaffected by soil type. However, those mammals that habitually live underground (fossorial) are more or less limited in distribution, as their digging ability is affected by soil texture. Three kinds of pocket gophers (Bailey's pocket gopher, *Thomomys baileyi*; the yellow-faced pocket gopher, *Cratogeomys castanops*; and the desert pocket gopher, *Geomys arenarius*) occur in the Trans-Pecos, and their distributions are determined primarily by soil conditions. *Thomomys* is extremely adaptable and occurs in soils ranging from loose sands to rocky and often clayey soils and in vegetation zones grading from dry deserts to montane meadows. Unlike *Thomomys*, *Cratogeomys* is partial to deep, sandy or silty soils that are relatively free from rocks. As a result, *Cratogeomys* is often partially isolated in small populations where suitable habitat is surrounded on one or more sides by the thin, rocky soils of uplands and mountains. *Geomys arenarius* seems to prefer sandy river-bottom habitats and cannot tolerate clayey or gravelly soils.

Perhaps the most striking example of mammalian preferences for specific types of substrate in the Trans-Pecos is found among the heteromyid rodents. In most desert localities no single species of heteromyid rodent occurs on all types of substrate, and some species are tightly restricted to a single type of soil. In the Big Bend area, the Chihuahuan Desert pocket mouse (*Chaetodipus eremicus*) prefers deep, sandy loams containing less than 5 percent gravel. The highland coarse-haired pocket mouse (*Chaetodipus collis*) reaches its peak abundance on shallow soils containing more than 70 percent gravel, large rocks, and boulders. The silky pocket mouse (*Perognathus flavus*) attains its maximum abundance on desert erosion pavement soils containing at least 50 percent rocks, few of which are as large as 7.5 cm (about 3 inches) in diameter (Porter 1962, 2011).

Fine, sandy soils are common throughout much of the Trans-Pecos, where they often mantle the coarser gravelly or rocky soils occurring on the foothills of desert mountain ranges. In such situations, soil type is often the critical factor in determining the species that will occur in a given area. Some species burrow into the soil, while others make their homes in rock fissures or beneath stones. The sand dwellers might have difficulty hopping through rock piles where there is little or no soil between closely packed boulders, and conversely, rock dwellers would be out of place on a sand dune. For mammals inhabiting sandy soils, the thin, rocky soils of uplands and mountains constitute a barrier. Mammals in the Trans-Pecos that commonly live in deep, sandy soils include the following:

Neotoma micropus	Southern Plains Woodrat
Onychomys leucogaster	Northern Grasshopper Mouse
Peromyscus labecula	Elliot's Deermouse
Sigmodon hispidus	Hispid Cotton Rat
Chaetodipus eremicus	Chihuahuan Desert Pocket Mouse
Dipodomys ordii	Ord's Kangaroo Rat
Cynomys ludovicianus	Black-tailed Prairie Dog
Xerospermophilus spilosoma	Spotted Ground Squirrel

Mammals that seem to prefer rocky or gravelly soils include the following:

Neotoma leucodon	White-toothed Woodrat
Onychomys arenicola	Mearns's Grasshopper Mouse
Sigmodon ochrognathus	Yellow-nosed Cotton Rat
Chaetodipus collis	Highland Coarse-haired Pocket Mouse
Chaetodipus intermedius	Rock Pocket Mouse
Dipodomys spectabilis	Banner-tailed Kangaroo Rat

A few kinds of mammals (e.g., the cactus deermouse, *Peromyscus eremicus*) are tolerant of both sandy and gravelly soils. Where clay soils predominate, *P. eremicus* is usually the only species of *Peromyscus* present. Merriam's kangaroo rat (*Dipodomys merriami*) lives on desert flats and in sandy areas but is equally abundant in coarser soils and rocky situations. The ability of this species to survive in most xeric situations probably accounts for its wide distribution within the Trans-Pecos.

Surface Water

For several mammals, surface water is necessary for survival in the desert, and freshwater streams and rivers are essential to their existence. The American beaver (*Castor canadensis*) and the common muskrat (*Ondatra zibethicus*) occur in the desert only along the Rio Grande and the Pecos River. These mammals must have access to a stream in order to build their safety retreats. The northern raccoon (*Procyon lotor*) occurs near most of the major streams, where the animal finds much of its food along the water's edge. Almost all bats require drinking water, and many species, especially in arid regions, congregate along streams and pools where water is available. The scattered occurrence of these watering places in the desert influences the local distribution of bats. Other mammals for which surface water may play a role in distribution include the common gray fox (*Urocyon cinereoargenteus*), white-tailed deer (*Odocoileus virginianus*), white-nosed coati (*Nasua narica*), and collared peccary (*Pecari tajacu*).

Food

Food is undoubtedly often a limiting factor in the distribution of mammals. However, if the preferred food items are not available, most species can shift to alternative food sources. Two mammals in the Trans-Pecos for which availability of a particular food item may influence distribution are the kit fox (*Vulpes macrotis*) and the black-footed ferret (*Mustela nigripes*). The kit fox generally lives in the open desert, where its presence and relative numbers are controlled primarily, under natural conditions, by the presence and abundance of nocturnally active rodents. In the Trans-Pecos the range of the kit fox coincides closely with that of kangaroo rats (*Dipodomys* sp.), which are its favored food items. The black-footed ferret, now thought to be extirpated in Texas, probably once occurred throughout the Trans-Pecos, where it lived in close association with the black-tailed prairie dog (*Cynomys ludovicianus*), on which it preyed.

The Influence of the Pecos River and Rio Grande

The Pecos River and Rio Grande form the eastern and southern boundaries, respectively, of the Trans-Pecos region, and both could serve as biogeographic barriers, or at least as a filter zone, to some of the faunal elements associated with the region. Biogeographic barriers are manifested primarily by limits on dispersal or the process by which species expand their ranges, including both the movement of individuals to a new location and their successful establishment there (Weins 2011). Thus, most so-called geographic barriers and physical barriers are simply areas of unsuitable habitat and are entirely organism specific. For example, in terrestrial environments, a river may be a barrier for some organisms but not others (e.g., especially for those organisms whose niche includes freshwater habitat). Mountains near river basins may limit dispersal in some cases, especially if they create zones of unsuitable habitat for the organisms in question. In the case of the Trans-Pecos, the Mountain Basin sector, which includes the mountain regions, is about 80 to 120 km from the Pecos River, but it abuts the Rio Grande directly. There can be other barriers to dispersal that are more subtle and more clearly related to habitat differences, which in turn may be influenced by abiotic factors, biotic factors, or some combination of the two (Weins 2011).

An estimate of a river's effectiveness as a physical barrier to dispersal can be made in two principal ways (Schmidly 1977a). One is to note the species whose distribution extends to the river without crossing over to the other side. The second is to examine, within a species with populations on both sides of the river, how much dissimilarity, if any, there is between individuals on opposite sides. Thus, the presence of different subspecies on the two sides of the river becomes a measure of its effectiveness as a barrier. But rivers can also constitute an ecological barrier to distribution if they delineate the boundary of different ecoregions with different habitats on opposite sides. For the mammals in the Trans-Pecos, all these factors are evident.

In table 2, we present a synopsis of the distributional limits of the native Trans-Pecos mammals that occur adjacent or proximal to the Pecos River (table 2a) and Rio Grande (table 2b). Because of their large size, mobility, and movement patterns, carnivores and artiodactyls are typically not especially useful in such an analysis, and the importance of bats is less than that of other small mammals because of their volant nature. Most bats and several species of deermice included in this table are montane species and are thus limited more by ecological than by physical barriers to dispersal.

PECOS RIVER

The Pecos River is the principal tributary of the lower Rio Grande. It enters Texas in Loving County at Red Bluff Lake and meanders in a general southeasterly course approximately 274 km through a narrow alluvial valley to Sheffield. From this point it continues in a southeasterly course 145 km through a deep canyon to its junction with the Rio Grande 16 km west of Comstock in Val Verde County (see fig. 15). Its principal tributaries are Toyah Creek, south of Pecos, which runs only intermittently; Comanche Creek, near Fort Stockton, which is now dry; and Independence Creek, south of Sheffield, which is still free flowing. Dams and the pumping of water for irrigation have significantly altered the natural flow of the Pecos.

Two previous studies of the distribution of mammals in the Pecos River basin have been conducted—one focused primarily on small mammals excluding bats (Hollander et al. 1990b); and a more recent analysis considering all the mammals in the river basin (C. Jones and Parish 2001). The following compilation has been derived from the tables in those publications, adjusted for recent taxonomic revisions and distributional reports, as well as a few errors in the tables.

FIG. 15. Lower Canyon of the Pecos River Valley near Langtry, Val Verde County. Photograph by Thomas R. Simpson

Eighty-five species of extant native mammals occur adjacent or proximal to the Pecos River and, of these, 66 (78 percent) occur on both sides of the river. All the carnivores and ungulates occur on both sides of the river. Eleven of the 85 species occur west of the river but their distributions do not cross to the east, and 8 species demonstrate the reverse (table 2a). Thus, the distributions of about 22 percent of the native mammal species in the Pecos River area may be influenced, to some degree, by the river. The lower canyon, which is both older and deeper than the upper valley, seems to have more influence on the geographic distribution of mammals in the region than has the northern section of the river (Hollander et al. 1990b).

Sixty-seven species of small mammals (defined as marsupials, armadillos, lagomorphs, shrews and moles, bats, and rodents) occupy the Pecos River area, and 19 of these (28 percent) reach distributional limits at or near the river. These 19 species include 8 eastern and northern species whose distributions approach the river but apparently do not cross it, and 11 western species that reach their eastern limits near the Pecos River but do not cross to the east. Of these 19 species, 8 are rodents, 10 are bats, and 1 is a shrew. The distributions of 3 other western species (cactus deermouse, *Peromyscus eremicus*; Chihuahuan Desert pocket mouse, *Chaetodipus eremicus*; and highland coarse-haired pocket

TABLE 2a. A summary of the extant native mammals known to occur adjacent or proximal to the Pecos River.

| Category | Total | Pecos River | | | |
		West side only	East side only	Both sides	Different subspecies
Number of species	85	11	8	66	12
Rodents	39	3[1]	5[2]	31	10
Bats	21	8[3]	2	11	0
Carnivores	14	0	0	14	1
Ungulates	4	0	0	4	0
Lagomorphs	3	0	0	3	0
Shrews and moles	2	0	1	1	1
Marsupials	1	0	0	1	0
Armadillos	1	0	0	1	0

Source: Information adapted and modified from C. Jones and Parish (2001).

[1] One of these (*Perognathus flavus*) occurs east of the Pecos River in Texas, but only in the Panhandle far disjunct from the Trans-Pecos population; thus, the population in this area is considered not to be linked biogeographically to the population across the Pecos River.

[2] One of these (*Perognathus flavescens*) occurs west of the Pecos River in Texas, but only in El Paso County in extreme western Trans-Pecos far disjunct from the eastern population; thus, the population in this area is considered not to be linked biogeographically to the population across the Pecos River.

[3] Five of these (*Myotis californicus, M. ciliolabrum, M. thysanodes, M. volans,* and *Eptesicus fuscus*) are known from east of the Pecos River in Texas, but from areas far disjunct from the Trans-Pecos (e.g., the Panhandle, Rolling Plains, East Texas); thus, the populations in these areas are considered not to be linked biogeographically to the populations across the Pecos River.

mouse, *Chaetodipus collis*) cross to the eastern side of the Pecos River but only by a short distance.

Eastern and northern species that are documented from near the Pecos River on its eastern side but whose distributions do not cross the river to the west include the least shrew (*Cryptotis parva*), hairy-legged vampire bat (*Diphylla ecaudata*), Seminole bat (*Lasiurus seminolus*), northern pygmy mouse (*Baiomys taylori*), Texas deermouse (*Peromyscus attwateri*), Jones's pocket gopher (*Geomys knoxjonesi*), Texas pocket gopher (*Geomys personatus*), and plains pocket mouse (*Perognathus flavescens*). Eight species of bats occur

TABLE 2b. A summary of the extant native mammals known to occur along the upper and lower canyons of the Big Bend region of the Rio Grande.

		Rio Grande			
Category	Total	North side only	South side only	Both sides	Different subspecies
Number of species	98	3	9	86	5
Rodents	40	2	5	33	4
Bats	28	1	3	24	0
Carnivores	15	0	0	15	0
Ungulates	6	0	0	6	0
Lagomorphs	4	0	0	4	0
Shrews and moles	3	0	1	2	1
Marsupials	1	0	0	1	0
Armadillos	1	0	0	1	0

Source: Information adapted and modified from personal communications with Bonnie McKinney and Gerardo Ceballos (2014).

near the river on the west side that have not been recorded immediately to the east; these include the Mexican long-nosed bat (*Leptonycteris nivalis*), California myotis (*Myotis californicus*), western small-footed myotis (*Myotis ciliolabrum*), fringed myotis (*Myotis thysanodes*), long-legged myotis (*Myotis volans*), big brown bat (*Eptesicus fuscus*), spotted bat (*Euderma maculatum*), and pocketed free-tailed bat (*Nyctinomops femorosaccus*). Rodents that occur near the Pecos on the west side but are not known from east of the river include the silky pocket mouse (*Perognathus flavus*), brush deermouse (*Peromyscus boylii*), and rock deermouse (*Peromyscus nasutus*).

Mammals with different subspecies on opposite sides of the river include the eastern mole (*Scalopus aquaticus*), 10 species of rodents (rock squirrel, *Otospermophilus variegatus*; black-tailed prairie dog, *Cynomys ludovicianus*; plains pocket mouse, *Perognathus flavescens*; hispid pocket mouse, *Chaetodipus hispidus*; Ord's kangaroo rat, *Dipodomys ordii*; American beaver, *Castor canadensis*; fulvous harvest mouse, *Reithrodontomys fulvescens*; plains harvest mouse, *Reithrodontomys montanus*; southern plains woodrat, *Neotoma micropus*; and northern grasshopper mouse, *Onychomys leucogaster*), and the long-tailed weasel (*Mustela frenata*). Collectively, the species with different subspecies on either side of the river make up about 14 percent of the extant mammalian fauna of the Pecos River area.

RIO GRANDE

The Rio Grande originates in the San Juan Mountains of southern Colorado and flows southward through New Mexico to the Gulf of Mexico. Its total length is approximately 3,050 km, including the southern border of Texas, with about half of that in the Trans-Pecos. The Rio Grande dwindles to zero flow above Presidio and does not flow again in earnest until water from the Rio Conchos of Mexico joins it at Presidio. Lower down, in the region of the Big Bend, the Rio Grande is wider and deeper and flows in a permanent channel, a steep-walled canyon (see fig. 16). For small, nonvolant mammals adapted for life in the desert, this part of the river can be an effective barrier to the interchange of individuals between populations on the two sides. Farther down in its lowest reaches of the Trans-Pecos, the river is shallow, with low banks that do not prevent the passage of mammals.

Far less is known about the distribution of mammals in the Rio Grande basin, particularly south of the river in northern Mexico (northern Coahuila and eastern Chihuahua). No detailed studies, similar to those for the Pecos

FIG. 16. Santa Elena Canyon, Rio Grande, Big Bend National Park, Brewster County. Photograph by Thomas R. Simpson

River basin, have been published, although a few articles have documented the distribution of some of the mammals (Easterla 1970a; Easterla and Baccus 1973; Baccus 1978). Bonnie McKinney recently compiled a checklist of mammals in the Sierra del Carmen region just south of the river in Coahuila, Mexico, and she has kindly allowed us to use her unpublished information. This source, in conjunction with distribution records from Hall (1981), a recently published account of Mexican mammals (Ceballos 2014), and accounts presented in chapter 6 of this book, provides sufficient information to compare the mammal fauna on both sides of the Rio Grande in the canyons of the Big Bend region. The results of this compilation are presented in table 2b.

Of the 98 species of extant native mammals in the Rio Grande basin, 86 (88 percent) occupy both sides of the river throughout most or all of its drainage in the Big Bend region, and all but five (95 percent) are represented by the same subspecies in the region. The five species represented by different subspecies on the two sides of the river are the eastern mole (*Scalopus aquaticus*), the Mexican woodrat (*Neotoma mexicana*), the yellow-faced pocket gopher (*Cratogeomys castanops*), Bailey's pocket gopher (*Thomomys baileyi*), and the spotted ground squirrel (*Xerospermophilus spilosoma*).

All the lagomorphs, carnivores, and artiodactyls occur on both sides of the river. The same is true for the Virginia opossum (*Didelphis virginiana*) and the nine-banded armadillo (*Dasypus novemcinctus*). There are only three species of small mammals (western red bat, *Lasiurus frantzii*; Lacey's white-ankled deermouse, *Peromyscus laceianus*; and tawny-bellied cotton rat, *Sigmodon fulviventer*) that occur north of the river and do not cross to the south side. Conversely, nine species occur right up to the southern edge of the Rio Grande but do not cross to the north side. Three of these are bats, the southwestern myotis (*Myotis auriculus*), Mexican big-eared bat (*Corynorhinus mexicana*), and Allen's big-eared bat (*Idionycteris phyllotis*), and six are small nonvolant mammals, namely the Carmen Mountain shrew (*Sorex milleri*), cliff chipmunk (*Tamias dorsalis*), Goldman's woodrat (*Neotoma goldmani*), Hooper's deermouse (*Peromyscus hooperi*), Mexican white-ankled deermouse (*Peromyscus pectoralis*), and Nelson's kangaroo rat (*Dipodomys nelsoni*).

Thus, the Rio Grande appears to be a more effective filter for inhibiting southern mammals from crossing to the north than for northern species moving southward. The river forms a distinct boundary separating two sister species of deermice (*Peromyscus laceianus* to the north and *Peromyscus pectoralis* south of the river; see Bradley et al. 2015), and two closely related species of kangaroo rats (*Dipodomys spectabilis* to the north and *Dipodomys nelsoni* to the south; see R. H. Baker 1956).

SUMMARY

The probable function of the Pecos River as a barrier to the dispersal of mammals is suggested, at least in part, by the differences in the mammalian faunas on either side of the river. The Rio Grande appears to have had far less of an impact, serving as a filter barrier for only a few of the mammals in the river basin. The impact of both river systems is reflected primarily in the distribution of some of the smaller mammals; neither river serves as a barrier for carnivores or artiodactyls. In the case of both river systems, it is the lower canyons that seem to have the greatest impact on mammalian distributions. Because the Pecos River serves as a boundary between diverse ecological regions, it appears to be a stronger biogeographic filter zone for mammalian faunas than is the Rio Grande. The Pecos River is the primary boundary between the Chihuahuan Desert ecological units of the Trans-Pecos to the west and the more grassland-like conditions of the Edwards Plateau to the east. This ecological barrier concept is particularly illustrated by the various species of bats recorded from only one side of the Pecos River, as well as the presence of different subspecies of the American beaver on opposite sides of the river. Volant bats and semiaquatic beavers are clearly capable of breaching the physical obstacles that the river might pose to small, nonvolant, terrestrial mammals (rodents, shrews, etc.) and thus refute the notion that variations in the mammalian fauna on opposite sides of the river are strictly the result of a physical barrier imposed by the river. The ecological conditions on both sides of the Rio Grande are generally similar.

This interpretation of biogeographic patterns suggests that the Trans-Pecos does not represent a natural biotic unit bounded on the south and east by the two river basins, but rather a mixture of at least three biotic components, namely the climax grassland habitats of the Great Plains, as developed on the High Plains and Edwards Plateau; the desert scrub and grasslands of the northern Chihuahuan Desert and Mexican Tableland; and a mixture of montane elements of the Rocky Mountains. The events that might have caused these patterns include cycles of expansion and contraction of montane woodlands during the Pleistocene associated with the more recent conversion of grasslands to desertlike conditions (discussed earlier in this chapter). Further support for the Trans-Pecos not being regarded as a distinct biogeographic region is provided by Escalante et al. (2013), who cited the lack of endemic mammals in the region as cause to exclude it as such.

Adaptations to Desert Conditions

As a desert, the Trans-Pecos is a harsh ecosystem. Water resources are inherently rare and desert ecosystems are unforgiving. But many mammals, both large and small, have adapted to live and do well in arid, hot desert areas where there is often little or no free water, little or no shade, extremely high temperatures during the daytime, a seemingly limited amount of food, and low humidity. Mammals overcome these problems in a variety of ways: many are nocturnal, some live belowground, many are seed eaters, some go into a state of torpor under adverse conditions, and some have other physiological, morphological, or behavioral adaptations (see Hoffmeister 1986). A few of these desert adaptations in representative mammals that live in the Trans-Pecos are described below.

Certain small rodents are adapted to exist with no free water and with relatively dry food throughout their life. Kangaroo rats, such as *Dipodomys merriami*, and pocket mice, such as *Perognathus flavus*, can live with no water, eating seeds and other dry plant materials. These are nocturnal mammals that do not go aboveground in the heat of the day, and they do not store quantities of water within their body. Rather, they store and consume seeds throughout the year and rely on water formed as a by-product of food oxidation when the seeds are metabolized. Most of the small amount of water produced by this process is conserved by efficient kidneys that produce urine with a urea concentration four times that of humans.

These small mammals reduce the loss of water from their body by having nest sites sufficiently deep that they remain cool. One meter below the desert floor, daily temperature fluctuation is about one-fifth that on the surface. The burrow leading to the nest is plugged with dirt much of the time to prevent the influx of warm air. Because of the unpredictable and infrequent but great production of seeds in a short time, desert rodents must store seeds for relatively long periods. Also, many desert rodents have cheek pouches to aid in gathering large quantities of seed (Hoffmeister 1986).

When small desert rodents are deprived of food and water, entering torpor or aestivation can reduce their metabolic rate, thus requiring less of these items. Research has shown that cactus deermice (*Peromyscus eremicus*), when deprived of food and water at any ambient temperature below 30°C, will go into torpor within 12 hours, with a lower metabolic rate that reduces their need for food and water (MacMillen 1965). There is a marked depression in their body temperature and oxygen consumption.

Most desert mammals are nocturnal, which aids in water conservation. However, ground squirrels, such as the spotted ground squirrel (*Xerospermophilus spilosoma*) and the Texas antelope squirrel (*Ammospermophilus interpres*), are diurnal despite living in some of the hottest deserts. Ground squirrels can store heat, if required, with their body temperature rising as high as 43°C with no ill effects. If body temperatures get too high, these ground squirrels retreat to their cooler burrows for short periods or flatten themselves against the ground in shaded places to thermoregulate. Texas antelope squirrels, and probably other desert species of ground squirrels, have kidneys that concentrate urine, resulting in a minimal loss of water. Furthermore, they can subsist on salty water, even saltier than that which kangaroo rats can tolerate. Without water, these squirrels in the laboratory will die within 15 to 35 days.

Black-tailed jackrabbits (*Lepus californicus*) and other long-eared mammals use their ears to radiate body heat. For example, a jackrabbit's two ears, with a surface area of 400 cm², can radiate about 5 kilocalories per hour, about one-third the heat production in a 3-kilogram rabbit (Schmidt-Nielson 1964). In addition, jackrabbits take advantage of any shade provided by bushes, trees, or cacti and rest for longer periods in slight depressions in the ground.

Larger mammals that must cope with desert conditions include several ungulates, such as the collared peccary (*Pecari tajacu*), pronghorn (*Antilocapra americana*), bighorn sheep (*Ovis canadensis*), and mule deer (*Odocoileus hemionus*); and medium-sized carnivores, such as the coyote (*Canis latrans*) and kit fox (*Vulpes macrotis*). Collared peccaries live in the prickly pear cactus deserts of the Big Bend region. In the summer peccaries usually forage at night to avoid high temperatures. During periods of dehydration, they can reduce evaporative loss up to 68 percent and urinary loss by 93 percent (Zervanos and Hadley 1973). When coyotes and foxes pant with open mouths while fluttering their throat region, air moving across the wet membranes carries off heat.

Bighorn sheep living on the rocky crags and slopes of mountain ranges within the desert can go up to two or three days, sometimes longer, without water. They may obtain considerable moisture from the cacti and agave on which they sometimes feed. Their thick, heavy fur coat may serve to prevent surface heat from reaching their skin. In large mammals much of the body heat acquired during the day can be stored and then dissipated during the cooler nights. This accounts in part for the greater activity of desert ungulates during the night and the twilight hours of evening and early morning (Hoffmeister 1986).

4

TWENTIETH- AND TWENTY-FIRST-CENTURY CHANGES IN LANDSCAPES AND THE MAMMAL FAUNA

A landmark study by the US government, conducted at the end of the nineteenth and beginning of the twentieth centuries, provides a detailed look at what the landscapes and the mammalian fauna of the region looked like during those times, and it provides a basis for assessing how things have changed during the past 100–125 years. Led by Vernon Bailey, a leading government naturalist (see Schmidly et al. 2016b; Schmidly 2018), the US Biological Survey sent teams of field agents across the state collecting animals and plants, describing landscapes and photographing them, and recording observations about land use. Bailey and the other federal agents visited 24 sites in the Trans-Pecos (Schmidly 2002). They were the first to survey the Big Bend of Texas in what is known today as BBNP. They also surveyed the Guadalupe Mountains, in western Texas and southern New Mexico, in the area where GMNP is today. During the time of their visits, most of the land was native range in large holdings, typically used for livestock grazing (cattle, sheep, mixed cattle and sheep, and some Angora goats). Cultivated areas were confined largely to the irrigated valleys.

In 1905 Bailey published the *Biological Survey of Texas*, a book that provides some of the earliest documentation of landscapes and the mammal fauna of the state, including the Trans-Pecos region. Many of the photographs taken by the federal naturalists provide a good portrayal of what this region was like a century or more ago. From their descriptions, it is obvious that landscape changes were already underway. A complete description of the Texas survey, including a reprint of its contents, is contained in two important twenty-first-century publications (Schmidly 2002; Schmidly et al. 2022).

Landscape Changes

The arrival of American settlers in the Trans-Pecos, with their dreams and plans for irrigated farms and well-stocked ranches, necessarily resulted in alteration of the region's land and water resources. These disturbances varied considerably in comparison to previous impacts, resulting in substantial modification of the habitats of the region. When naturalists first explored the region in the early and mid-nineteenth century, prior to increased human settlement, the region surrounding the mountains supported an extensive grassland interspersed with a few shrubs. Subsequently, however, over the next 125 years, with the advent of modern settlement and development, the vegetation and wildlife changed more rapidly in composition, abundance, and distribution than at any other time in recorded history.

Over time, grasslands decreased and gradually gave way to increases in woody plant abundance. Causal factors included overgrazing by livestock, suppression of grassland fires, droughts of short and long duration in association with plant competition, changes in climate, and erosion of topsoil in areas where vegetation was removed (Richardson 2003). By the time that Bailey and the other federal naturalists began to explore the region, the impact of overgrazing on the rangelands was evident, although the riparian habitats were relatively unaltered compared to today. Desert scrub was already expanding at the expense of grassland habitat, lumbering of the "islets" of coniferous and hardwood timber at the summits of desert mountains was underway, and freestanding natural water was beginning to become rare throughout the region.

Extensive ranching in the Trans-Pecos began in the early 1870s when the first Anglo-Americans settled in the Big Bend region. Livestock numbers peaked in the late 1880s soon after completion of the Texas and Pacific Railway in 1883. But it was not long before drought and severe winters (1885–1895) drastically reduced the herds. The early Anglo settlers were not aware of the brutal droughts that frequently occurred in the region, and they did not comprehend the critical role of periodic natural fires in maintaining the health and integrity of the grassland systems. They also failed to appreciate that the carrying capacity of an ecosystem maintained by frequent drought, periodic fire, and low numbers of grazing animals was incapable of supporting high numbers of grazing animals on a continuous or long-term basis without rangeland reduction. The high stock densities during the 1880s and 1890s had an impact on vegetation and on rangeland productivity, including soil erosion (Richardson 2003).

Historically, fire played a major role in shaping and maintaining the Trans-Pecos grasslands. Lightning-caused fires were known to sweep 60 to 80 km across the country (Warnock and Loomis 2002). Fire kept the grasslands healthy and prevented shrubs from dominating the land. Shrubs and trees were often killed by fire and only roots were left, giving the Chihuahuan Desert the famous description "the land where you must dig for wood and climb for water" (Downie 1978). In the absence of fire, grasslands gradually revert to dominance by woody plants. If woody plants are allowed to increase their deep, spreading root systems, they will eventually outcompete grasses with the interacting effect of repeated droughts. When the soil surface is not covered by grasses and forbs and is thus exposed to the elements (wind and rainfall), soil erosion is inevitable. The long-term effect of this process is soil loss, which reduces the capability of the land to support vegetation and permanently decreases the carrying capacity of the land for livestock and wildlife (Richardson 2003).

Fire suppression and heavy grazing pressure in the twentieth century resulted in a gradual shift in species composition from herbaceous plants to invasive desert communities, a process known as desertification that continues today. The combination of reduced fire frequency and continued topsoil erosion on these marginal grazing lands has sustained a land cover in which much of the remaining grassland is being converted to desert scrubland. In fact, it is difficult today to precisely delineate the extent of the Chihuahuan Desert in western Texas because of overgrazing and erosion. Desert indicator plants have encroached so much into other types of vegetation in the Trans-Pecos that the boundaries between the desert and adjacent communities have been blurred. The most pristine example of Chihuahuan Desert vegetation today can be found surrounding the base of the Chisos Mountains in southern Brewster County.

The best examples of healthy grassland savannas today in the Trans-Pecos exist on sites where wildfires have occurred or where prescribed burning is practiced, as well as on ranches that have been conservatively grazed and properly managed for decades. Most of these healthy grassland savannas occur at moderate to high elevations (with cooler temperatures and greater average rainfall) in Hudspeth, Jeff Davis, Presidio, and Brewster Counties.

When the first Euro-Americans traversed the Stockton Plateau, just west of the Pecos River, grasses also dominated the vegetation and numerous freshwater springs occurred in the region. But after years of overstocking and intensive grazing (particularly by sheep and goats) in the last decades of the nineteenth century, erosion thinned the soils and the grasslands gave way to a shrub-savanna association dominated by woody plants and many kinds of desert plants. Many of the once numerous springs in this area were eliminated by the lack of runoff percolating into aquifers (Chapman and Bolen 2018).

Spring-fed streams are rare in the Trans-Pecos, and they have been heavily impacted in the past century and a half. The loss of grass cover and increasing loss of water through runoff (caused by reduced percolation into the water table) is one of the primary reasons for their demise (Chapman and Bolen 2018). Virtually every natural spring in Pecos County was pumped dry for irrigation purposes (Brune 1975, 2002). The most significant of these was Comanche Springs, a series of six large springs that formed Comanche Creek, a stream that flowed for 48 km near Fort Stockton and supported an atypical desert fauna that included common muskrats (*Ondatra zibethicus*). The establishment of Fort Stockton in 1859, however, with the subsequent use of the springs' supply for the fort, a gin (1904), and irrigation (from 1875 on), lowered the water table, and discharge began to decline by 1947. Because of poor land management and overirrigation, the spring was dried up by March 1961 (Brune 1975).

A similar fate befell many other named springs that fed the Pecos and Devils Rivers, significantly contributing to the reduced flow in the creeks and rivers of the Rio Grande watershed. Historically, more than 50 springs flowed throughout much of Reeves and Loving Counties, but no more than 8 of these remain active. The result has been devastation to the local fauna that relied on these springs for water and associated habitat.

Where springs form shallow saline wetlands, or ciénagas, marsh vegetation develops and these wetland habitats often contain endemic species. One of the most extensive ciénagas today in the Trans-Pecos is San Solomon Springs, near Balmorhea, which was destroyed years ago when the spring water was diverted, or "channelized," into concrete-lined irrigation canals. After Comanche Springs at nearby Fort Stockton went dry in 1961, the residents of Balmorhea noted the economic impacts on agriculture and tourism and formed a group of local citizens and supporters, which, with help from state universities and TPWD personnel, took steps to establish a new wetland at San Solomon Springs. This is now a permanent wetland that looks and functions like a natural ciénaga and provides habitat for several endangered species.

In desert canyons and arroyos where water flows occasionally, tinajas became another vital source for stored surface water. Sand and gravel carried by floodwaters cascading over waterfalls scour out deep pockets in the bedrock below, leaving pools of standing water in the well-like depressions. Sheltered from the heat and wind at the base of cliffs or ledges, tinajas epitomize miniature oases of lush vegetation and centers of wildlife activity, especially for many of the bat species that occupy the region (Chapman and Bolen 2018).

The major aquatic habitats of the Trans-Pecos today include the large riparian areas that are prominent on the banks of the only two rivers—the Pecos River and the Rio Grande—that drain the region as well as along the few permanent and many ephemeral watercourses in the region. Although less than 5 percent of the Trans-Pecos land area consists of riparian zones, these areas support an amazing diversity of animals and serve as nesting habitats for birds as well as movement corridors for many mammals, and they may be the most valuable natural communities in the Trans-Pecos.

Dense groves of cottonwoods, willows, desert willows, screwbean, and two cane grasses (common reed and giant reed) crowd the banks of the Rio Grande wherever a floodplain exists. Cottonwoods were once abundant, but most fell to woodcutters in the late nineteenth century. Today, the flow of the river has been greatly reduced from its historical rates, and floods, which once occurred annually, no longer displace rocks and sand bars or scour stream-bank vegetation. As a consequence, introduced species, especially salt cedar, Russian olive, and giant reed, have invaded the bottomlands bordering long stretches of the river. These processes produced a number of major changes along the river during the twentieth century.

Major hydrological changes from Fort Quitman to Presidio, where the river is now dry except for occasional flash floods, have resulted in a filling and narrowing of the channel, as well as deposition of fill at the mouths of arroyos. The Rio Conchos of Mexico joins the Rio Grande at Presidio and subsequently flows into the canyon lands of BBNP. A 310 km strip of the American bank, known as the Rio Grande Wild and Scenic River, begins in the national park and runs downstream to the Terrell–Val Verde county line. This lower section of the river is fed by several major tributaries, including Terlingua Creek, Tornillo Creek, the Pecos River, and the Devils River, that contribute important volumes of water to this portion of the Rio Grande. Although changes have occurred, this segment of the river has not changed as drastically as the Fort Quitman to Presidio section (see Schmidly and Jones 2001).

Photographs and descriptions of the Rio Grande from the turn of the twentieth century show a mosaic of floodplain habitats consisting of open grasslands, mesquite thickets, willow thickets, cottonwood forests, and wetlands (see Schmidly 2002). The introduction and rapid spread of salt cedar (*Tamarix gallica*) has dramatically altered this landscape by replacing native willows and cottonwoods, as depicted in recent photographs of the river. Since its introduction, salt cedar has proved to be very resilient and difficult to control.

Another human-induced change to the riparian habitats along the Rio Grande has been the construction of reservoirs and subsequent flooding. Prior to construction of Amistad Dam, the region between Langtry and Del Rio consisted primarily of steep river canyons and dry creek beds bordered by limestone cliffs. Since the construction of the reservoir in 1968, over 161 km of this limestone canyon country have been flooded.

The original shoreline of the river and its associated riparian vegetation of cattails, reeds, and tree tobacco were eliminated when the reservoir was flooded. Two disjunct populations of trees (the Texas pistache and Chisos oak) were extirpated by the reservoir inundation. The riparian association was replaced with a new, lakeshore association composed primarily of willow, some cattails, annual weeds, and a variety of brush and cacti that are periodically killed by the rising and falling lake levels.

Several major springs and their associated habitats were also inundated by Amistad Reservoir. Goodenough Springs, an artesian spring originally located 19 km southwest of Comstock, was the third largest spring in Texas before it was covered by the waters of the reservoir (Brune 1975).

By the mid-1970s, concern about human impacts on the riparian habitats of BBNP and ANRA prompted the NPS to conduct studies in those areas. Recreational activities and livestock grazing were monitored for their effect on terrestrial rodents and vegetation in riparian habitats (see Boeer and Schmidly 1977). The results of these studies indicated that overgrazing by domestic livestock had a more devastating effect on vegetation and rodent populations than did recreational impacts. Trespass livestock from Mexico were attracted to the riparian zone by the quality and variety of forage available. Further, the impact of livestock was more generalized over the area, whereas human impacts were localized more around campsites and along well-defined trails. Site impacts had occurred as a result of recreational use, but not to the point where ecological conditions, as indicated by the biological health of the mammal fauna, were in jeopardy (Schmidly et al. 1979).

Changes in the Mammal Fauna

We are fortunate to have a good historical perspective on the mammal fauna of the Trans-Pecos, beginning with Vernon Bailey's *Biological Survey of Texas*, published in 1905, and several more recent works (see Schmidly 1977b, 2002; Schmidly et al. 2022). From these sources it is possible to chronicle the changes in the mammal fauna during the twentieth century, including both a loss and a gain of species since 1900. Among the significant trends, four are particularly important and are discussed here. In addition to these biological changes, two important philosophical changes have impacted how mammalian species are defined and what happens when they come under conservation pressure.

SPECIES EXTIRPATIONS

Species extirpations increased dramatically during the twentieth century. When Bailey published his work in 1905, the only extirpated mammals in the Trans-Pecos were the American bison (*Bos bison*) and the elk (*Cervus canadensis*). During the course of the century, the grizzly bear (*Ursus arctos*), gray wolf (*Canis lupus*), black-footed ferret (*Mustela nigripes*), jaguar (*Panthera onca*), ocelot (*Leopardus pardalis*), and bighorn sheep (*Ovis canadensis*) all joined the list of extirpated species. A variety of factors can cause extirpation, and in chapter 6 we discuss the particular circumstances for the disappearance of each of these species from the Trans-Pecos. Thanks to the efforts of the TPWD and committed landowners, bighorns and elk have been reestablished in the region, although the stock used in the reintroductions were not of the native subspecies.

DECLINES IN GEOGRAPHIC DISTRIBUTION AND POPULATION ABUNDANCE

At least three Trans-Pecos mammals have undergone drastic range reductions and today occupy a scant portion of their former range. These include the pronghorn (*Antilocapra americana*), black-tailed prairie dog (*Cynomys ludovicianus*), and common muskrat (*Ondatra zibethicus*). Their current status in the Trans-Pecos region is discussed in the species accounts for each of these mammals.

ADDITIONS AND RECENT INVASIONS OF MAMMALS

Since the beginning of the twentieth century, a number of mammals have expanded their range into the Trans-Pecos. These include mammals such as the Virginia opossum (*Didelphis virginiana*) and nine-banded armadillo (*Dasypus novemcinctus*), which are now present but were not recorded by Bailey (1905) when he worked in the area. Also, a few new faunal elements have been added since the 1977 account by Schmidly (1977b). This latter category includes five species of bats (American perimyotis, *Perimyotis subflavus*; evening bat, *Nycticeius humeralis*; western yellow bat, *Dasypterus xanthinus*; Mexican long-tongued bat, *Choeronycteris mexicana*; and lesser long-nosed bat, *Leptonycteris yerbabuenae*) along with a single rodent species (tawny-bellied cotton rat, *Sigmodon fulviventer*). Detailed explanations for the discovery of these species in the Trans-Pecos are provided in the various species accounts.

INTRODUCTIONS OF NONNATIVE MAMMALS

During the twentieth century, at least 10 nonnative species of mammals were introduced into the Trans-Pecos, including one carnivore, five artiodactyl ungulates, and four rodents. Several other feral species have also become established in the wild and are creating conservation challenges. These include feral cats, feral dogs, feral hogs, feral burros, and wild horses. The character and composition of the fauna have changed somewhat as a result of these biological invaders. The details and challenges associated with their presence are discussed in chapter 6.

CHANGES IN CLASSIFICATION AND TAXONOMY OF MAMMALS

The twentieth century also saw a dramatic shift in the philosophy for classifying mammals for the purpose of taxonomic designation (see Schmidly and Bradley 2016 for an extensive discussion). The most significant change comes from the use of DNA sequence data and the reemergence of an old species concept, the genetic species concept. The major taxonomic changes for mammals in the Trans-Pecos region since the publication of Schmidly's 1977 account are summarized in the various species accounts in chapter 6.

INCREASE IN THE NUMBER OF THREATENED, ENDANGERED, AND RARE SPECIES

Beginning in the 1960s and continuing today, state and federal agencies as well as private organizations have developed lists of rare and endangered mammals that seemingly require protection. The USFWS publishes a list of endangered and threatened species in the *Federal Register* that includes mammals assigned to these categories based on the ESA, and the TPWD maintains a list of protected nongame wildlife (established in 1973). These are the official lists governed by federal and state laws, statutes, and regulations. Additionally, the IUCN publishes the *IUCN Red List of Threatened Species (Red List)* (IUCN 2022), which is the most comprehensive inventory of the global conservation status of biological species.

Seventeen Trans-Pecos mammals, about 16 percent of the total native mammal fauna, appear on one or more of these lists (table 3). These include a rabbit, five bat species, seven carnivore species, one ungulate, and three rodent species. Almost half the species on these lists are already extirpated in the Trans-Pecos, including five of the carnivore species and the single ungulate species. Overhunting and trapping for predator control in the twentieth century caused most of these extirpations.

There is a great deal of concern and even resentment among many landowners in the Trans-Pecos over the application of regulations, particularly the

TABLE 3. Trans-Pecos mammals with critical status as defined by the Texas Parks and Wildlife Department (TPWD), US Fish and Wildlife Service (USFWS), and International Union for Conservation of Nature (IUCN)

Common name	TPWD	USFWS	IUCN	Status in Trans-Pecos
Holzner's Mountain Cottontail	NL	NL	VU	Rare, Restricted
Mexican Long-tongued Bat*	NL	NL	NT	Rare
Mexican Long-nosed Bat	E	E	NT	Endangered
Lesser Long-nosed Bat	NL	E	NT	Extralimital
American perimyotis (tricolored bat)	NL	NL**	VU	Rare
Spotted Bat	T	NL	LC	Rare
Gray Wolf	NL	E	LC	Extirpated
Ocelot	E	E	LC	Extirpated
Jaguar	NL	E	NT	Extirpated
Black-footed Ferret	NL	E	E	Extirpated
White-nosed Coati	T	NL	LC	Rare
American Black Bear	T	NL	LC	Rare
Grizzly Bear	NL	T	LC	Extirpated
American Bison	NL	NL	NT	Extirpated
Desert Pocket Gopher	NL	NL	NT	Common, Restricted
Banner-tailed Kangaroo Rat	NL	NL	NT	Uncommon, Declining
Tawny-bellied Cotton Rat	T	NL	LC	Enigmatic

Note: E = endangered, T = threatened, NT = near threatened, VU = vulnerable, LC = least concern, NL = not listed.

*Listed as endangered in Mexico.

**Currently under consideration for listing as endangered.

ESA, to enforce habitat protection for these species on private lands. The concerns center around the impact of enforcement and how it potentially impinges on private landowner rights. These conflicts have made habitat conservation for many rare and threatened species difficult. But, as more and more habitats are altered and fragmented by development, it seems likely that additional protection of species will be required to stem the tide of extirpations.

5

KEY TO THE ORDERS OF TRANS-PECOS MAMMALS

The first step in identifying mammals is to determine the taxonomic order to which the animal belongs. The key provided below facilitates this process. For those orders with multiple taxa, keys to the families and/or species are provided in chapter 6, in some cases with illustrations to aid in identification. Distribution maps are provided for each species to indicate its projected occurrence within the Trans-Pecos region. However, a note of caution is required in interpreting these maps. Although often depicted as continuous in field guides, the distribution of a species is often not uniform throughout its geographic range; rather, it is a mosaic of the distributions of the local populations. Neither are individuals in the population equally abundant in all regions. The natural world is heterogeneous, resulting in a patchwork distribution of habitats of varying quality.

1. Bony plates covering back; almost no hair anywhere on back
   . . . . ORDER CINGULATA, armadillos
   Back without bony plates; at least part of back (usually entire back)
   covered with hair. 2
2. Forelimbs modified for flight. . . . . ORDER CHIROPTERA, bats
   Forelimbs not modified for flight 3
3. Hooves present on one or more toes of each foot 4
   Toes armed with claws, not hooves 5
4. A single large hoof on each foot . . . .
   . . . . ORDER PERISSODACTYLA, odd-toed ungulates
   Two subequal and conspicuous hooves on each foot . . . .
   . . . . ORDER ARTIODACTYLA, even-toed ungulates
5. Canines present; no pronounced diastema (space) between incisors
   and cheek teeth 6

Canines absent; diastema present between incisors and cheek teeth;
 cheek teeth pronounced 8
6. Incisors 5/4; hallux (first toe of hindfoot) opposable but without
 a claw; tail prehensile . . . .
 . . . . .Order Didelphimorphia, marsupials
 Incisors 3/3 or fewer; hallux not opposable; all toes armed with claws;
 tail not prehensile 7
7. Canines not markedly longer than adjacent teeth; size small, total
 length less than 150 mm; eyes very small or hidden . . . .
 . . . . . Order Eulipotyphla, shrews, moles, and relatives
 Canines markedly larger than adjacent teeth; size medium to large;
 eyes normal . . . . .Order Carnivora, carnivores
8. Incisors 2/1 . . . . Order Lagomorpha, rabbits and hares
 Incisors 1/1 . . . . .Order Rodentia, rodents

The dichotomous keys to the species within each order are generally easy
to follow and will produce reliable identifications. However, for some of the
small mammals there are groups of similar species that can be difficult to iden-
tify other than by a trained and experienced expert. Examples include bats of
the genus *Myotis*, species of deermice (genus *Peromyscus*), and species of pocket
mice (genera *Perognathus* and *Chaetodipus*). In these cases, we have included
stylized drawings that illustrate the salient identification features described in
the keys to assist the reader.

PLATE 1. Desert cottontail, *Sylvilagus audubonii*. Photograph by Franklin D. Yancey II

PLATE 2. Holzner's mountain cottontail, *Sylvilagus holzneri*. Photograph by Mark W. Lockwood

PLATE 3. Black-tailed jackrabbit, *Lepus californicus.* Photograph by Franklin D. Yancey II

PLATE 4. Big free-tailed bat, *Nyctinomops macrotis.* Photograph by Mark W. Lockwood

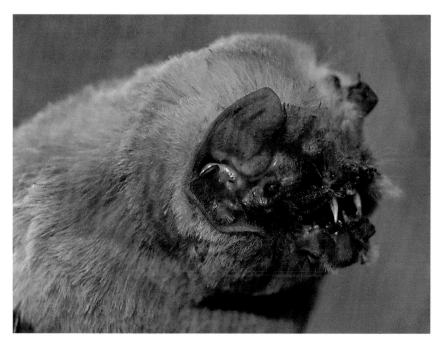

PLATE 5. Ghost-faced bat, *Mormoops megalophylla*. Photograph by Mark W. Lockwood

PLATE 6.
Fringed myotis,
Myotis thysanodes.
Photograph by
Mark W. Lockwood

PLATE 7. Silver-haired bat, *Lasionycteris noctivagans*. Photograph by Richard W. Manning

PLATE 8. Townsend's big-eared bat, *Corynorhinus townsendii*. Photograph by Mark W. Lockwood

PLATE 9.   Coyote, *Canis latrans*. Photograph by Mark W. Lockwood

PLATE 10.   Common gray fox, *Urocyon cinereoargenteus*. Photograph by
Mark W. Lockwood

PLATE 11. Kit fox, *Vulpes macrotis*. Photograph by Maryann M. Eastman

PLATE 12. Bobcat, *Lynx rufus*. Photograph by Mark W. Lockwood

PLATE 13. Mountain lion, *Puma concolor*. Photograph by Mark W. Lockwood

PLATE 14. Striped skunk, *Mephitis mephitis*. Photograph by Mark W. Lockwood

PLATE 15. Northern raccoon, *Procyon lotor*. Photograph by Mark W. Lockwood

PLATE 16. American black bear, *Ursus americanus*. Photograph by Mark W. Lockwood

PLATE 17. Pronghorn, *Antilocapra americana*. Photograph by Franklin D. Yancey II

PLATE 18. Bighorn sheep, *Ovis canadensis*. Photograph by Mark W. Lockwood

PLATE 19. Elk, *Cervus canadensis*. Photograph by Mark W. Lockwood

PLATE 20. Mule deer, *Odocoileus hemionus*. Photograph by Mark W. Lockwood

PLATE 21. White-tailed deer, *Odocoileus virginianus*. Photograph by Mark W. Lockwood

PLATE 22. Collared peccary, *Pecari tajacu*. Photograph by Franklin D. Yancey II

PLATE 23. Texas antelope squirrel, *Ammospermophilus interpres*. Photograph by Mark W. Lockwood

PLATE 24.
Rock squirrel,
*Otospermophilus
variegatus*.
Photograph by
Mark W. Lockwood

PLATE 25. Bailey's pocket gopher, *Thomomys baileyi*. Photograph by Franklin D. Yancey II

PLATE 26. Silky pocket mouse, *Perognathus flavus/Perognathus merriami*. Photograph by Franklin D. Yancey II

PLATE 27. Rock pocket mouse, *Chaetodipus intermedius*. Photograph by Franklin D. Yancey II

PLATE 28. Merriam's kangaroo rat, *Dipodomys merriami*. Photograph by Franklin D. Yancey II

PLATE 29.   Elliot's deermouse, *Peromyscus labecula*. Photograph by Mark W. Lockwood

PLATE 30.   Yellow-nosed cotton rat, *Sigmodon ochrognathus*. Photograph by Franklin D. Yancey II

PLATE 31. White-toothed woodrat, *Neotoma leucodon*. Photograph by Franklin D. Yancey II

PLATE 32. Nutria, *Myocastor coypus*. Photograph by Mark W. Lockwood

6

ACCOUNTS OF SPECIES

ORDER DIDELPHIMORPHIA (Opossums and Allies)

Didelphimorphia is the only order within the mammalian infraclass Metatheria (marsupials) that occurs in North America. Metatherians are characterized by the presence of a marsupium (external pouch) that develops in the abdominal area of females. The young are born in a relatively undeveloped stage and in most species complete their development in this pouch. Australia and South America are centers of radiation of marsupials.

Family Didelphidae
OPOSSUMS
Virginia Opossum
Didelphis virginiana Kerr, 1792

Virginia opossums are about the size of a domestic cat and may be recognized by their long, scantily haired, prehensile tail; the opposable or thumb-like "big toe" of the hindfoot; and the external, fur-lined abdominal pouch of females (fig. 17). Average external measurements (in mm) for males and females, respectively, are total length 782, 710; tail 324, 320; hindfoot 64, 63; ear is about 50 for both sexes. Weight ranges from 1.8 to 4.5 kg, with males typically weighing more than females.

The Virginia opossum has a spotty distribution and is one of the rarest mammals in the Trans-Pecos. It has been documented only from BBNP (Brewster County), El Paso (El Paso County), the Stockton Plateau near Sheffield (Terrell County), near the southern edge of the Davis Mountains (Jeff Davis County), and Balmorhea State Park in Reeves County (Hollander and Hogan 1992; Yancey and Lockwood 2017) (map 4). It resides in a wide

FIG. 17. Virginia opossum, *Didelphis virginiana*

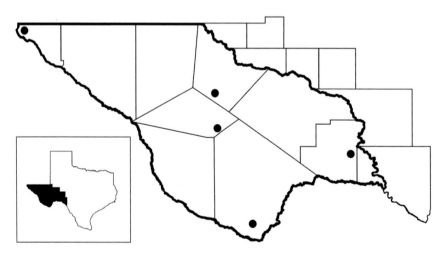

MAP 4. Distribution of the Virginia opossum, *Didelphis virginiana*

variety of habitats and can be found in forests, grasslands, and agricultural and urban situations; however, streamside riparian habitats and other wetlands are preferred. The paucity of these moist habitats in the Trans-Pecos lends to its scarcity in the region.

Didelphis virginiana does not hibernate and is active throughout the year. Individuals are typically nocturnal, although they may rarely be encountered during daylight. Virginia opossums are true omnivores and typically rely on the most readily available food items. They commonly consume a variety of plant material, including fruits, nuts, seeds, tubers, grasses, and other green vegetation. Animal matter eaten includes earthworms, insects, and a host of other invertebrates, as well as small reptiles, amphibians, birds and their eggs, and mammals. Carrion is consumed frequently, and cannibalism is not uncommon.

Breeding typically begins in January or occasionally December and may continue through June or July. One or two, rarely three, litters per year of up to 14 or more young each are produced. The young are tiny at birth, weighing only about 0.13 g each. Upon birth, young immediately grasp hairs on the mother's abdomen and crawl toward the pouch. Upon reaching the pouch, neonates attach to a teat, where they remain for about 60 days. The number of surviving young is limited to the number of functional teats on the mother (usually 13).

Predators include large snakes, raptors, domestic dogs, coyotes, and bobcats. Caution should be exercised when handling opossums because they are known carriers of rabies, tuberculosis, relapsing fever, and tularemia.

Because of a paucity of museum voucher specimens from the Trans-Pecos, the subspecific status of *D. virginiana* in the region is ambiguous. Gardner (1973) acknowledged a single subspecies, *D. v. virginiana*, whereas Hall (1981) recognized two, *D. v. virginiana* in the north and *D. v. californica* in the southeast. The documentation of the latter was based solely on "a few" secondhand and thirdhand verbal accounts of individuals taken by trappers in the Big Bend area in the 1920s and 1930s (Borell and Bryant 1942). Because of the absence of any confirmed *californica* specimens from the region, we have taken a conservative approach and, pending future specimen acquisitions, recognize a single subspecies in the region, *D. v. virginiana*. This species is in serious need of taxonomic revision in the Trans-Pecos.

ORDER CINGULATA (Armadillos and Allies)

This order originated in South America and only recently invaded North America. Its members, the armadillos, are highly specialized with a protective armored appearance. This bony carapace is unique among mammals.

Family Dasypodidae
ARMADILLOS
Nine-banded Armadillo
Dasypus novemcinctus Linnaeus, 1758

The nine-banded armadillo is an unusual-looking mammal about the size of a terrier. Its upperparts are encased in a bony carapace, with large shields on the shoulders and rump and nine bands in between (fig. 18). The front feet have four toes and the hindfeet five, and all are provided with large, strong claws. They have small, peg-like teeth and a protrusible tongue. Average external measurements (in mm) are total length 760; tail 345; hindfoot 85; ear 40; weight is 5–8 kg.

Nine-banded armadillos have been recorded from Reeves, Pecos, Terrell, and Jeff Davis Counties in the Trans-Pecos (map 5). They prefer areas near water with deeper, softer soils that are easier to burrow into, such as the alluvial stream bottoms of the Stockton Plateau. They avoid the hard and rocky soils of desert or grassland areas. This is one of the rarest mammals in the Trans-Pecos. Specimen records are from east of Fort Stockton in Pecos County, along Tunis Creek, which is a tributary of the Pecos River (Schmidly 1977b); and north of

FIG. 18. Nine-banded armadillo, *Dasypus novemcinctus*

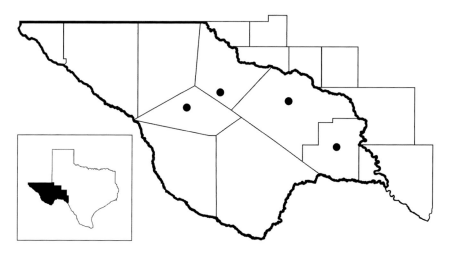

MAP 5. Distribution of the nine-banded armadillo, *Dasypus novemcinctus*

Fort Davis, Jeff Davis County (Kennedy and Jones 2006). Sightings have been reported from a live-oak association in Terrell County (Hermann 1950), and game wardens have reported them from the edge of the Davis Mountains in Reeves County (Cleveland 1970).

Their diet is more than 75 percent insects and other arthropods, but individuals will readily feed on snails, worms, small vertebrates, bird eggs, carrion, and berries. They are nocturnal and crepuscular, although they may be active during the daytime on colder days. They do not hibernate but may reduce their activity during rainy and cold periods. When alarmed, armadillos will jump straight up or may arch their back, a behavior that allows them to wedge into a burrow.

Their reproduction is unusual, as it involves delayed implantation and polyembryony. Breeding occurs in July–August, with copulation occurring with the female on her back. A fertilized egg (zygote) develops for five to seven days, forming a blastocyst that passes into the uterus, where development ceases. About four months later, the blastocyst implants into the uterus, where it divides, then divides again, leaving four embryonic growth centers that induce a placenta with four separate umbilical cords. After gestation of about 120 days, four genetically identical young are born in a burrow during March or April.

Predators include mountain lions and humans. Armadillos are known to carry leprosy, apparently from the same strain of the microbe (*Mycobacterium leprae*) that infects humans. *Armadillo* means "little armored one" in Spanish. The Aztec name was translated to "turtle-rabbit." The subspecies in the Trans-Pecos is *D. n. mexicanus*

Order Lagomorpha (Rabbits, Hares, and Pikas)

Lagomorphs are medium-sized terrestrial mammals characterized by a short, well-furred tail and long ears. Locomotion is by hopping or leaping, and the hindfeet are longer than the front. Rabbits and hares have four upper and only two lower incisors. The first pair of upper incisors is large; the second pair is small, peg-like, and located directly behind the first. Young of hares (jackrabbits) are born precocial, furred, and with their eyes open; they have no nest prepared for them. Young rabbits (cottontails) are born altricial (blind, helpless, and hairless), in a fur-lined nest constructed especially for them. Four species, one hare and three rabbits, occur in the Trans-Pecos, and all belong to the family Leporidae.

1. Length of ear from notch > 100 mm; tail with black dorsal stripe; interparietal bone of cranium indistinct or absent.........
    .... **Black-tailed Jackrabbit**, *Lepus californicus*
    Length of ear from notch usually < 100 mm; tail cotton-white below, brownish above with no black dorsal stripe; interparietal bone of cranium distinct 2

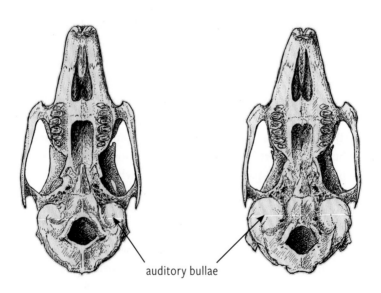

auditory bullae

FIG. 19. Ventral view of the skulls of cottontail rabbits, genus *Sylvilagus*, showing relative sizes of the auditory bullae: left, *S. floridanus*; right, *S. audubonii*

2. Auditory bullae large, with a rough-textured surface (fig. 19); nuchal patch behind ears typically dull gray to rusty orange; hindfoot usually < 90 mm; ear usually > 60 mm . . . .
. . . . **Desert Cottontail**, *Sylvilagus audubonii*
Auditory bullae small, with a smooth surface (fig. 19); nuchal patch behind ears typically a rich reddish orange; hindfoot usually
> 90 mm 3

3. Hindfoot near 90 mm; ear near 55 mm; premolar enamel patterns simple, with the enamel plates even and smooth . . . .
. . . . **Eastern Cottontail**, *Sylvilagus floridanus*
Hindfoot near 100 mm; ear near 70 mm; premolar enamel patterns complex, with the enamel plates having intricate folds . . . . . .
. . . . **Holzner's Mountain Cottontail**, *Sylvilagus holzneri*

Family Leporidae
RABBITS AND HARES
Desert Cottontail
Sylvilagus audubonii (Baird, 1858)

Desert cottontails are medium-sized rabbits (average weight 750–800 g) with relatively long ears (not approaching the size of those of the jackrabbit, however), long hind legs, buff-brown upperparts, and a short but conspicuous tail that is white on the underside (fig. 20). The auditory bullae are large and rough in appearance (see fig. 19). Average external measurements vary according to subspecies (see below).

Desert cottontails are common and widespread in all counties in the Trans-Pecos (map 6). These rabbits favor desert scrub habitats, but they will also occupy riparian woodlands, grasslands, and occasionally juniper woodlands. They seem more at home in open country and less dependent on dense cover for protection than *S. floridanus*.

Their diet includes a variety of grasses, forbs, and shrubs, with seasonal availability influencing their dietary preference. They are known to utilize a variety of green vegetation, including grasses, leaves, green pods and leaves of mesquite, pads of prickly pear cactus, and bark and small branches of shrubs.

Desert cottontails are most active during twilight hours and at night but may also be active on cooler days. During hot days, they squat in a scratched-out or existing depression (called a form) in the shade. When alarmed, they often sit motionless and then run to the nearest cover or into a burrow dug by another mammal.

FIG. 20. Desert cottontail, *Sylvilagus audubonii*

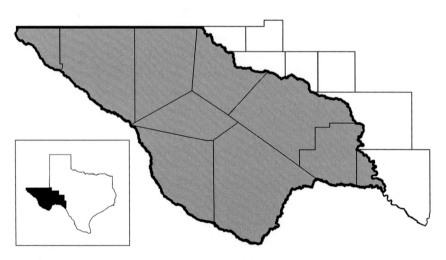

MAP 6. Distribution of the desert cottontail, *Sylvilagus audubonii*

In the Trans-Pecos, they breed throughout much of the year, with gravid females having been recorded from March to August (Davis 1940b; Stangl et al. 1994; Yancey 1997; Yancey et al. 2006). Gestation is about 28–30 days, followed by birth of two to four blind, altricial young. A pregnant female taken on June 26 in the Guadalupe Mountains carried three embryos (Davis 1940b).

Two pregnant females were examined from BBRSP—one taken on April 29 had three embryos, and another on June 3 contained two embryos (Yancey 1997). Grass and fur-lined nests are formed in shallowly excavated depressions that the female will crouch over to nurse the kittens.

Predators include coyotes, gray foxes, kit foxes, bobcats, raccoons, striped skunks, owls, hawks, and snakes. A large western diamondback rattlesnake (*Crotalus atrox*) killed in the Guadalupe Mountains had two juvenile desert cottontails in its stomach (Genoways et al. 1979).

The desert cottontail is a reservoir for tularemia, or rabbit fever. Rabbit hemorrhagic disease (RHDV2), a highly contagious viral disease, has been identified in desert cottontails from Hudspeth and El Paso Counties. Should this disease spread across the Trans-Pecos, it would be devastating to populations.

Two subspecies occur in the Trans-Pecos, *S. a. minor* in the western and southern parts and *S. a. neomexicanus* in the northeastern section, which includes almost all of Culberson, Reeves, and Pecos Counties. The two subspecies can be distinguished based on coloration and average external measurements. The upperparts of *S. a. minor* are pale gray sprinkled with black, with a dull gray to buff nuchal patch, whereas those of *S. a. neomexicanus* are buffy brown heavily lined with black, with a rusty orange nuchal patch (Bailey 1931). Average external measurements (in mm) for *S. a. neomexicanus* are total length 367; tail 44; hindfoot 80; ear 65 (Stangl et al. 1994), and, respectively, 347, 40, 88, and 69 for *S. a. minor* (Yancey 1997).

Eastern Cottontail
Sylvilagus floridanus (J. A. Allen, 1890)

The eastern cottontail is a moderately large rabbit with relatively short ears; large hind legs and feet; short front legs and feet; a soft pelage; and a short, fluffy tail that is white beneath (fig. 21). *Silvilagus floridanus* differs from *S. audubonii*, with which its range overlaps, in having shorter ears and small, smoothly rounded auditory bullae, as opposed to large and rough bullae in *S. audubonii* (fig. 19). Average external measurements (in mm) are total length 380; hindfoot 85; ear 50; tail 56; weight is 1–2 kg.

Eastern cottontails occur throughout most of the Trans-Pecos but are absent from the far western part of the region (El Paso County and most of Hudspeth County) and parts of the north-central Trans-Pecos (eastern Culberson County and most of Reeves County) (map 7). These cottontails are usually found inhabiting brushy lowlands at low to moderate elevations. Specimens have been obtained from dense thickets of mesquite, sumac, and

FIG. 21. Eastern cottontail, *Sylvilagus floridanus*

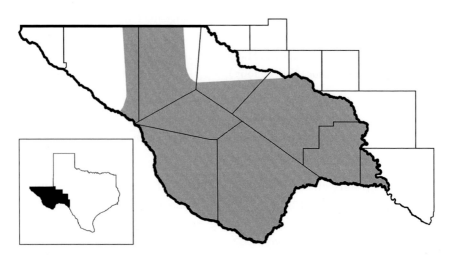

MAP 7. Distribution of the eastern cottontail, *Sylvilagus floridanus*

condalia in Terrell County (Hermann 1950). They are usually most active around dawn and dusk, and during moonlit nights.

Their diet varies seasonally, with grasses, forbs, and other herbaceous vegetation consumed during the growing season and stems and bark of woody plants added during the dormant season. Most feeding occurs around sunrise and sunset. As with other species of *Sylvilagus*, coprophagy is practiced, in which soft, green caecal pellets composed of not fully digested plant material are excreted and then reconsumed, sometimes directly from the anus (Kirkpatrick 1956).

In the Trans-Pecos, eastern cottontails produce young annually, usually in the spring. Several young of the year have been observed in June, and a female collected in the Chisos Mountains on April 6 was lactating and, based on uterine scars, had just given birth to two young (Davis 1974). Natural predators include owls, foxes, coyotes, bobcats, and eagles.

Eastern cottontails are prime reservoirs for tularemia (rabbit fever), especially when population densities become high. Rabbit hemorrhagic disease (RHDV2), a highly contagious viral disease, has been documented in jackrabbits and desert cottontails in Hudspeth and El Paso Counties but has not yet been identified in eastern cottontails.

In the first edition of this book (Schmidly 1977b), *S. floridanus* in the Trans-Pecos included two distinct subspecies, *S. f. chapmani* and *S. f. robustus*. Subsequently, *S. f. robustus* was elevated to the status of a separate species, *S. robustus* (see Ruedas 1998). However, a recent taxonomic study (Diersing and Wilson 2021) has now placed *robustus* as a subspecies of *S. holzneri* (see next account for an explanation).

Holzner's Mountain Cottontail
Sylvilagus holzneri (Mearns, 1896)

Holzner's mountain cottontail is a larger, paler version of the eastern cottontail with longer ears (> 60 mm) and hindfeet (near 100 mm) (fig. 22). Upperparts are pale buff gray integrated with black hairs; hips and rump are a slightly darker gray; front and sides of forelegs are rusty orange; top of tail is dull brownish; and nape of neck (nuchal patch) is rufous. The underparts are clear white with a cotton-white tail, and the underside of the neck is grayish or brownish. Complicated folds and creases in the enamel walls of the second and third upper premolars and the third lower premolar distinguish this rabbit from both *S. audubonii* and *S. floridanus* (Ruedas 1998). Average external measurements (in mm) are total length 416; tail 53; hindfoot 98; ear 71; weight is 1.3–1.8 kg.

Holzner's mountain cottontail is restricted to the central core of mountains (Guadalupe, Davis, and Chisos) of the Trans-Pecos (map 8). These cottontails inhabit brushy and forested areas above 1,800 m. They are associated with sumac, mountain mahogany, scrub oak, and piñon-oak-juniper woodlands, and they use crevices and gaps in rocks for refuge. In GMNP, the species was

FIG. 22. Holzner's mountain cottontail, *Sylvilagus holzneri*

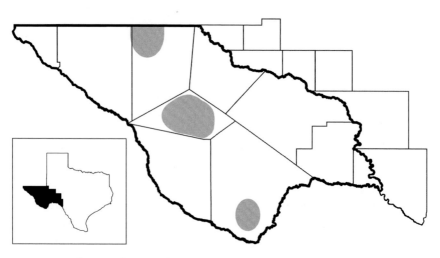

MAP 8. Distribution of Holzner's mountain cottontail, *Sylvilagus holzneri*

observed using dense stands of Douglas fir and ponderosa pine (Genoways et al. 1979). *Sylvilagus holzneri* is irregularly sympatric with *S. floridanus* in the Davis Mountains.

Their diet consists of grasses, leaves of mountain mahogany, whitebrush, and various herbs, as well as piñon pine twigs (Davis 1940b). They normally rest in brushy thickets and emerge to feed mostly in late evening and early morning. They are not common, and when disturbed, they usually run short distances, stop behind brush, and eye the disturbance.

Breeding occurs in spring and early summer. Several young of the year have been observed in June (Davis 1940b; Genoways et al. 1979). In the Chisos Mountains in April, a female was lactating and birthed two young. Another had three young and was lactating in June. Natural predators are probably similar to those of the eastern cottontail—namely, owls, hawks, eagles, foxes, coyotes, and bobcats.

Concerns have been expressed about the conservation status of *S. holzneri* in the Trans-Pecos. Based on the most recent assessment of the taxon in 2018 (not as *S. holzneri* but under its former name *S. robustus*; see the following paragraph), this rabbit is currently listed as vulnerable on the IUCN Redlist (IUCN 2022). It does not, however, appear on USFWS or TPWD lists of threatened and endangered species. The Guadalupe and Chisos Mountains populations have been severely reduced, with few specimens collected from those areas in the last 30 years. Mammalogists from Texas Tech University recently obtained several specimens from the Davis Mountains, and it appears that healthy populations remain in that area. The narrow elevation range where the species occurs (1,400–2,347 m) suggests it could be sensitive to climate change. Careful monitoring of this species is needed throughout the isolated mountain ranges of the Trans-Pecos to determine its overall conservation status in the region (Schmidly et al. 2022). If rabbit hemorrhagic disease (RHDV2) spreads into Jeff Davis County, it could potentially decimate populations of this rare rabbit.

The taxonomic status of this rabbit has been bantered back and forth over the past 75 years. It was originally described as a distinct species and named *S. robustus* (Nelson 1909). It was later relegated to a subspecies of *S. floridanus* (Hall and Kelson 1951), although some taxonomists (e.g., Davis 1960, 1966, 1974) continued to recognize it as a distinct species. Ruedas (1998) demonstrated trenchant morphological differences between *robustus* and other subspecies of *S. floridanus*, and subsequent molecular genetic studies (Vestal 2005; D. Lee et al. 2010; Nalls et al. 2012) seemed to confirm its status as a separate species. Diersing and Wilson (2021), in a morphological study of specimens from mountain populations in the southwestern United States and northern Mexico, arranged *robustus* as a subspecies of *Sylvilagus holzneri*. Unfortunately, these authors did not have any genetic data to confirm this taxonomic interpretation. Therefore, while we tentatively accept Diersing and Wilson's recent conclusion, we suggest that additional investigation is necessary to fully resolve the status of this rabbit. *Sylvilagus holzneri robustus* is considered to be the subspecies that occurs in the Trans-Pecos.

Black-tailed Jackrabbit
Lepus californicus Gray, 1837

Black-tailed jackrabbits are large, long-eared hares. Ears are > 100 mm, nearly the length of the hindfeet, and practically naked, with a black patch at the tip (fig. 23). The top of the tail has a black stripe that extends over the rump. Upperparts are light brownish to pale buff gray, heavily peppered with black. Underparts are pale, with individual hairs gray at the bases. There may be a dark pectoral band, or mane, across the chest. Average external measurements (in mm) are total length 533; tail 66; hindfoot 125; ear 138; weight is 1.5–4 kg.

These jackrabbits have been recorded from all counties in the Trans-Pecos (map 9). They prefer open habitat where visibility is not obscured. They are common in hot, dry desert scrub and open shortgrass prairie, but during the warmer months they are known to occupy montane meadows and woodlands above 1,800 m in the Davis Mountains (Blair 1940) and 2,100 m in the Guadalupe Mountains (Genoways et al. 1979). At BBRSP, they have been taken in desert scrub and grassland habitats, but not in riparian areas. Apparently, they avoid such wooded areas because their primary defense

FIG. 23. Black-tailed jackrabbit, *Lepus californicus*

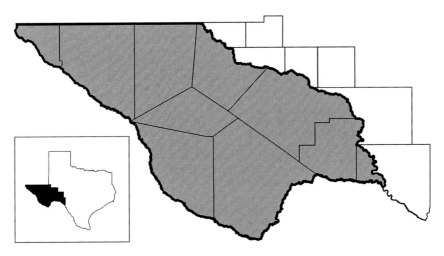

MAP 9. Distribution of the black-tailed jackrabbit, *Lepus californicus*

mechanisms of exceptional eyesight and escape speed are rendered ineffective (J. Jones et al. 1985).

Their diet consists mainly of grasses and forbs, although any green vegetation may be consumed. In periods of drought and during winter, they subsist on prickly pear pads and fruits, and the bark of mesquite and other desert shrubs. Water is acquired from consumed plants.

Jackrabbits are active throughout the year. They are most active at night, when they are commonly seen along roadsides and highways, but they can be active during morning and evening hours and on winter days. During the hotter part of the day, they usually crouch close to the ground, sleeping in a scratched-out depression (called a form) under shade. A similar shelter is employed during cold winter days where vegetation shields against prevailing winds. These behaviors, together with their long ears for radiating heat, are excellent adaptations for surviving in the desert.

The abundance of jackrabbits varies with different seasons and localities but seems to follow a wavelike pattern. After increasing for a few years until they are extremely numerous, they decline rather suddenly, are unusually scarce for a few years, and then gradually increase again. An unusually "high" population of jackrabbits was reported from the Trans-Pecos in May 1947 (Taylor 1948). A quantitative check of highways revealed that dead jackrabbits occurred at the rate of 3.8 per km in Brewster and Pecos Counties between US 90 and US 290 on US 67. Farther east in Pecos County (US 290), dead jackrabbits occurred at the rate of 5.2 per km

The breeding season extends throughout the year. During breeding, males will frequently fight by rearing up on their hind legs and "boxing" with their forefeet, and individuals may bite one another (Hoffmeister 1986). Two to six litters of up to six young per litter are produced annually. Two females taken in June in the Guadalupe Mountains contained five and three embryos, respectively (Genoways et al. 1979). A female examined on July 3 in BBRSP carried two embryos. Gestation is about 41–47 days. No nest is prepared for the young at birth; they are born precocial with open eyes, full fur, and distinctive black tails. Young are immediately mobile but uncoordinated for the first few days.

Known predators include coyotes, gray foxes, bobcats, badgers, owls, hawks, and eagles. They use speed and leaping agility to escape an attack. Rabbit hemorrhagic disease (RHDV2), a highly contagious viral disease, has recently been documented in jackrabbits in Hudspeth and El Paso Counties, and should this spread throughout the Trans-Pecos it could be devastating to populations. The subspecies that occurs in the Trans-Pecos is *L. c. texianus*.

Order Eulipotyphla (Shrews, Moles, and Relatives)

North American representatives of this group are relatively small animals with long, pointed noses and tiny, bead-like eyes. Most feed on insects and other invertebrates and lack a caecum. Only one species of shrew and one species of mole have been recorded within the boundaries of the Trans-Pecos, and they are among the rarer mammals of the area.

1.   Front feet broad and paddle shaped; eyes nonfunctional; ears not
            visible; total length > 150 mm . . . .
            . . . . **Eastern Mole,** *Scalopus aquaticus*
      Front feet normal, not paddle shaped; eyes small, but functional;
            total length < 90 mm . . . . **Crawford's Desert Shrew,**
            Notiosorex crawfordi

Family Soricidae
SHREWS
Crawford's Desert Shrew
Notiosorex crawfordi (Coues, 1877)

Crawford's desert shrew is a small grayish mammal with a long, pointed snout and tiny eyes (fig. 24). The ears are conspicuous and the tail is relatively long, more than two times as long as the hindfoot. Average external measurements

FIG. 24. Crawford's desert shrew, *Notiosorex crawfordi*

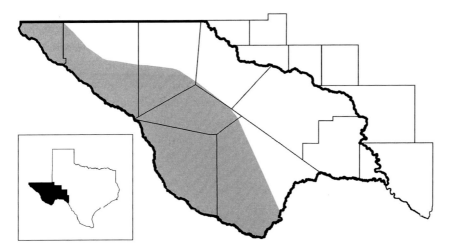

MAP 10. Distribution of Crawford's desert shrew, *Notiosorex crawfordi*

(in mm) for males and females, respectively, are total length 83.4, 86.5; tail 26.3, 27.5; hindfoot 10.1, 10.7; ear 6.4, 5.2 (Punzo 2003b).

Crawford's desert shrews have been recorded from western and southwestern parts of the Trans-Pecos, but they have yet to be recorded in the eastern and northern counties (map 10). They favor desert scrub vegetation and probably occur throughout the Trans-Pecos in this type of habitat. Although they are considered uncommon because they are difficult to collect by conventional methods, in the Big Bend region they are easily captured in pitfall traps used to obtain amphibians and reptiles (Punzo 2003b). Interestingly, they have not been recorded from El Paso County since the type specimen for the species was obtained there over 160 years ago.

Their diet is predominantly arthropods and their larvae. In BBNP, food items include crickets, grasshoppers, moths, beetles, and spiders, and, to a lesser degree, scorpions, solifuges, millipedes, centipedes, termites, ants, wasps, and true bugs and on rare occasions small vertebrates (Punzo 2003a, 2003b). Toxins in the saliva have been experimentally confirmed (Camargo and Álvarez-Castañeda 2019).

These shrews may be found in depressions under rocks, within rock piles, in association with decayed yucca, agave, and sotol, beneath brush and logs, and within woodrat middens. In BBNP, nests were constructed with dried grasses and leaves in decaying yucca and agave plants and in depressions under rocks (Punzo 2003a), and in Presidio County under decaying leaves of yucca and lechuguilla (Punzo and Lopez 2003). Nests may be found in both abandoned and occupied woodrat nests.

Most behavioral information is limited to captive shrews. They are known to construct and use latrines away from their sleeping nests (Hoffmeister and Goodpaster 1962), and they practice frequent dust bathing (Punzo 2003b). When Crawford's desert shrews are together, they possibly form a dominant-submissive hierarchy.

Breeding extends throughout the warmer months of the year, with typical litters of three to five young. Gestation ranges from 19 to 22 days (Punzo 2003a). Litters for 20 pairs of captives from BBNP averaged 2.9 young, with four females producing a second litter of 1.5 young (Punzo 2003a). At two sites in Presidio County, 12 females averaged 3.8 young and 8 females averaged 1.9 young (Punzo and Lopez 2003). Sexual maturity occurs between 43 and 46 days for males and between 42 and 44 days for females (Punzo 2003a).

Great horned owls and barn owls are known predators, and regurgitated owl pellets are one of the best sources of cranial material for records of these secretive shrews. *Notiosorex crawfordi* is a monotypic species, with no subspecies recognized (see Schmidly et al. 2023).

Family Talpidae
MOLES
Eastern Mole
Scalopus aquaticus (Linnaeus, 1758)

Eastern moles are small, fossorial mammals with a sharp-pointed nose and broadened, shovel-like front feet that are webbed to the base of the claws (fig. 25). The eyes and ears are not visible, and the tail is short and sparsely haired. Average external measurements (in mm) are total length 165; tail 29; hindfoot 22; weight is 60–90 g. The same measurements from the dried skin of the single specimen from the Trans-Pecos are 124, 24, and 16.5, respectively (Allen 1891).

FIG. 25. Eastern mole, *Scalopus aquaticus*

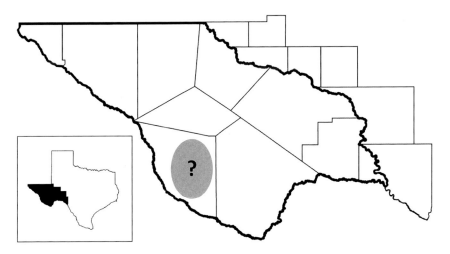

MAP 11. Distribution of the eastern mole, *Scalopus aquaticus*

The distribution of this species is difficult to explain with any degree of biological certainty (hence its designation in the checklist as "enigmatic"). Only a single specimen of *S. aquaticus*, obtained in 1887, has been reported from the Trans-Pecos, and the exact locality where it was obtained is unknown (map 11). The specimen was noted as being collected in Presidio County (Allen 1891, 1893), which at that time included the present counties of Brewster, Jeff Davis, and Presidio (Blair 1940). The lack of additional records suggests that if moles still occur in the Trans-Pecos, they are extremely limited in numbers. Because of their preference for moist, loamy or sandy soils, they would most likely occur in the alluvial soils associated with the Rio Grande, which are more suitable for the construction of their underground burrows. They are usually scarce or absent from heavy clayey, stony, or gravelly soils.

Interestingly, two specimens of eastern moles, belonging to the subspecies *S. a. montanus* (Yates and Schmidly 1977), have been discovered in the Sierra del Carmen south of the Trans-Pecos in Coahuila, Mexico (R. H. Baker 1951; McKinney 2012). These are the only other records of eastern moles from the Chihuahuan Desert.

Moles spend more than 99 percent of their life belowground in constructed tunnels. The eastern mole has a voracious appetite and consumes from 25 to 100 percent of its weight in food daily. Its diet consists of earthworms, grubs, and insects, although vegetable matter is occasionally eaten. This mole has but one litter of two to five young annually. Because of their underground lifestyle, eastern moles are somewhat protected from predators, but they may occasionally be captured by coyotes, foxes, badgers, and venomous snakes. The single specimen from the Trans-Pecos is referred to the subspecies *S. a. texanus*.

ORDER CHIROPTERA (Bats)

Bats are second only to rodents in numbers of species worldwide, and they are among the most common mammals in the Trans-Pecos. On warm evenings they can be seen in flight almost anywhere, but they are most often found feeding over a pond or stream, along a forest edge, beside a cliff or ravine, or among buildings. Many species seem to be attracted by water, especially in arid desert areas.

There are two rather broad categories of bats in the Trans-Pecos, "cavity-roosting" and "foliage-roosting" bats. Cavity-roosting bats tend to have drab brown coloration and seek refuge in dark (often aphotic) caves, tunnels, and other structures where colors are not advantageous, and many of them are migratory. Foliage-roosting bats often have variously colored pelage (red, yellow, tan, gray, often with whitish-tipped hairs or overtones) and roost in trees where colors and a "frosted" appearance help camouflage them among leaves, branches, and bark. Foliage-roosting bats may be hibernators or migrate, depending on the species.

Because of the great variety of species that have been recorded there, the Trans-Pecos is described as a "bat-rich" area. It is home to the highest number of bat species recorded in the United States, and 82 percent of the bat species that occur in Texas have been documented in the region. Four families, 19 genera, 27 species, and possibly 5 other species (see checklist in chapter 2) occur in the Trans-Pecos. The key below includes the 27 species documented from the region. Much of the natural history information has been adopted from

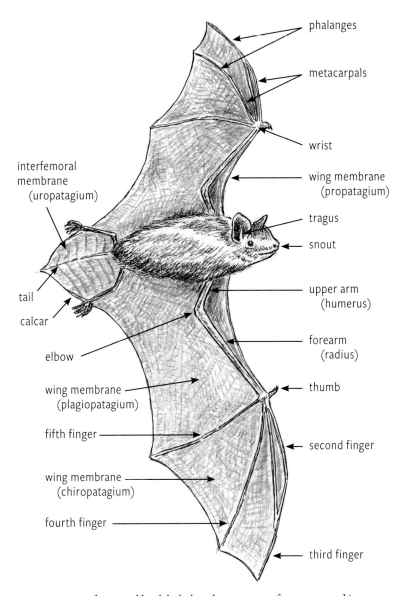

phalanges

metacarpals

wrist

wing membrane
(propatagium)

tragus

snout

upper arm
(humerus)

forearm
(radius)

thumb

second finger

third finger

interfemoral
membrane
(uropatagium)

tail

calcar

elbow

wing membrane
(plagiopatagium)

fifth finger

wing membrane
(chiropatagium)

fourth finger

FIG. 26. Anatomy of a typical bat labeled to show names of parts, as used in text

the two editions of *Bats of Texas* (Schmidly 1991; Ammerman et al. 2012), from David Easterla's 1973 publication on the bats of BBNP, and the book by Harvey, Altenbach, and Best (2011) about US and Canadian bats. The anatomy of a typical bat is presented in fig. 26 to illustrate the names of parts that are used in the text. Tables 4 and 5 contain information about the size variation and dental formulas, respectively, of all Trans-Pecos bats.

TABLE 4. Average body weight and wing measurements for Trans-Pecos bats

Species	Weight (grams)	Forearm length (mm)	Wingspan (mm)	Wing area (cm²)
Tadarida brasiliensis	11–14	36–46	308	110
Nyctinomops femorosaccus	10–18	44–50	345	132
Nyctinomops macrotis	22–30	58–64	426	178
Eumops perotis	65	72–82	550	322
Mormoops megalophylla	15–16	46–59	395	188
Choeronycteris mexicana	25	43–49	345	166
Leptonycteris nivalis	24	56–59	410	223
Leptonycteris yerbabuenae	25–27	55–60	375	158
Myotis occultus	7–9	36–41	260	86
Myotis yumanensis	4–8	32–38	240	77
Myotis velifer	12–15	39–46	298	131
Myotis thysanodes	6–11	40–45	282	125
Myotis volans	12–15	35–41	260	108
Myotis californicus	3–5	29–35	230	77
Myotis ciliolabrum	4–5	29–35	235	83
Lasionycteris noctivagans	8–12	37–44	290	104
Lasiurus borealis	10–15	35–45	315	134
Lasiurus frantzii	10–15	35–40	315	110
Aeorestes cinereus	20–35	46–58	395	180
Dasypterus xanthinus	12–19	45–48	345	203
Parastrellus hesperus	3–6	27–33	205	60
Perimyotis subflavus	3–6	31–35	235	81
Eptesicus fuscus	13–20	42–51	335	162
Nycticeius humeralis	5–8	33–39	270	107
Euderma maculatum	16–20	48–51	335	181
Corynorhinus townsendii	7–12	39–48	305	136
Antrozous pallidus	12–17	46–60	375	228

Source: Adapted from Schmidly (1991) and Ammerman et al. 2012; specimens used to determine figures obtained from throughout a species' range.

TABLE 5. Dental formulas for Trans-Pecos bats

Species	Upper teeth				Lower teeth				Total
	I	C	PM	M	I	C	PM	M	(×2)
Tadarida brasiliensis	1	1	2	3	3	1	2	3	32
Nyctinomops femorosaccus	1	1	2	3	2	1	2	3	30
Nyctinomops macrotis	1	1	2	3	2	1	2	3	30
Eumops perotis	1	1	2	3	2	1	2	3	30
Mormoops megalophylla	2	1	2	3	2	1	3	3	34
Choeronycteris mexicana	2	1	2	3	0	1	3	3	30
Leptonycteris nivalis	2	1	2	2	2	1	3	2	30
Leptonycteris yerbabuenae	2	1	2	2	2	1	3	2	30
Myotis occultus	2	1	2/3*	3	3	1	3	3	36/38*
Myotis yumanensis	2	1	3	3	3	1	3	3	38
Myotis velifer	2	1	3	3	3	1	3	3	38
Myotis thysanodes	2	1	3	3	3	1	3	3	38
Myotis volans	2	1	3	3	3	1	3	3	38
Myotis californicus	2	1	3	3	3	1	3	3	38
Myotis cIllolabrum	2	1	3	3	3	1	3	3	38
Lasionycteris noctivagans	2	1	2	3	3	1	3	3	36
Lasiurus borealis	1	1	2	3	3	1	2	3	32
Lasiurus frantzii	1	1	2	3	3	1	2	3	32
Aeorestes cinereus	1	1	2	3	3	1	2	3	32
Dasypterus xanthinus	1	1	1	3	3	1	2	3	30
Parastrellus hesperus	2	1	2	3	3	1	2	3	34
Perimyotis subflavus	2	1	2	3	3	1	2	3	34
Eptesicus fuscus	2	1	1	3	3	1	2	3	32
Nycticeius humeralis	1	1	1	3	3	1	2	3	30
Euderma maculatum	2	1	2	3	3	1	2	3	34
Corynorhinus townsendii	2	1	2	3	3	1	3	3	36
Antrozous pallidus	1	1	1	3	2	1	2	3	28

Source: Adapted from Schmidly (1991).

Note: Numbers are the number of teeth on each side of jaw. I = incisors, C = canines, PM = premolars, M = molars.

*Second premolar (PM) usually absent or minute; if present, crowded medially and not visible in lateral view.

1. Prominent triangular nose leaf freely projecting upward over nostrils; snout elongated; Phyllostomidae (New World leaf-nosed bats) 2

   No triangular nose leaf; snout not elongated. 4

2. Tail extending only 6–10 mm into interfemoral membrane; distance from eye to ear opening about half of distance from eye to nose leaf; forearm < 48 mm; rostrum of skull extremely long; no lower incisors . . . . **Mexican Long-tongued Bat,** *Choeronycteris mexicana*

   Tail not visible, reduced to a few vertebrae; interfemoral membrane constricted at midline of body; eye about equidistant between ear opening and nose leaf; forearm > 48 mm 3

3. Long, fluffy, grayish pelage; forearm > 55 mm; third finger > 104 mm; condylobasal length of skull > 27 mm . . . . **Mexican Long-nosed Bat,** *Leptonycteris nivalis*

   Short, dense, brownish pelage; forearm < 55 mm; third finger < 104 mm; condylobasal length of skull < 27 mm . . . . **Lesser Long-nosed Bat,** *Leptonycteris yerbabuenae*

4. Conspicuous grooves and flaps on chin; distal third of tail protruding from dorsal surface of interfemoral membrane; cranium extremely inflated; rostrum abruptly upturned; Mormoopidae (leaf-chinned bats) . . . . **Ghost-faced Bat,** *Mormoops megalophylla*

   No substantial grooves or flaps on chin; tail extending to or beyond edge of interfemoral membrane, not protruding from dorsal surface; cranium not inflated; rostrum not upturned 5

5. Distal third or more of tail free of interfemoral membrane, extending conspicuously beyond edge of membrane; Molossidae (free-tailed bats) 6

   Tail extending to edge of interfemoral membrane, or just barely beyond; Vespertilionidae (vesper bats) 9

6. Forearm > 70 mm; upper lips without vertical grooves; no premaxillary gap in bony palate; greatest length of skull > 30 mm . . . . **Western Bonneted Bat,** *Eumops perotis*

   Forearm < 70 mm; upper lips with distinct, deep vertical grooves; premaxillary gap in bony palate; greatest length of skull < 30 mm 7

7. Forearm > 52 mm; greatest length of skull > 21 mm . . . . **Big Free-tailed Bat,** *Nyctinomops macrotis*

   Forearm < 52 mm; greatest length of skull < 21 mm 8

ears not joined at base   ears joined at base

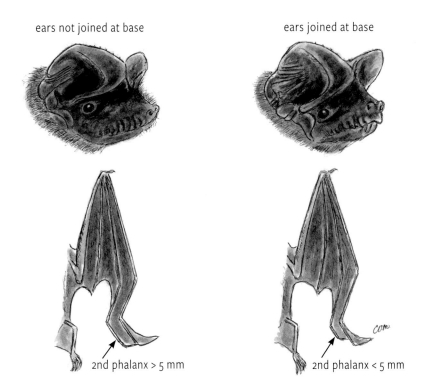

2nd phalanx > 5 mm   2nd phalanx < 5 mm

FIG. 27. Facial views (upper) and second phalanx of fourth finger (lower) of the wing of two free-tailed bats: left, *Tadarida brasiliensis*; right, *Nyctinomops femorosaccus*

8. Ears connected at base (fig. 27); second phalanx of fourth finger
       < 5 mm; greatest length of skull > 18 mm; incisors 1/2 ......
       .... **Pocketed Free-tailed Bat**, *Nyctinomops femorosaccus*
    Ears not connected at base (fig. 27); second phalanx of fourth finger
       > 5 mm; greatest length of skull < 18 mm; incisors 1/3 ....
       .... **Brazilian Free-tailed Bat**, *Tadarida brasiliensis*
9. Ears disproportionately large, > 25 mm from notch to edge 10
    Ears not extremely large, < 25 mm from notch to edge 12
10. Pelage black with three large white spots, one behind each shoulder
       and one on rump at base of tail ....
       .... **Spotted Bat**, *Euderma maculatum*
    Pelage without large white spots, color variable but not black 11
11. Distinctive glands (large bulges) wrapped around each side of snout
       between eyes and nose; dorsal color pale to dark brown; two upper
       incisors and two upper premolars on each side ....
       .... **Townsend's Big-eared Bat**, *Corynorhinus townsendii*

No distinctive glands present; dorsal color pale yellow; one
   upper incisor and one upper premolar on each side . . . .
   . . . . **Pallid Bat,** *Antrozous pallidus*

12. Anterior (or proximal) half (or more) of dorsal surface of
   interfemoral membrane well furred. 13
   Dorsal surface of interfemoral membrane naked, scantily haired,
   or at most lightly furred on anterior third 17

13. Pelage black overlaid with abundant silver-tipped hairs, giving
   upperparts a frosted appearance. . . . .
   . . . . **Silver-haired Bat,** *Lasionycteris noctivagans*
   Pelage color variable but never uniformly black 14

14. Pelage pale yellow; total teeth 30; one upper premolar on each side
   . . . . **Western Yellow Bat,** *Dasypterus xanthinus*
   Pelage reddish, brownish, or grayish; total teeth 32; two upper
   premolars on each side 15

15. Pelage wood brown and heavily frosted with white; forearm
   > 45 mm; greatest length of skull > 15 mm; lacrimal ridge absent
   . . . . **Hoary Bat,** *Aeorestes cinereus*
   Upperparts reddish to brownish with some to no frosted white;
   forearm < 45 mm; greatest length of skull < 15 mm; lacrimal
   ridge present 16

16. Pelage reddish to brick red with a frosted appearance resulting from
   numerous white-tipped hairs; band of reddish hairs on neck,
   chest, and abdomen below white tips; interfemoral membrane
   fully haired; greatest length of skull usually > 12.8 mm in males,
   > 13.0 mm in females . . . . **Eastern Red Bat,** *Lasiurus borealis*
   Pelage rusty red to brownish with little to no frosted appearance;
   band of dark brown to black hairs on neck, chest, and abdomen
   below lighter tips; posterior third of interfemoral membrane
   naked or only scantily haired; greatest length of skull usually
   < 12.8 mm in males, < 13.0 mm in females . . . .
   . . . . **Western (Desert) Red Bat,** *Lasiurus frantzii*

17. Tragus (fleshy projection within base of ear) short, blunt, and
   curved (fig. 28) 18
   Tragus long, straight, and pointed, or nearly so (fig. 28). 20

18. Forearm > 40 mm. . . . . **Big Brown Bat,** *Eptesicus fuscus*
   Forearm < 40 mm. 19

19. Forearm > 32 mm; interfemoral membrane naked; color brown
   . . . . **Evening Bat,** *Nycticeius humeralis*

broad
round
tragus

curved
blunt
tragus

straight
pointed
tragus

FIG. 28. Characters of three genera of vespertilionid bats, showing the tragus: left, *Eptesicus*; center, *Parastrellus*; right, *Myotis*

Forearm < 32 mm; anterior third of interfemoral membrane sparsely furred on dorsal surface; color drab yellow buff to smoke gray .... **American Parastrelle (Canyon Bat)**, *Parastrellus hesperus*

20. Dorsal pelage tricolored when parted (black at base, wide band of light yellowish brown in middle, tips darker); leading edge of wing membrane obviously paler than remaining membrane .... **American Perimyotis (Tricolored Bat)**, *Perimyotis subflavus*

Dorsal pelage bicolored or unicolored with no light band in middle; color of leading edge of wing membrane similar to remaining membrane 21

21. Calcar (cartilaginous heel spur) with well-marked keel (fig. 29) 22

Calcar without well-marked keel (fig. 29) 24

22. Forearm > 36 mm; foot > 8 mm long; underside of wing furred to elbow; pelage dark brown ....**Long-legged Myotis**, *Myotis volans*

Forearm < 36 mm; foot < 8 mm long; underside of wing not furred to elbow; pelage light brown to buff brown 23

23. Pelage on back with long, glossy, brownish tips; black mask usually noticeable; thumb > 4 mm long (including wrist, 8–8.5 mm); naked part of snout about 1.5 times width of nostrils when viewed from above; tail extending up to 4 mm beyond posterior edge of interfemoral membrane; gradual slope from rostrum to braincase; skull usually > 13.6 mm .... **Western Small-footed Myotis**, *Myotis ciliolabrum*

Hairs on back with dull reddish-brown tips; black mask indistinct; thumb < 4 mm long (including wrist, 6–7.5 mm); naked part of snout about same length as width of nostrils; tail not extending beyond posterior edge of interfemoral membrane; abrupt upward slope from rostrum to braincase; skull usually < 13.6 mm ... . **California Myotis**, *Myotis californicus*

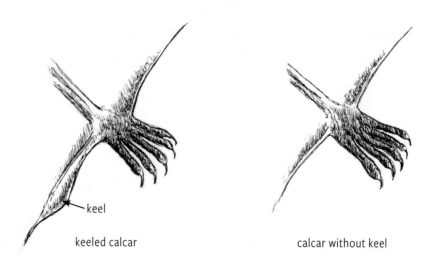

keel

keeled calcar   calcar without keel

FIG. 29. Characters of the calcar in vespertilionid bats

24. Forearm > 40 mm 25
    Forearm usually < 40 mm 26
25. Conspicuous fringe of stiff hairs on posterior edge of interfemoral
    membrane; sagittal crest not well defined; breadth across upper
    canines < 4.3 mm . . . . **Fringed Myotis,** *Myotis thysanodes*
    No conspicuous fringe of stiff hairs on posterior edge of interfemoral
    membrane; sagittal crest well defined; breadth across upper canines
    > 4.3 mm . . . . **Cave Myotis,** *Myotis velifer*
26. Dorsal pelage usually with a slight sheen; forearm > 36 mm; total
    length > 80 mm; second upper premolar minute and crowded
    inward, or often missing; gradual slope from rostrum to braincase
    . . . . **Southwestern Little Brown Myotis,** *Myotis occultus*
    Dorsal pelage usually lacking a sheen; forearm < 36 mm; total length
    < 80 mm; abrupt upward slope from rostrum to braincase
    . . . .**Yuma Myotis,** *Myotis yumanensis*

Family Molossidae
FREE-TAILED BATS
Brazilian Free-tailed Bat
Tadarida brasiliensis (I. Geoffroy Saint-Hilaire, 1824)

This medium-sized bat (forearm 36–46 mm; wingspan 290–325 mm) has broad
ears that do not join at the midline of the head, vertical wrinkles on the lips

along the muzzle, and dark brown to dark gray pelage (fig. 27). The individual hairs of the pelage are uniform in coloration, not bicolored. As in all molossid bats, the terminal half of its tail is free from the interfemoral membrane. In *Tadarida* the incisors are 1/3; other free-tails (*Nyctinomops* and *Eumops*) known from West Texas have incisors that are 1/2. Average external measurements (in mm) are total length 95; tail 34; hindfoot 9; ear 18; weight is 11–14 g. *Tadarida brasiliensis* is the smallest of the free-tailed bats in the Trans-Pecos (fig. 30) and may be confused with *N. femorosaccus*, from which it may be distinguished by its lack of a junction between the ears (fig. 27). Also, *T. brasiliensis* is the only free-tailed bat in which the hair is of uniform color.

Brazilian free-tailed bats are abundant statewide, and they have been recorded from every county in the Trans-Pecos (map 12). They are especially abundant in BBNP, where they seem to prefer riparian floodplain areas (Higginbotham and Ammerman 2002). They are not as abundant in BBRSP, CMSNA, or GMNP (Genoways et al. 1979; Yancey 1997; C. Jones et al. 2011), probably because there is less water and riparian floodplain habitat in these areas.

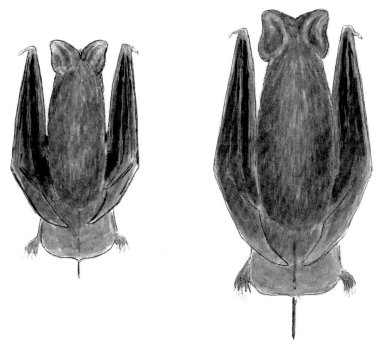

FIG. 30. Two species of free-tailed bats: left, *Tadarida brasiliensis*; right, *Nyctinomops macrotis*

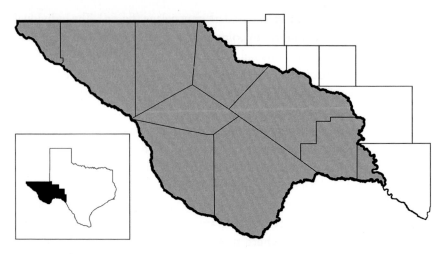

MAP 12. Distribution of the Brazilian free-tailed bat, *Tadarida brasiliensis*

In the Hill Country of Texas, these bats inhabit caves (referred to as "guano caves") and form huge maternity colonies that may number into the millions. Guano caves are absent from the Trans-Pecos, and these bats have been found primarily in abandoned railroad tunnels, crevices, culverts, bridges, and other human-made structures. A large colony of several hundred bats has been documented in a cement block barn near Canutillo, Texas, along the Rio Grande (Dooley 1974). Bats were located in numerous cracks and inside the walls of the barn, where they had established a roost. Another colony of note is at the Fort Leaton State Historic Site, east of Presidio off Farm to Market Road 170. Personnel at the historic site have been trying to eliminate the bats for some time; they are considered a nuisance because of their guano deposits and urine smell.

Most Brazilian free-tailed bats in the Trans-Pecos migrate southward in autumn and spend the winter in Mexico, although a few remain and are year-round residents. There are winter records of individuals, or small aggregates of these bats, hibernating in buildings or other structures, as well as a recurring, relatively large (several thousand individuals) colony that overwinters at the Fresno Creek Bridge in BBRSP (Kasper and Yancey 2018b). In addition, there are winter records from several other bridges in the Trans-Pecos (Stevens et al. 2021). Migratory bats returning from Mexico usually begin arriving in the Trans-Pecos in March or April and become absent after November.

Their reproductive biology in the Trans-Pecos is poorly understood. A few pregnant females have been taken in BBNP in June and from late April to early June in BBRSP, but they are thought to represent late migrants from Mexico. A maternity colony of several thousand individuals occurs at the Fresno Creek Bridge, where young are born and mature annually. This colony is unique in that adults migrate while the young of the year remain and over-winter at the bridge (Kasper and Yancey 2018b). Females typically produce a single young annually.

Their diet is made up of moths, ground beetles, weevils, flying ants, leaf chafers, leaf beetles, green blowflies, and other insect groups. They often feed on swarms of insects and are estimated to consume thousands of tons of insects per year in Texas. Known predators include raccoons, ringtails, opossums, striped and hog-nosed skunks, house cats, and a variety of snakes, owls, hawks, and falcons (Yancey et al. 1996; Roberts et al. 1997; Sparks et al. 2000). Freeman (1981) reviewed the systematics of the family Molossidae and proposed the tax-onomic arrangement used today (see account of *N. femorosaccus*). The subspe-cies in the Trans-Pecos is *T. b. mexicana*.

Pocketed Free-tailed Bat
Nyctinomops femorosaccus (Merriam, 1889)

This is a medium-sized free-tailed bat (forearm 44–50 mm; wingspan 330–360 mm) with long, narrow wings, vertical wrinkles on the lips along the muz-zle, and broad ears that are joined basally at the midline of the forehead (see fig. 27). Dorsal pelage varies from dark brown to gray, with bicolored individual hairs that are darkish above and whitish near the base. The common and sci-entific names of this bat refer to a shallow fold of skin on the underside of the uropatagium near the knee, which forms a pocket-like area. Average external measurements (in mm) are total length 113; tail 42; hindfoot 10; ear 20; weight is 10–18 g.

This bat was first reported from BBNP (Brewster County) in 1967 (Easterla 1968) and was thought to be rare, but further study has shown that it is a year-round resident and may be locally common at some sites in the park (Higginbotham and Ammerman 2002). The species has now been doc-umented in two additional counties of the Trans-Pecos, Presidio and Terrell (Ammerman et al. 2002; Higginbotham et al. 2002) (map 13).

Pocketed free-tailed bats inhabit both lowland and upland areas in the Chihuahuan Desert. In BBNP they have been netted in desert scrub and river

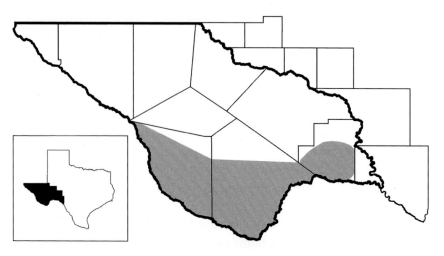

MAP 13. Distribution of the pocketed free-tailed bat, *Nyctinomops femorosaccus*

and floodplain arroyos as well as in habitats that feature rugged cliffs, caves, and rocky outcrops with many crevices, and occasionally in human-made structures (Higginbotham and Ammerman 2002). In BBRSP, they have occasionally been collected in desert scrub and riparian cottonwood associations (Higginbotham et al. 2002).

Pocketed free-tailed bats are colonial like other free-tailed bats, but they usually form aggregations of fewer than 100 individuals. Both males and females are thought to roost together, but in BBNP females have been found to outnumber males by a factor of two to one. These bats have been obtained in BBNP from January through November, suggesting they are year-round residents of the park. Individuals are mostly sedentary during winter, but activity inside the roost has been observed. During warmer months, bats typically leave the roost at or slightly before sunset, although there are reports of them leaving the roost much later in the evening, well after dark.

Females usually produce a single young during the summer months. Pregnant and lactating females have been captured from May through August, and volant young are known from August to November in BBNP. Their diet includes not only moths but crickets, flying ants, stinkbugs, beetles, froghoppers and leafhoppers, lacewings, beetles, and other insects (Easterla and Whitaker 1972). This bat is probably preyed on by a variety of snakes, raptors, and small mammals.

A comprehensive taxonomic study of the Molossidae revealed that the New World *Tadarida* species (excluding *T. brasiliensis*) are morphologically distinct from Old World species of *Tadarida* (Freeman 1981). Consequently, the generic name *Nyctinomops* has since been used for New World species formerly classified in the genus *Tadarida*. *Nyctinomops femorosaccus* is monotypic, and no subspecies are recognized.

Big Free-tailed Bat
Nyctinomops macrotis (Gray, 1839)

This is the largest species (forearm 58–64 mm; wingspan 417–436 mm) in the genus *Nyctinomops*, with adults weighing from 22 to 30 g (fig. 30). It is characterized by large, broad ears that are joined basally at the midline of the head and extend beyond the tip of the snout when laid forward, and vertical wrinkles on the lips along the muzzle (fig. 31); pelage is reddish brown to dark brown or gray and somewhat glossy. Individual hairs are bicolored, with basal portions white. Average external measurements (in mm) are total length 130; tail 54; hindfoot 12; ear 28; and weight 24–30 g.

Nyctinomops macrotis is an uncommon species that has been recorded from a few locations throughout the Trans-Pecos except for Terrell and Val Verde Counties in the southeast (map 14). It has been encountered most often in the mountainous regions of the Front Range (GMNP, Davis Mountains, and Chisos Mountains). Most other records are represented by single captures (e.g., Pecos, Reeves County; Tornillo, El Paso County; and BBRSP, Presidio County).

Big free-tailed bats inhabit primarily rugged, rocky country, where they apparently roost in crevices in high, rocky cliffs in both lowland and highland

 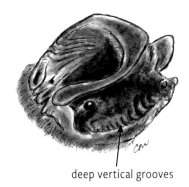

no vertical grooves   deep vertical grooves

FIG. 31. Facial view of the two largest free-tailed bats: left, *Eumops perotis*; right, *Nyctinomops macrotis*

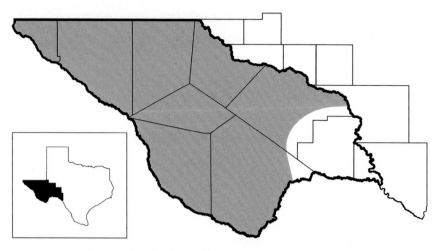

MAP 14. Distribution of the big free-tailed bat, *Nyctinomops macrotis*

habitats. In the Guadalupe Mountains this species was most common at elevations above 1,500 m (Genoways et al. 1979), whereas in BBNP it has been found in both lowland and upland areas. The records from the Davis Mountains are from upland sites with conifer trees and riparian areas dominated by cottonwoods (Bradley et al. 1999). An individual captured in the Fine Arts Building of Sul Ross State University in Alpine demonstrates that these bats will occasionally roost in buildings (Axtell 1961).

Big Bend National Park is the only place in the Trans-Pecos where this bat appears to be abundant. A nursery colony of about 130 bats was located on May 7, 1937, in the Chisos Mountains (Borell and Bryant 1942). The bats occupied a horizontal crevice about 600 cm long and 150 mm wide on the side of a sheer rock cliff about 12 m above a talus slope. Big free-tailed bats were also recorded from April through October at three separate locations in the river floodplain of the park (Higginbotham and Ammerman 2002). There are no winter records, and it is not known whether this species hibernates or migrates to some other area. These bats feed almost entirely on large moths, but crickets and long-horned grasshoppers are occasionally taken (Easterla and Whitaker 1972).

Little is known about their reproduction. Females typically produce a single young; pregnant females have been taken in June in the Guadalupe Mountains (Genoways et al. 1979), in May in BBRSP (Yancey 1997), and from May through September in BBNP (Higginbotham and Ammerman 2002). Few adult males have been taken, and evidently the sexes are segregated during summer when the young are being raised. Of 85 captures in BBNP, only 6 were males (Higginbotham and Ammerman 2002). Specific predators have not been reported, but this bat is most likely hunted by a variety of snakes, raptors, and small mammals.

The big free-tailed bat was formerly placed in the genus *Tadarida* but was later moved to the genus *Nyctinomops*, as explained in the previous account. The big free-tailed bat is monotypic, and no subspecies are recognized.

Western Bonneted Bat
Eumops perotis (Schinz, 1821)

The western bonneted bat is the largest bat in the United States (forearm 72–82 mm; wingspan 530–570 mm). Its large ears are united across its forehead and project about 10 mm beyond its snout (fig. 31). The lips are smooth along the muzzle and not wrinkled vertically. Pelage is dark gray dorsally and slightly paler below; individual hairs are bicolored, being nearly white at the base. Average external measurements (in mm) are total length 167; tail 57; hindfoot 17; ear 40; weight is 60–70 g.

These rare bats have been recorded in the Trans-Pecos only from Val Verde, Terrell, Brewster, and Presidio Counties and only at localities close to the Rio Grande (map 15). There is also an extralimital record from Midland County, just to the north of the region. Western bonneted bats are residents of arid habitats at lower elevations, usually near water. They prefer rugged, rocky canyons and cliffs where crevices provide their favored daytime retreats. Like other free-tailed bats, these bats are colonial, although their colonies are small, usually

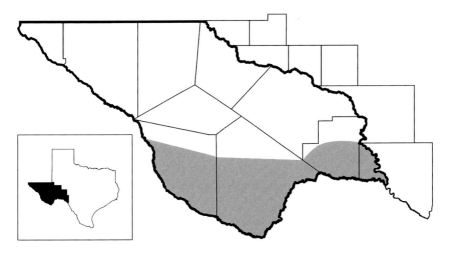

MAP 15. Distribution of the western bonneted bat, *Eumops perotis*

numbering fewer than 100 individuals. A single individual was collected at Fort Leaton State Historic Site in Presidio County (C. Jones and Lockwood 2008).

Harry Ohlendorf located and observed a colony of 71 of these bats in Presidio County. They were roosting in crevices in vertical cliffs on the western face of the Sierra Vieja at an elevation of approximately 1,130 m (Ohlendorf 1972). The entrance to the roost site was horizontal and faced downward so that the bats could leave the roost by simply dropping out. These bats leave their day roosts late and are not thought to use night roosts. Upon exiting, they presumably begin to forage, going as far as 10–25 km from roosting sites. Their diet is strictly insectivorous and includes moths, crickets, grasshoppers, dragonflies, bees, leaf bugs, beetles, and cicadas (Easterla and Whitaker 1972).

They are nonmigratory, nonhibernating, year-round residents of the Big Bend region. In BBNP, bats have been netted in floodplain habitat from the middle of March through the end of November (Higginbotham and Ammerman 2002). Female captures outnumbered males by a factor of three to one, and pregnant females were taken from late May to early August. In BBRSP, an adult male was captured in desert scrub habitat, and three individuals were obtained at Arroyo Segundo (Yancey 1997). Females usually produce a single young, but twins are known. Volant young have been recorded from August and September. Known predators include barn owls, peregrine falcons, American kestrels, and red-tailed hawks (Sparks et al. 2000). The subspecies in the Trans-Pecos is *E. p. californicus*.

Family Mormoopidae
LEAF-CHINNED BATS
Ghost-faced Bat
Mormoops megalophylla Peters, 1864

The ghost-faced bat is not likely to be confused with any other bat in the Trans-Pecos because of its unique body features and facial pattern. It is large (forearm 46–59 mm; wingspan 380–410 mm), with a pale brown to reddish to dark brown pelage and a pug face with leaflike appendages that stretch from ear to ear and are fused below the chin (fig. 32). Its tail is short and protrudes from the upper side of the interfemoral membrane. Average external measurements (in mm) are total length 90; tail 26; hindfoot 10; ear 14; weight is 15–16 g. Females tend to be slightly larger than males.

Mormoops megalophylla was first documented in the Trans-Pecos from BBNP in 1959 (Constantine 1961). Since then, it has been confirmed from the Apache (Stangl et al. 1994), Davis (DeBaca and Jones 2002), Chinati (C. Jones

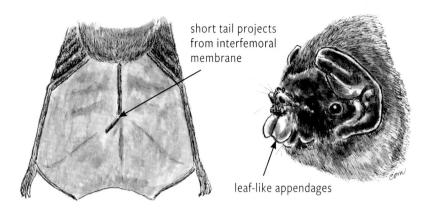

short tail projects from interfemoral membrane

leaf-like appendages

FIG. 32. Tail and face of the ghost-faced bat, *Mormoops megalophylla*

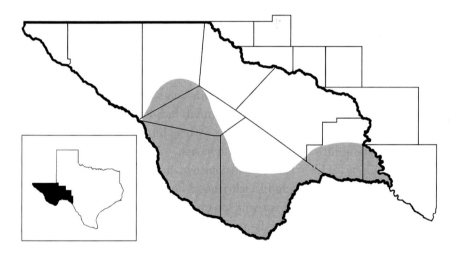

MAP 16. Distribution of the ghost-faced bat, *Mormoops megalophylla*

et al. 2011), and Chisos Mountains (Higginbotham and Ammerman 2002); EMWMA (Bradley et al. 1999); BBRSP (Yancey 1997); and other localities adjacent to the Rio Grande in Brewster, Presidio, and Terrell Counties (map 16).

Ghost-faced bats may be found in both lowland areas (flatlands) and upland areas (canyonlands), but they appear to be most common in desert scrub and riparian floodplain habitats with some water and little or no vegetation. Density of vegetation and physical topography appear to be important factors in determining their foraging sites (Yancey 2016). They are abundant in both BBNP and BBRSP in situations such as these (Higginbotham and Ammerman 2002; Yancey et al. 2006). In the Davis Mountains, they were netted at higher elevations along Limpia Creek in oak-juniper, montane grass habitat (DeBaca and Jones 2002).

This bat is a highly migratory, colonial, cave-dwelling species. Individuals have been reported only during the spring, summer, and autumn months. The species is apparently absent during winter, and females are far more prevalent in the region than males (Yancey 1997, 2016; Higginbotham and Ammerman 2002). Individuals have been seen foraging over slow-moving water, with peak foraging activity about two hours after sunset (Yancey 2016). Moths make up the majority of their diet along with beetles, flies, true bugs, and net-winged insects (Yancey 2016).

Pregnant ghost-faced bats arrive in the Trans-Pecos and set up nursery colonies in the spring. At BBNP pregnant females are known from April to July. Females give birth to a single young, with lactation and nursing reported from June to September. Volant young have been recorded in early September (Yancey 1997). No information is available on their specific predators, although they most likely include snakes, raptors, and small mammalian carnivores. One of us (FDY) observed two ringtails high on a sheer cliff scampering in and out of a cave that supported a large maternity colony of *M. megalophylla*. It is presumed the ringtails were foraging on debilitated juvenile bats that had fallen from their elevated roost. The subspecies in the Trans-Pecos is *M. m. megalophylla*.

Family Phyllostomidae
NEW WORLD LEAF-NOSED BATS
Mexican Long-tongued Bat
Choeronycteris mexicana Tschudi, 1844

This is a sooty-gray to brownish, relatively large bat (forearm 43–49 mm; wingspan 330–360 mm) with a long, slender muzzle, a prominent nose leaf, and a long, protrusible tongue. A minute tail is present and extends less than halfway to the edge of the interfemoral membrane (fig. 33). The rostrum is longer than the length of the braincase, and the skull lacks zygomatic arches. Average external measurements (in mm) are total length 85; tail 10; hindfoot 14; ear 16; weight is about 25 g. Females average slightly larger than males.

This is a predominantly Mexican species. Its range in the United States extends from extreme southern California eastward through southeastern Arizona to extreme western Texas. It is known from the Trans-Pecos based on a single specimen from El Paso County (Balin 2009). (map 17).

These bats often form small colonies (fewer than 50 individuals) in areas associated with steep canyons and caves. They occur in a variety of habitats

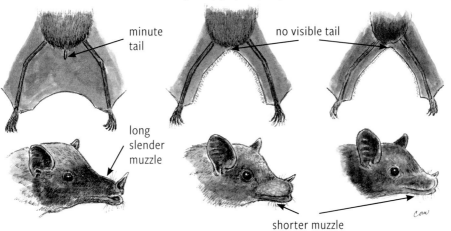

long fluffy grayish pelage

short dense brownish pelage

minute tail

no visible tail

long slender muzzle

shorter muzzle

FIG. 33. Interfemoral membranes and facial views of three species of long-nosed bats: left, *Choeronycteris mexicana*; center, *Leptonycteris nivalis*; right, *Leptonycteris yerbabuenae*

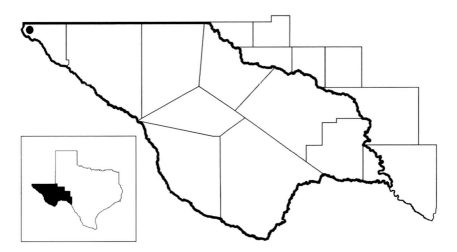

MAP 17. Distribution of the Mexican long-tongued bat, *Choeronycteris mexicana*

ranging from desert scrub to oak-juniper woodlands. They typically roost in dimly lit areas near the mouth of caves and abandoned mines. If disturbed, they retreat to the dark, or aphotic, zone of the cave.

Their diet includes nectar, pollen, some fruits, and incidental numbers of small insects. It is thought that they migrate seasonally along corridors with a rich diversity of agave and cacti, which are their primary food source (Burke et al. 2019). They can hover while feeding and are known to aid in pollination of night-blooming plants (Findley 1987).

Little is known about their reproduction. A single offspring is usually produced. In Arizona and New Mexico parturition occurs in June and July. There are observations of a female giving birth in Arizona. The event lasted about 15 minutes as the female hung her head down from a vertical surface. The neonate was well furred and appeared to be in an advanced stage of development. Females are known to carry their rather large young while foraging.

The Mexican long-tongued bat is rare in the Trans-Pecos. It does not appear on federal or state lists of species of concern in the United States, but it is listed as endangered in Mexico. This is a monotypic species, and subspecies are not recognized.

Mexican Long-nosed Bat
Leptonycteris nivalis (Saussure, 1860)

This is a relatively large (forearm 56–59 mm; wingspan 410 mm), sooty-brown bat with a long muzzle, prominent nose leaf, and long, protrusible tongue. The tail is minute and appears to be lacking but actually consists of three vertebrae, and the interfemoral membrane exists as a narrow fringe along the inside of each leg (fig. 33). *Choeronycteris mexicana* is similar in appearance but has a broader interfemoral membrane, a distinct tail, and a longer, narrower muzzle. Average external measurements (in mm) are total length 83; tail 0; hindfoot 17; ear 15; weight is about 24 g.

This is a predominantly Mexican species. In the Trans-Pecos it is known to occur only in Brewster and Presidio Counties in the Big Bend region, where it is a summer and spring resident (map 18). It was first reported in the Trans-Pecos from BBNP (Borell and Bryant 1942). Later, it was documented from the Chinati Mountains in Presidio County (Mollhagen 1973). The only known nursery colony is in the Chisos Mountains (Emory Peak Cave) in BBNP, and numbers at the colony are known to vary between years.

This species typically forms small colonies (fewer than 500 individuals) in caves and abandoned mines. It is found at midelevations (between 1,000 and 2,000 m) and inhabits deserts, pine-oak forests, and areas in between. This bat is relatively rare throughout its range, and there are indications of substantial population decline. Several caves in central Mexico that housed considerable numbers of these bats in the past now contain only small colonies or lack bats altogether. This species is considered endangered and is on both federal and state lists of species of concern in the United States.

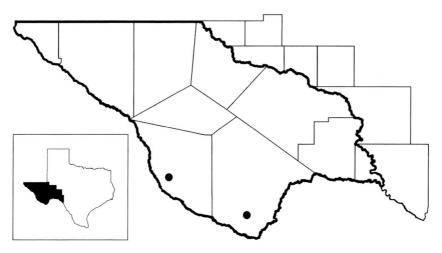

MAP 18. Distribution of the Mexican long-nosed bat, *Leptonycteris nivalis*

Its diet is primarily nectar and pollen of night-blooming plants such as century plants, some morning-glories, silk trees, columnar cacti, and agaves (Burke et al. 2019). These bats can hover while feeding, like hummingbirds, and are known to aid in pollination of night-blooming plants.

The breeding season is April to June, and females give birth to a single pup. Parturition usually occurs in Mexico before this migratory species enters the Trans-Pecos, but two pregnant females have been obtained in BBNP on April 25 at Glenn Springs, a desert scrub site at the southern edge of the Chisos foothills. During July, adults of both sexes and young of the year have been reported from Emory Peak Cave at BBNP. This is a monotypic species, and subspecies are not recognized.

Lesser Long-nosed Bat
Leptonycteris yerbabuenae (Martínez and Villa-R., 1940)

This large leaf-nosed bat (forearm 55–60 mm; wingspan 370–380 mm) is similar in appearance to *L. nivalis* but differs in having a brownish dorsal pelage that is shorter and denser (fig. 33). The head and body length is approximately 10 percent less than in *L. nivalis*. Average external measurements (in mm) are total length 81; tail 0; hindfoot 16; ear 18; weight is 25–27 g.

The lesser long-nosed bat was unrecorded from Texas until an adult female was received on October 19, 2010, by the DSHS in El Paso, El Paso County, for rabies testing (map 19). The specimen, a female that tested negative, was

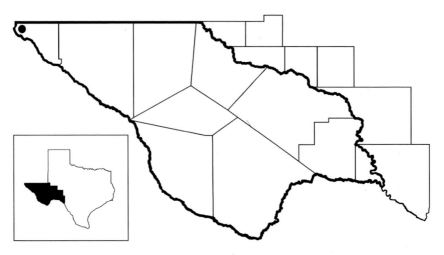

MAP 19. Distribution of the lesser long-nosed bat, *Leptonycteris yerbabuenae*

placed in the mammal collection at the Museum of Arid Land Biology at the University of Texas at El Paso until it was discovered and reported in the scientific literature 10 years later (Krejsa et al. 2020). At the present time, this is the only known specimen of this rare bat from Texas. It is thought to represent an extralimital distribution record and is not indicative of an established population (Krejsa et al. 2020).

This is a migratory species and only a seasonal resident of southwestern New Mexico, southern Arizona, and southern California, where it is found in semiarid to arid grasslands, scrubland, and oak woodlands. These bats winter in Mexico, where they occupy arid grasslands, tropical thorn scrub, and deciduous and/or pine-oak forests. Their primary food is nectar from agaves and columnar cacti, but they seem to specialize on the latter (Burke et al. 2019). They roost in large, densely packed colonies, typically with several thousand individuals. Females produce a single young each year, with parturition and lactation in June–July.

Its diet is primarily nectar and pollen of night-blooming plants such as century plants, some morning-glories, silk trees, columnar cacti, and agaves (Burke et al. 2019). These bats can hover while feeding, like hummingbirds, and are known to aid in pollination of night-blooming plants.

Until recently, *L. yerbabuenae* was listed as "endangered" across its range in the United States and Mexico. However, thanks to the efforts of a Mexican mammalogist, Rodrigo Medellin, and some tequila agave growers, the species has made a remarkable recovery and was recently delisted as an endangered species. This is a monotypic species, and subspecies are not recognized.

Family Vespertilionidae
VESPER BATS
Southwestern Little Brown Myotis
Myotis occultus Hollister, 1909

This is a medium-sized *Myotis* (forearm 36–41 mm; wingspan 245–275 mm) with large feet and relatively small ears. The pelage has a slight sheen and is tan to dark brown dorsally, with somewhat paler underparts (fig. 34). The calcar is not keeled. The ears when laid forward extend > 2 mm beyond the nose. The skull appears low and flattened in profile, with a broad, enlarged rostrum that

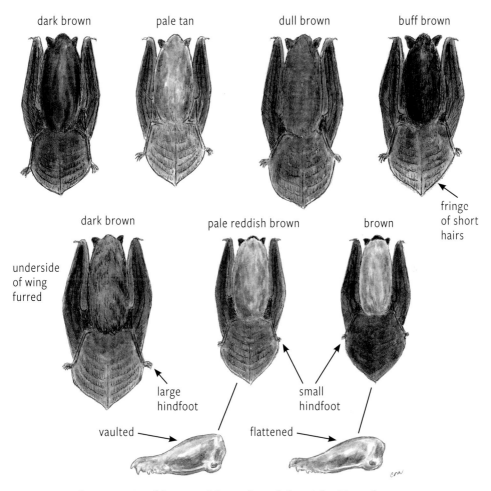

FIG. 34. Seven species of the genus *Myotis*: above, left to right, *M. occultus*, *M. yumanensis*, *M. velifer*, and *M. thysanodes*; below, left to right, *M. volans*, *M. californicus*, and *M. ciliolabrum*

TABLE 6. Trenchant morphological characters useful in distinguishing Trans-Pecos species of *Myotis*

Species	General size (forearm length, mm)	Pelage coloration	Ear size (typical length, mm)	Hindfoot (typical length, mm)	Unique hair on body	Keel on calcar	Elevation from rostrum to braincase	Sagittal crest
M. occultus	Medium (34–41)	Shiny, dark brown	Small (11–15)	Large (8–10)	N/A	None	Gradual	Distinct
M. yumanensis	Small (32–38)	Light, buff brown	Small (12–14)	Large (9–11)	N/A	None	Abrupt	None
M. velifer	Large (39–47)	Dark, dull	Intermediate (12–17)	Large (8–11)	Bare patch on back between scapulae	None	Moderate	Distinct
M. thysanodes	Large (40–45)	Buff brown	Large (16–20)	Large (9–12)	Thick fringe of hair on trailing edge of tail membrane	None	Moderate	Slight
M. volans	Medium (35–41)	Dark brown	Small (12–15)	Large (8–10)	Underside of wing furred to elbow	Well developed	Abrupt	None
M. californicus	Small (29–36)	Bright pale reddish brown	Small (9–13)	Small (6–8)	N/A	Well developed	Abrupt	None
M. ciliolabrum	Small (29–36)	Buff brown	Small (12–15)	Small (6–8)	N/A	Well developed	Gradual	None

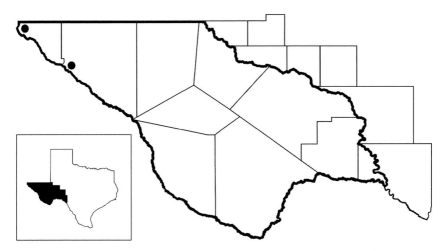

MAP 20. Distribution of the southwestern little brown myotis, *Myotis occultus*

gradually rises to the braincase. The molar teeth are large and the premolars show a tendency toward crowding, with most individuals possessing only two upper premolars. The second upper premolar is often missing, and if present it is reduced and crowded out of the tooth row so that it is usually not visible in side view. The lower second premolar is usually present but crowded inwardly. Average external measurements (in mm) are total length 85; tail 35; hindfoot 9; ear 13; weight is 7–9 g. Females are slightly larger than males.

This bat may be confused with the Yuma myotis (*M. yumanensis*), which is a smaller species with pale brown to yellowish pelage and small, lighter-colored ears (table 6). The cave myotis (*M. velifer*) is larger and usually darker than *M. occultus* (fig. 34). The California myotis (*M. californicus*) and the western small-footed myotis (*M. ciliolabrum*) are both smaller and have a keeled calcar. The long-legged myotis (*M. volans*) is also similar but has a keeled calcar, and the underside of the wing is furred to the elbow and knee.

Myotis occultus is known from only two specimens in far West Texas (map 20). It was first recorded in 1893 from Fort Hancock in Hudspeth County. A second record has been reported from El Paso County (Krejsa et al. 2020). The identity of this specimen, a female obtained in August 2011, was confirmed using molecular analysis.

The southwestern little brown myotis is considered an extralimital species in the Trans-Pecos. It is possible these bats are simply migrants passing through the area, and it seems doubtful there is an established resident population in the region. *Myotis occultus* is primarily a cavity-roosting bat, using crevices in rocks, caves, and tunnels and beneath tree bark for shelter. This bat is most often associated with high-elevation montane habitats composed of mixed conifers in New Mexico and Arizona. Like other myotis species, *M. occultus* is an aerial insectivore that feeds on a variety of insects including moths, flies, mosquitoes, and true bugs.

This bat has a complex taxonomic history. Most recently it was a subspecies of the little brown myotis (*Myotis lucifugus*; Findley and Jones 1967), and it has gone back and forth either as a subspecies of *M. lucifugus* or as a separate species (Findley and Jones 1967; Hoffmeister 1986). A molecular analysis of its taxonomic status has confirmed *M. occultus* as a distinct species (Piaggio et al. 2002). It is monotypic and there are no described subspecies.

Yuma Myotis
Myotis yumanensis (H. Allen, 1864)

This is a small (forearm 32–38 mm; wingspan 235–245 mm), pale brown myotis with relatively large feet, short ears that extend < 2 mm beyond the nose when laid forward, and an unkeeled calcar (fig. 34). The skull lacks a well-developed sagittal crest, and the short rostrum rises abruptly to the braincase when viewed in profile. Other myotis species sympatric with it include *M. velifer*, which is larger and darker in coloration; *M. volans*, *M. californicus*, and *M. ciliolabrum*, all of which have keeled calcars; and *M. occultus*, which has tan to dark brown, glossy fur (table 6). Average external measurements (in mm) are total length 80; tail 34; hindfoot 9; ear 13; weight is 4–8 g. Females are larger than males.

The Yuma myotis is a common summer resident along the southern tier of counties in the Rio Grande corridor, but it has not yet been recorded from the northern part of the Trans-Pecos (map 21). Its winter status is poorly documented. Two individuals were netted on November 26 at BBNP, suggesting they possibly hibernate in the park (Higginbotham and Ammerman 2002). In addition to caves, they are known to roost in and use human-made structures such as buildings, bridges, and tunnels. These bats have been taken in both upland and lowland habitats but seem to prefer lower-elevation areas near water. At BBRSP they were captured in small numbers from mid-June to mid-August. Two bats were netted over water in riparian habitat with cottonwood and willows nearby, and another was found in a day roost in an abandoned building near Whitroy Mine in desert scrub–creosote bush habitat (Yancey 1997). Multiple captures are known from BBNP, in river floodplain habitat over open water and in areas with crowded vegetation and canyon walls (Higginbotham and Ammerman 2002).

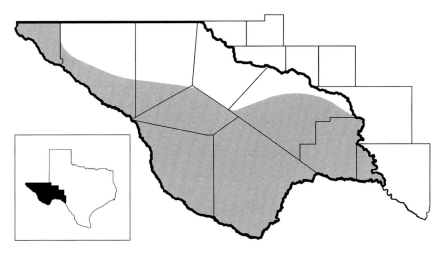

MAP 21. Distribution of the Yuma myotis, *Myotis yumanensis*

This species is highly colonial and, like many other bats, forms nursery colonies. In June 1965, one such colony was located in an old railroad tunnel on the east bank of the Pecos River in Val Verde County. Young bats as well as lactating adult females were found roosting in small clusters of 5 to 25 individuals on the walls of the tunnel (Schmidly 1977b). Two males were found roosting within a large cluster of Brazilian free-tailed bats under the Fresno Creek Bridge in BBRSP (Kasper and Yancey 2018b).

Their reproductive biology is poorly known. They are thought to mate in autumn, undergo delayed fertilization, and produce a single young in the spring. No pregnant or lactating females were caught during a baseline inventory of bats at BBRSP (Yancey 1997); however, during the nighttime on May 12 two of us (FDY and SK) observed three unattended infants estimated to be less than one week old clinging to the underside of the Madera Creek Bridge in the park. The Yuma myotis is a food generalist, with moths and beetles forming the bulk of its diet. Reported predators include bobcats and house cats (Sparks et al. 2000). The subspecies in the Trans-Pecos is *M. y. yumanensis*.

Cave Myotis
Myotis velifer (J. A. Allen, 1890)

This is the largest species of *Myotis* in the Trans-Pecos (forearm 39–46 mm; wingspan 280–315 mm). It occurs sympatrically with several other species of *Myotis*, but its larger size; dark, dull brown coloration; and lack of a keeled calcar serve to distinguish it from the others (fig. 31; table 6). The skull has a well-developed

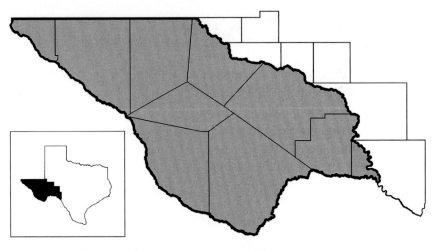

MAP 22. Distribution of the cave myotis, *Myotis velifer*

sagittal crest and a short rostrum that rises moderately to the braincase when viewed in profile. Average external measurements (in mm) are total length 94; tail 42; hindfoot 10; ear 15; weight is 12–15 g. Females are larger than males.

The cave myotis is a common and widespread resident of the Trans-Pecos (map 22). Previously thought to be absent from Pecos County, it has now been recorded from there as well as from Ward County just east of the Pecos River (Halsey et al. 2018). There is only one report from the winter months—a single individual captured in Presidio County on February 26 (Yancey and Jones 1996). Throughout its range, *Myotis velifer* is a colonial cave-dwelling species with colonies that often number into the thousands. These bats also are known to roost in and use human-made structures, such as buildings and tunnels, in addition to caves. They are common inhabitants of bridges in the Trans-Pecos during nonwinter months (Stevens et al. 2021). At BBRSP, the cave myotis was netted over water near thick vegetation between May and September and is considered uncommon rare in the state park (Yancey 1997), whereas it appears to be more abundant in CMSNA, where several individuals were captured in ciénaga wetland habitat and within the porch area of an old cabin that bats were using as a night roost (C. Jones et al. 2011). At BBNP, it has been recorded at five localities, most commonly in canyons containing confined water in desert scrub habitats (Higginbotham and Ammerman 2002).

They are thought to mate in autumn or winter, with copulation occurring during periods of arousal from hibernation. Following delayed fertilization, a single young is produced in the spring or early summer. In BBNP, pregnant females have been recorded in May and June (Higginbotham and Ammerman

2002). Lactating females have been observed in June and July in BBNP and BBRSP, respectively (Yancey 1997; Higginbotham and Ammerman 2002).

Myotis velifer is an aerial insectivore with two separate feeding periods each night—one begins shortly after sunset and the second just prior to sunrise before the bats enter their daytime roost. Information on specific food items is lacking, but in Arizona moths are a major part of their diet. Reported predators include barn owls, rough-legged hawks, American kestrels, gray foxes, ringtails, striped skunks, raccoons, and southern plains woodrats (Sparks et al. 2000). The subspecies in the Trans-Pecos is *M. v. incautus*.

Fringed Myotis
Myotis thysanodes G. S. Miller, 1897

This medium-sized *Myotis* (forearm 40–45 mm; wingspan 265–300 mm) is the most easily distinguished species of *Myotis* in the Trans-Pecos. A conspicuous fringe of short hairs lines the free edge of the interfemoral membrane, from which the species gets its common name (fig. 34). Pelage is buff brown above and dull white below, and, compared with those of other myotis species, the ears are long (extending 3–5 mm beyond the nose when laid forward) and the feet are large. The calcar is robust but lacks a distinct keel. The skull is delicate, with a short rostrum that rises moderately to the braincase when viewed in profile (table 6). Average external measurements (in mm) are total length 86; tail 35; hindfoot 9; ear 17; weight is 6–11 g. Females average larger than males.

The fringed myotis is a common spring, summer, and autumn resident of the Trans-Pecos, where it occurs across most of the western half of the region (map 23). It has been documented in El Paso, Hudspeth, Culberson, Jeff Davis, Presidio, and Brewster Counties but is apparently absent from the lowlands of the Pecos River drainage in Reeves, Pecos, Terrell, and Val Verde Counties. Specific areas where it has been found include BBNP (Higginbotham and Ammerman 2002), GMNP (Genoways et al. 1979), CMSNA (C. Jones et al. 2011), and BBRSP (Yancey 1997).

This is a cavity-roosting bat that occupies crevices in rocks and caves, beneath tree bark, and occasionally in human-made structures such as buildings, tunnels, highway culverts, and bridges (Stevens et al. 2021). It is known primarily from grasslands at intermediate elevations, but it can also be found in oak, piñon-juniper, pine, and Douglas fir habitats at higher elevations as well as in desert scrub lowland areas. Only two fringed bats were obtained (one each in June and July) in BBRSP (Yancey 1997); both were netted over small pools of water in canyons.

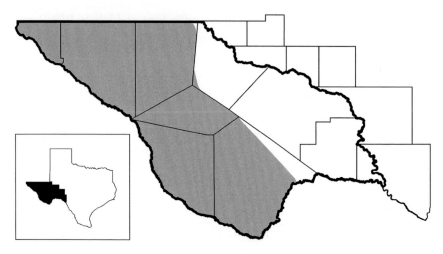

MAP 23. Distribution of the fringed myotis, *Myotis thysanodes*

Mating occurs during autumn in temperate regions, and sperm is stored in the female reproductive tract over winter. Females may form maternity colonies with hundreds of individuals the following spring. Ovulation, fertilization, gestation, and parturition occur in the spring. Females typically give birth to one young during May or June, and young bats are volant by mid-July. *Myotis thysanodes* is an aerial insectivore thought to feed on a diet of mostly beetles, although other flying insects have been reported in its diet as well. The subspecies in the Trans-Pecos is *M. t. thysanodes*.

Long-legged Myotis
Myotis volans (H. Allen, 1866)

The long-legged myotis is a rather large brown myotis (forearm 35–41 mm; wingspan 250–270 mm) that can be distinguished by a combination of features including a keeled calcar, relatively long tail, short ears that just reach the nose when laid forward, and large feet (fig. 34; table 6). Most importantly, the underside of the wing membrane is lightly furred to an imaginary line connecting the elbow with the knee. The skull has a short rostrum that rises abruptly to the braincase when viewed in profile, giving the face a "pug-nose" look. Average external measurements (in mm) are total length 94; tail 42; hindfoot 10; ear 15; weight is 12–15 g.

This is a common spring, summer, and autumn resident of the Trans-Pecos, with most records from the central part of the region in the Guadalupe (Genoways et al. 1979), Davis (Bradley et al. 1999), and Chisos Mountains (Higginbotham and Ammerman 2002) (map 24). These bats prefer forests and

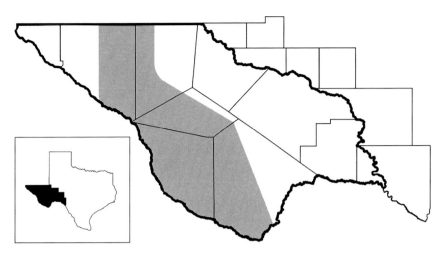

MAP 24. Distribution of the long-legged myotis, *Myotis volans*

open woodlands on rugged terrain at higher elevations (Higginbotham et al. 2002; C. Jones et al. 2011). Because of the lack of suitable montane habitat, this species has not been recorded in BBRSP (Yancey 1997). Interestingly, a record from the Indio Mountains, obtained over a small pool surrounded by cattails, was from 1,280 m, which is low for a species that normally occurs at elevations between 2,000 and 3,000 m (Higginbotham et al. 2002).

These bats are known to roost in and use human-made structures such as buildings, tunnels, and bridges. Apparently, they use caves as roosting sites at night, but not during the daytime. They are thought to hibernate in caves and tunnels, but there are no winter records from the region. Parturition occurs in early summer, with females normally producing a single young. *Myotis volans* is an insectivorous bat and likely feeds on moths and other flying insects. Some bat biologists affectionately refer to the long-legged myotis as the "woolly bear" because of its hairy underwings that extend from elbow to knee. The subspecies in the Trans-Pecos is *M. v. interior*.

California Myotis
Myotis californicus (Audubon and Bachman, 1842)

This is one of the smallest (forearm 29–36 mm; wingspan 215–245 mm) species of *Myotis* in the Trans-Pecos. Coloration of its upperparts is bright and light red-dish brown in contrast to its black ears. A distinctive dark facial mask extends across the eyes and rostrum from ear to ear. The calcar is keeled, and compared with other species of *Myotis*, the feet are tiny and the tail and ears relatively

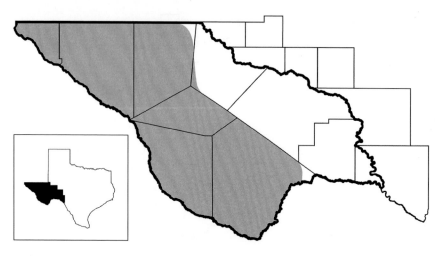

MAP 25. Distribution of the California myotis, *Myotis californicus*

long (table 6). This bat is easily confused with the western small-footed myotis (*M. ciliolabrum*), from which it differs in having a sharply rising braincase, in contrast to the flattened skull of *M. ciliolabrum* (fig. 34). This cranial feature gives the California myotis a more prominent forehead than is evident in the western small-footed myotis (Bogan 1974). This species differs from *M. volans* in not having ventral surfaces of the membranes furred to the level of a line joining the elbow and knee (table 6). Females tend to be slightly larger than males. Average external measurements (in mm) are total length 82; tail 36; hindfoot 6; ear 13; weight is 3–5 g.

The California myotis is a common year-round resident of the western half of the Trans-Pecos in El Paso, Hudspeth, Culberson, Jeff Davis, Presidio, and Brewster Counties (map 25). It is primarily a cavity-roosting bat (often occupying crevices in rocks and beneath tree bark), and it may occasionally occupy human-made structures such as buildings, tunnels, and construction features (e.g., culverts). This bat is known from deserts, grasslands, and wooded areas (Higginbotham et al. 2002). At BBRSP, it was most frequently captured near creeks and springs in closed-canopy vegetation associated with riparian areas as well as open desert scrub or grassland areas (Yancey 1997). The single record from the Indio Mountains was netted over a small pool surrounded by dense cattails on June 18 (Higginbotham et al. 2002). Records from BBNP are from April through August (Higginbotham and Ammerman 2002).

Females may form small maternity colonies (usually fewer than 25 individuals) during the breeding season. In temperate regions mating occurs in autumn, and sperm is stored in the female reproductive tract over the winter. Ovulation, fertilization, gestation, and parturition occur in the spring. In BBNP, pregnant females have been recorded in April and May; lactating females from June and July; and young volant bats from July and August (Higginbotham and Ammerman 2002). Reproductive data from BBRSP revealed the capture of pregnant females on May 29 and June 4; lactating females on June 6, 7, and 9, July 15 and 25, and August 2 and 3; and volant young on August 3 (Yancey 1997).

This bat is an aerial insectivore thought to feed on a variety of insects including moths, beetles, flies, and true bugs. It is one of the best "desert-adapted" bat species because its urine-concentrating ability conserves water, an important adaptation for any metabolically active tiny bat (Geluso 1978). The subspecies in the Trans-Pecos is *M. c. californicus*.

Western Small-footed Myotis
Myotis ciliolabrum (Merriam, 1886)

This small myotis (forearm 29–35 mm; wingspan 225–245 mm) has buff-brown pelage and black ears that extend about 1 mm beyond the nose when laid forward. It lacks the reddish tone of *M. californicus*. The calcar is keeled, and a dark facial mask is present across the rostrum and eyes from ear to ear (table 6). This bat is easily confused with *M. californicus*, from which it differs as described in the previous account (fig. 34). Average external measurements (in mm) are total length 79; tail 37; hindfoot 7; ear 13; weight is 4–5 g. Females are slightly larger than males.

This myotis is an uncommon summer resident of the Trans-Pecos. It has been recorded from El Paso, Culberson, Jeff Davis, Presidio, and Brewster Counties and is biogeographically projected to range throughout most of the western and southern parts of the region (map 26). It is a migratory, cavity-roosting species that occupies caves, rock crevices, and locations beneath tree bark. It has been recorded from montane as well as lowland wooded areas (Higginbotham et al. 2002) and occasionally in human-made structures such as buildings. At BBRSP, a single individual was taken over water in a small canyon in an area with sparse vegetation, but the species is uncommon in the state park because of the lack of suitable montane habitat (Yancey 1997).

Females usually breed in the autumn, store sperm in the reproductive tract over winter, and then ovulate and become gravid in the spring. Females give birth to a single young from May to July. The diet is thought to consist of moths, flies, beetles, true bugs, and other insects. In areas where it occurs with *M. californicus*, the two species segregate while feeding, with *M. ciliolabrum*

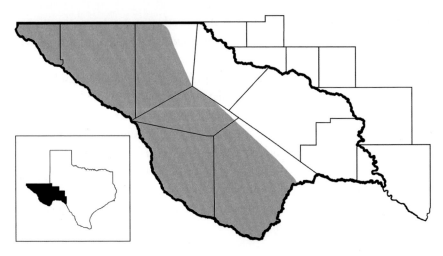

MAP 26. Distribution of the western small-footed myotis, *Myotis ciliolabrum*

foraging over rougher, desertlike arroyos and on different prey species than *M. californicus*.

This bat has a rather confusing and convoluted taxonomic history. For years it was classified as a subspecies of *M. subulatus*, then aligned with *M. leibii*, and finally recognized as a full species, *M. ciliolabrum*. The subspecies in the Trans-Pecos is *M. c. ciliolabrum*.

Silver-haired Bat
Lasionycteris noctivagans (Le Conte, 1831)

This is a medium-sized bat (forearm 37–44 mm; wingspan 270–310 mm) that is easily recognized by its striking black fur tipped with silver, giving the pelage a frosted appearance, and by the dense fur on the dorsal surface of the inter-femoral membrane (fig. 35). Other bats having similar "frosted" pelage are the hoary bat (*Aeorestes cinereus*), which is much larger and grayer, and the eastern (*Lasiurus borealis*) and western (*L. frantzii*) red bats, which are reddish rather than black. Average external measurements (in mm) are total length 100; tail 40; hindfoot 8; ear 16; weight is 8–12 g. Females are slightly larger than males.

Silver-haired bats are seasonal residents in the Trans-Pecos, present only during spring and autumn migration periods. They have been recorded in El Paso, Culberson, Jeff Davis, Presidio, Brewster, and Terrell Counties, and biogeographical patterns suggest they range throughout the western and southern Trans-Pecos (map 27).

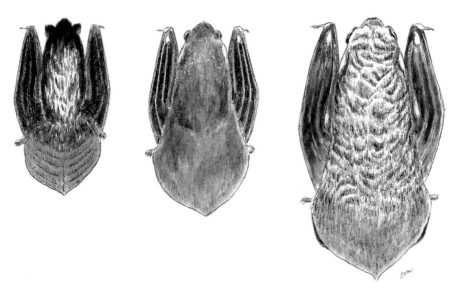

FIG. 35. Three species of tree bats: left, *Lasionycteris noctivagans*; center, *Lasiurus borealis*; right, *Aeorestes cinereus*

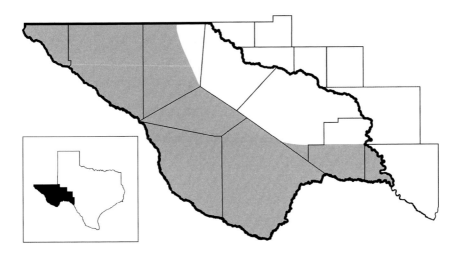

MAP 27. Distribution of the silver-haired bat, *Lasionycteris noctivagans*

Silver-haired bats are highly migratory. Females are thought to migrate earlier in the year, and there is some evidence that the sexes segregate during the summer months. As foliage-roosting bats, they favor woodland habitats across their distribution. In GMNP, 18 males were netted over water in both montane and riparian habitats between May (no specified date) and June 26, leading to speculation that a resident summer population of these bats existed in the national park (Genoways et al. 1979). If so, this would be the only known summer population of silver-haired bats in Texas and the Trans-Pecos. In BBRSP, two females were taken on March 13 along Terneros Creek in riparian habitat lined with mature cottonwood trees (Brant et al. 2002). Two males taken along a drainage of Boot Spring in BBNP (Chisos Mountains) in late May represent the southernmost known record of these bats in the Trans-Pecos (Ammerman 2005).

There are no records of these bats breeding in the Trans-Pecos. Elsewhere they are thought to mate in the fall and undergo delayed fertilization until spring. Parturition occurs in late June or July following a gestation period of about 50–60 days. Females typically produce two (occasionally one) young per year, and it is thought that young bats can fly in about three weeks. Knowledge about their winter ecology is incomplete, but they are not known to hibernate in the Trans-Pecos. These bats consume a variety of prey items including small moths, flies, beetles, midges, and true bugs. Owls, particularly great horned owls, and striped skunks are reported predators (Sparks et al. 2000). *Lasionycteris noctivagans* is a monotypic species, and there are no recognized subspecies.

Eastern Red Bat
Lasiurus borealis (Müller, 1776)

This medium-sized (forearm 35–45 mm; wingspan 290–335 mm), distinctly reddish bat has short, broad, rounded ears and a densely furred interfemoral membrane (fig. 35). In females the reddish upperparts are tipped with white, producing a frosted appearance; males usually lack the white-tipped hairs and are much redder. The tragus is triangular and has a slight forward bend to the tip. *Lasiurus borealis* is similar in appearance to *L. frantzii*. The latter species is rusty red to brownish and lacks the frosted appearance of the former species. Also, *L. frantzii* is significantly smaller in most cranial measurements, and the posterior third of the interfemoral membrane is only sparsely haired. Average external measurements (in mm) are total length 108; tail 48; hindfoot 9; ear 12; weight is 10–15 g. Females are slightly larger than males.

Eastern red bats are uncommon summer residents across much of the Trans-Pecos. They have been recorded in the western counties of El Paso, Jeff Davis (Davis Mountains), Presidio (Chinati Mountains, Sierra Vieja, and BBRSP), and Brewster (Chisos Mountains), and, based on biogeographic

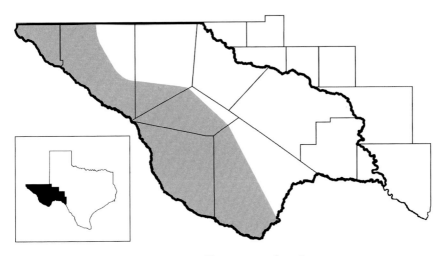

MAP 28. Distribution of the eastern red bat, *Lasiurus borealis*

factors, their range is projected to traverse Hudspeth and Culberson Counties. They apparently do not occur in the eastern part of the region (map 28). They are known primarily from montane habitats, although they have also been taken in riparian areas with dense stands of cottonwoods (C. Jones and Bradley 1999; Higginbotham et al. 2002). During summer they tend to be solitary, and in autumn and winter they migrate from the Trans-Pecos. Because they are better insulated with thicker, denser fur than cave-roosting species, they can enter deep torpor and withstand cold ambient temperatures.

Mating occurs in August and September, and sperm is stored in the female reproductive tract over winter. Ovulation, fertilization, gestation (80–90 days), and parturition occur in the spring. Females give birth to one to four pups in May or June. Females have four teats and can raise multiple young. Three pregnant female bats, taken on May 22, 2001, have been reported from BBRSP; each of these females carried three embryos with crown-rump lengths of between 12 and 20 mm (Higginbotham et al. 2002).

Lasiurus borealis is an aerial insectivore that feeds on a variety of insects including moths, scarab beetles, plant hoppers, flying ants, leafhoppers, ground beetles, and assassin bugs. This bat also routinely forages for insects around artificial lights (for example, floodlights over streets or parking areas). Known predators include bullfrogs, short-eared owls, sharp-shinned hawks, merlins, American kestrels, greater roadrunners, opossums, and ringtails (Sparks et al. 2000). This is a monotypic species, and there are no recognized subspecies.

Western Red Bat or Desert Red Bat
Lasiurus frantzii (Peters, 1871)

This medium-sized bat (forearm 35–40 mm; wingspan 290–335 mm) is similar in overall appearance to the eastern red bat (*Lasiurus borealis*), thus causing confusion in identification. It has short, rounded ears and a relatively long tail. Pelage is rusty red to brownish and lacks the white-tipped hairs (fig. 36) that give the frosted appearance so characteristic of *L. borealis*; other differences are described in the account of *L. borealis*. Average external measurements (in mm), which overlap those of the eastern red bat, are total length 103; tail 49; hindfoot 10; ear 13; weight is 10–15 g.

The distribution of the western red bat is difficult to explain, and for that reason it is listed as enigmatic in the species checklist. It is known from a single record from ZH Canyon in the Sierra Vieja of Presidio County, taken on July 15 (Genoways and Baker 1988) (map 29). The eastern red bat is also known from Presidio County, and it is possible these two red bats occur there in sympatry. In Arizona and New Mexico, *L. frantzii* is most often reported from lowland riparian habitats characterized by cottonwoods, oaks, sycamores, and walnuts. This type of habitat is abundant in the Big Bend region, and more records of this species should be expected.

Their reproductive biology is similar to that of the eastern red bat. Mating occurs in August and September, followed by delayed fertilization and

rusty red-brownish pelage   pale yellow pelage

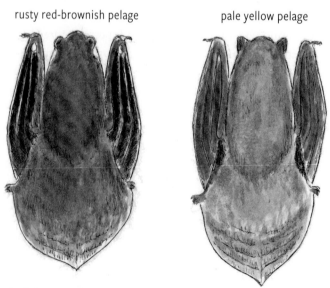

FIG. 36. Left, *Lasiurus frantzii*; right, *Dasypterus xanthinus*

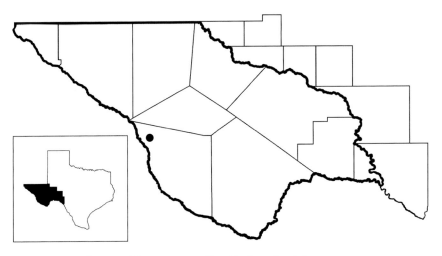

MAP 29. Distribution of the western red bat, *Lasiurus frantzii*

implantation in the spring, and gestation and parturition in May or June. In New Mexico, pregnant western red bats were captured on May 15 and June 15, 25, 26, and 29, and lactating females from June 13 to July 12 (Findley et al. 1975). *Lasiurus franztii* is an aerial insectivore that feeds on moths and several other types of insects. Virtually nothing is known about its predators.

The western red bat has an interesting taxonomic history. Previously it was arranged as one of three subspecies of the eastern red bat: *L. b. borealis* (eastern US subspecies), *L. b. teliotis* (western US subspecies), and *L. b. frantzii* (Mexican and Central American subspecies). Based on the results of a genetic analysis, two of these subspecies (*L. b. teliotis* and *L. b. frantzii*) were combined to form a taxonomic unit that was subsequently elevated to specific status as *Lasiurus blossevillii* (R. J. Baker et al. 1988). A later taxonomic review of the genus then recognized *L. frantzii* as distinct from *L. blossevillii* and elevated it to independent species status (Baird et al. 2015). *Lasiurus frantzii* is monotypic, with no subspecies recognized.

Hoary Bat
Aeorestes cinereus (Palisot de Beauvois, 1796)

This is a large (forearm 46–58 mm; wingspan 380–410 mm), dark brown to grayish bat with a frosting of white on the tips of the fur giving it a hoary appearance (fig. 35). The face, throat, and hairs on the underside of the wing are yellowish; the fur on the back and on the tail membrane, beneath the white

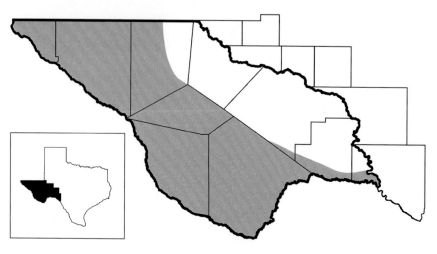

MAP 30. Distribution of the hoary bat, *Aeorestes cinereus*

tips, is mahogany brown. Other features include large eyes; large, sharp teeth; and short, rounded, well-furred ears with a black rim. Average external measurements (in mm) are total length 128; tail 60; hindfoot 12; ear 18; weight is 20–35 g. Many biologists consider the hoary bat to be one of the most beautiful bats in North America.

Aeorestes cinereus is a highly migratory bat that has been recorded throughout the Trans-Pecos except on the flat plains in the eastern part of the region (map 30). It has been recorded in all types of habitat, including lowlands (floodplain) and montane areas. There are reports of hoary bats being restricted to wooded montane habitats in the Trans-Pecos (Schmidly 1977b), but in BBRSP they were captured in floodplain habitat among cottonwood and willow trees (Yancey 1997). In BBNP they have been reported from six sites between March 18 and October 11 in river floodplain and desert scrub habitats (Higginbotham and Ammerman 2002).

Hoary bats are solitary and not known to form large aggregates (colonies) as described for some other species. They roost in leaves (foliage) of trees (both deciduous hardwoods and conifers), rather than in caves and mines, and they are highly migratory. Trans-Pecos records are from March through October; thus, *A. cinereus* appears to be only a spring, summer, and fall resident of the area. Apparently, the sexes are segregated throughout most of the summer. Females migrate through the area in March, April, and May and again in September and October. Males appear in May and are common in wooded regions from June through September in BBNP. However, at BBRSP the only record of a male is a single individual captured in September, and there are no summer records of males in the park (Yancey 1997). At CMSNA, a single female captured in April has been reported (C. Jones et al. 2011).

Females generally produce twins, with births typically occurring from May to early July. Food items include moths (a preferred item), beetles, flies, true bugs, grasshoppers, termites, dragonflies, and wasps (Black 1972). There are no accounts of their predators in the Trans-Pecos, although they are probably taken by birds of prey.

Hoary bats were formerly included in the genus *Lasiurus*, but recent evidence from molecular genetic studies has resulted in their placement in a separate genus, *Aeorestes* (Baird et al. 2015). This is a monotypic species, and subspecies are not recognized.

Western Yellow Bat
Dasypterus xanthinus Thomas, 1897

This is a medium-sized bat (forearm 45–48 mm; wingspan 335–355 mm) with a pale yellow pelage. The interfemoral membrane is furred for about half its length, with bright yellow hairs that contrast with the rest of the pelage (fig. 36). The flight membranes are dark, and there is no dark face mask. The ears are of medium size and rounded. Average external measurements (in mm) are total

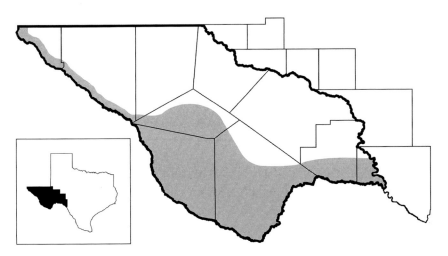

MAP 31. Distribution of the western yellow bat, *Dasypterus xanthinus*

length 105; tail 50; hindfoot 10; ear 13; weight is 12–19 g. Females average slightly larger than males.

This bat has only recently been recorded in the Trans-Pecos. The first specimen was documented in Brewster County in 1996 (Higginbotham et al. 1999), and since then it has been recorded in the Trans-Pecos from El Paso and Jeff Davis Counties, as well as from just east of the region in Val Verde County (Tipps et al. 2011; Schmidly and Bradley 2016; Decker et al. 2020). Based on biogeographic factors, its distribution is projected to extend across the southern Trans-Pecos (map 31). These relatively recent records suggest it may be extending its range to the north, but currently it remains rare in the Trans-Pecos.

Dasypterus xanthinus is a solitary, migratory bat that typically roosts in large trees along waterways. It is also known to roost in giant dagger yucca at BBNP. Four bats were captured from 1996 to 1998 at BBNP, and all were taken over a spring-fed pool in riparian habitat between early September and late November (Higginbotham and Ammerman 2002). There are other records from March, June, July, September, October, and November. Records of lactating females from Jeff Davis (C. Jones et al. 1999) and Brewster (Bradley et al. 1999) Counties suggest that at least some of these bats are reproducing in the Trans-Pecos.

Dasypterus xanthinus tends to be a relatively late flier, often not netted until three to four hours after sunset. Food items analyzed from one male bat from Brewster County included several types of insect prey such as small moths, true bugs, flies, wasps, crickets, and beetles (Higginbotham et al. 1999). The predators of these bats are unknown.

The western yellow bat was formerly classified as a subspecies of the eastern yellow bat, *Lasiurus ega xanthinus*, but subsequently it was elevated to specific status and moved from the genus *Lasiurus* to *Dasypterus* (R. J. Baker et al. 1988; Baird et al. 2015). This is a monotypic species, and there are no described subspecies.

American Parastrelle or Canyon Bat
Parastrellus hesperus (H. Allen, 1864)

This is the smallest bat in the Trans-Pecos (forearm 27–33 mm; wingspan 190–215 mm). In addition to its small size, its drab gray or smoked gray coloration, leathery face mask, and curved, blunt tragus may be used for identification (fig. 37). The flight membranes are thin and delicate, and the proximal fourth of the interfemoral membrane is slightly furred on top. The calcar is keeled. Average external measurements (in mm) are total length 78; tail 32; hindfoot 6; ear 11; weight is 3–6 g. Males are typically smaller than females of the same age.

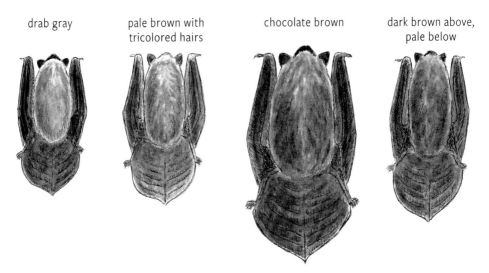

drab gray

pale brown with
tricolored hairs

chocolate brown

dark brown above,
pale below

FIG. 37. Four species of vespertilionid bats: left to right, *Parastrellus hesperus*;
Perimyotis subflavus; *Eptesicus fuscus*; *Nycticeius humeralis*

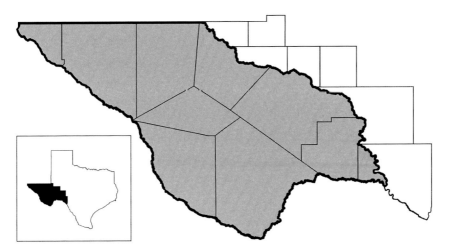

MAP 32. Distribution of the American parastrelle, *Parastrellus hesperus*

    Parastrellus hesperus may be confused with the California myotis (*Myotis californicus*) and the western small-footed myotis (*M. ciliolabrum*), both of which are larger and have the narrow, straight, and sharply pointed tragus characteristic of the genus *Myotis*. The American perimyotis (*Perimyotis subflavus*) is also similar in appearance to *Parastrellus hesperus*, and the ranges of these two bats can overlap. Characters distinguishing the two species are given in the account for *P. subflavus*.

The American parastrelle is the most ubiquitous bat in the Trans-Pecos. It has been recorded in all counties except Reeves, and because of the presence of continuous suitable habitat it undoubtedly occurs throughout the region (map 32). If you are in the Chihuahuan Desert early in the evening, chances are these bats will be flying about. *Parastrellus hesperus* is most common in low, often treeless, desert scrub habitats in rocky canyons, but it also occurs at mid-elevations around mesquite and other small trees. In BBNP, these bats have been taken from March through November, but there are no winter records. This is the most common bat in BBRSP, where it is active all year, although less so in winter months (Yancey 1997).

They are active year-round when temperatures are favorable. They are a sedentary, nonmigratory, cavity-roosting species that will occupy caves, abandoned mines, rock crevices, and human-made structures such as buildings and bridges. They often use rocky outcrops for roosting and as foraging sites for insects. Their flight, described as slow, erratic, fluttering, flitting, or zigzagging, has a "butterfly-like" motion. They are often seen well before sunset hawking insects over vegetation, canyons, or water sources.

They are thought to hibernate in small groups during the winter months, but they will occasionally arouse during warm periods and leave the roost. Males are more active in cooler weather than females. Mating occurs in autumn, with fertilization in spring. The gestation period is about 40 days. In BBNP pregnant females have been observed from late April to mid-July, and lactating females from mid-June to mid-July (Higginbotham and Ammerman 2002). Females typically give birth to twins, which is unusual for such a small bat. Maternity roosts have been discovered in caves, mines, and human-made buildings and structures.

Parastrellus hesperus feeds on flies, moths, beetles, mosquitoes, true bugs, and several other types of insects. Of 11 desert bats tested in physiological experiments, *P. hesperus* had the highest mean maximum urine-concentrating ability (Geluso 1978). It was also able to maintain a positive water balance and retain weight in "water-never" experiments, illustrating that it truly is a desert-adapted specialist. Known predators include largemouth bass, Cooper's hawks, great horned owls, and short-eared owls (Sparks et al. 2000).

This bat was formerly classified in the genus *Pipistrellus* and called the western pipistrelle. Its taxonomic status and phylogenetic relationships were clarified using molecular genetics, and it was subsequently placed in the genus *Parastrellus* (Hoofer and Van Den Bussche 2003; Hoofer et al. 2006). The subspecies in the Trans-Pecos is *P. h. maximus*.

American Perimyotis or Tricolored Bat
Perimyotis subflavus (F. Cuvier, 1832)

This is a small (forearm 31–35 mm; wingspan 210–260 mm), pale yellowish-brown bat characterized by unique "tricolored" hairs in which the base of each individual hair is dark, the middle band lighter, and the tip dark; and by the leading edge of the wing membrane being noticeably paler than the rest of the membrane (fig. 37). The calcar is not keeled, and the tragus is straight and slightly curved forward. It can be confused with the American parastrelle (*Parastrellus hesperus*), from which it is distinguished by its unique, tricolored fur; unkeeled calcar; and paler wing membranes. Average external measurements (in mm) are total length 77; tail 35; hindfoot 9; ear 12; weight is 3–6 g. Females are slightly larger than males.

The American perimyotis was first reported from the Trans-Pecos from BBRSP, Presidio County, in 1995 (Yancey et al. 1995b). Subsequently, other specimens have been obtained from the Chisos Mountains in BBNP in Brewster County (Ammerman 2005) and from ZH Canyon in the Sierra Vieja of Presidio County (Geluso et al. 2005); recent biogeographic patterns project this bat to range across the southern Trans-Pecos (map 33). Interestingly, a recent record from GMNP in Culberson County (Hanttula and Valdez 2021) suggests the species may be dispersing into some northern parts of the Trans-Pecos as well. It roosts in caves, tunnels, rock crevices, and human-made structures and uses similar places as hibernacula in the winter months. Mating is

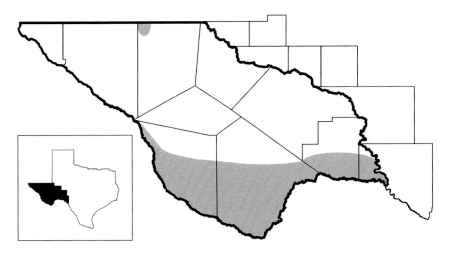

MAP 33. Distribution of the American perimyotis, *Perimyotis subflavus*

thought to occur in the autumn, with parturition the following spring. Females normally give birth to twins, which is somewhat unusual for such a small bat. Like many other North American bats, they feed on a variety of small aerial insects, including moths, beetles, and true bugs (Hanttula and Valdez 2021). They are early fliers and are commonly seen near, or at, sunset flying around the crown or canopy of trees and under low-hanging branches in pursuit of insect prey. Known predators include house cats and hoary bats (Sparks et al. 2000).

The American perimyotis is a "rare" bat in the Trans-Pecos. Because of its susceptibility to white-nose syndrome, this bat has been proposed for listing as an endangered species by the USFWS. Historically this species was included in the genus *Pipistrellus* (along with *Parastrellus*) until molecular genetic studies warranted placement in a separate genus (Hoofer and Van Den Bussche 2003; Hoofer et al. 2006). The subspecies that occurs in the region is *P. s. subflavus*.

Big Brown Bat

Eptesicus fuscus (Palisot de Beauvois, 1796)

This bat is moderately large (forearm 42–51 mm; wingspan 325–350 mm), with a broad nose; a broad, rounded tragus; broad wings; and a keeled calcar. The ears are relatively small, leathery, and black. Coloration varies from pale brown to chocolate or deep, rich brown (fig. 37). The skull has a distinctive sagittal crest. Average external measurements (in mm) are total length 114; tail 46; hindfoot 11; ear 17; weight is 13–20 g. Females are slightly larger than males.

These nonmigratory, cavity-roosting bats are known only from the western counties of the Trans-Pecos, where they commonly occur year-round in caves, abandoned mines, and rock crevices, beneath the bark of trees, and in human-made structures such as buildings and bridges. They appear to be absent from the arid lowlands adjacent to the Pecos River in the four eastern counties of the region (Reeves, Pecos, Terrell, and Val Verde) (map 34). They have been captured predominantly in woodlands and montane habitats. During summer months they tend to be solitary, but in winter females form small aggregates and use the same habitats for hibernacula that were used as night roosts during the summer. Males tend to hibernate singly and apart from females.

Big brown bats are habitat generalists found evenly distributed among lowland and highland sites. In BBNP these bats apparently segregate by sex during the summer months, with males occupying habitat at relatively higher elevations and females forming small nursery colonies at lower elevations. Big brown bats are common in GMNP (Genoways et al. 1979), and they are broadly distributed, especially in lowland areas, in BBNP (Higginbotham and

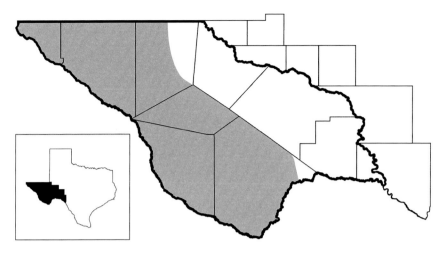

MAP 34. Distribution of the big brown bat, *Eptesicus fuscus*

Ammerman 2002). At BBRSP, they have been captured between April 11 and September 4, most frequently over water near riparian habitats surrounded by cottonwoods and willows (Yancey 1997).

Mating occurs between autumn and spring, with ovulation and fertilization delayed until after hibernation is complete. Pregnant females have been captured between April 26 and May 24 in BBNP. At BBRSP, a pregnant female was captured carrying a single embryo (crown-rump length 9 mm) on May 5, and lactating females were recorded between June 23 and July 16 (Yancey 1997).

Big brown bats feed on a variety of insects such as flies, stoneflies, bees, mayflies, true bugs, scorpion flies, caddisflies, cockroaches, and especially beetles. Curiously, moths seem to be missing from their diet. Bullfrogs, great horned owls, peregrine falcons, American kestrels, common grackles, long-tailed weasels, raccoons, house cats, and black rats are reported predators (Sparks et al. 2000). The subspecies in the Trans-Pecos is *E. f. pallidus*.

Evening Bat
Nycticeius humeralis (Rafinesque, 1818)

This is a small (forearm 33–39 mm; wingspan 260–280 mm), rather nondescript bat with pelage that is dark brown dorsally and paler below. Superficially it resembles a "scaled-down" version of the big brown bat (*Eptesicus fuscus*) (fig. 37). The wings are short and narrow, and the ears are small. The rostrum has well-developed facial (scent) glands. The wing and tail membranes are dark and leathery, as are the ears. The calcar is not keeled, and the tragus is short,

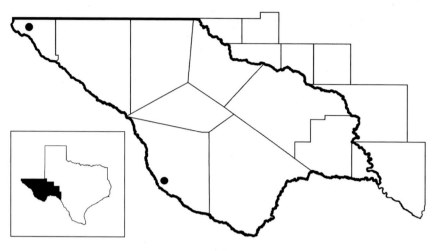

MAP 35. Distribution of the evening bat, *Nycticeius humeralis*

broad, and blunt. The skull is relatively large, but with a short, broad rostrum and low profile. Average external measurements (in mm) are total length 87; tail 32; hindfoot 8; ear 12; weight is 5–8 g. Females are slightly larger than males.

The evening bat is common and broadly distributed in the deciduous forests of eastern Texas, but it is "rare" and a recent invader of the Trans-Pecos. The first record was of a young female collected at the Ocotillo Unit of Las Palomas Wildlife Management Area, 3 km northwest of Ruidosa in Presidio County (Dowler et al. 1999) (map 35). This bat was netted over a grassy area near the Rio Grande, not too distant from a stand of salt cedar. A second specimen, a male from El Paso County, represents the westernmost record of this species in Texas (Krejsa et al. 2020). Evening bats have recently been recorded in New Mexico, suggesting they could be expanding their distribution westward in the United States (Yancey and Jones 2006; Andersen et al. 2017).

This migratory species is known to roost in caves, tunnels, rock and tree crevices, and human-made structures. Males and females segregate by sex during the summer months, but their winter ecology is not well known. Mating is thought to occur in the autumn, with parturition the following spring. Females normally give birth to twins, and young bats are volant in about 20 days.

Evening bats feed on a variety of small aerial insects including moths, true bugs, beetles, and flying ants. They are early fliers and may commonly be seen in flight at, or near, sunset. They are known to have two separate feeding periods, one at dusk and a second just prior to sunrise. Reported predators include short-eared owls, house cats, and raccoons (Sparks et al. 2000). The subspecies in the Trans-Pecos is *N. h. humeralis*.

Spotted Bat
Euderma maculatum (J. A. Allen, 1891)

The spotted bat is a moderately large bat (forearm 48–51 mm; wingspan 325–350 mm) with extremely large ears (about three-fourths as long as the forearm) and a conspicuous dorsal color pattern with three large white spots, one on each shoulder and one on the rump, on a black background (fig. 38). Average external measurements (in mm) are total length 124; tail 51; hindfoot 12; ear 42; weight is 16–20 g. Females are slightly larger than males.

At the present time, the spotted bat is known to occur in the Trans-Pecos only in BBNP in Brewster County, where it was first encountered in 1967 (Easterla 1970b) (map 36). It has been captured at several locations in diverse habitats at both high elevations (piñon pine–juniper associations) and in lowland areas with open scrub vegetation (Higginbotham and Ammerman 2002). Several individuals have been netted during the summer months at Ernst Tinaja Canyon (Easterla 1973) and in Mariscal Canyon (Higginbotham

FIG. 38. Spotted bat, *Euderma maculatum*

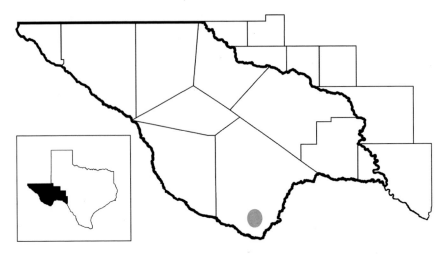

MAP 36. Distribution of the spotted bat, *Euderma maculatum*

and Ammerman 2002) in BBNP. They seem to be more dependent on suitable roosting sites (high vertical cliffs with rocky crevices) than on vegetation type. Virtually nothing is known about their winter ecology. They are thought to hibernate, but there is no evidence of this in the Trans-Pecos. *Euderma maculatum* is a late-flying insectivorous bat that feeds almost exclusively on moths.

Little is known about their reproductive biology. Females are thought to produce a single young each year, with parturition probably occurring in mid-June (Higginbotham and Ammerman 2002). A pregnant female that gave birth to a single altricial young, weighing 4 g, on June 11 in BBNP has been documented (Easterla 1971), and lactating females have been recorded in New Mexico on June 23 and 30 and July 1 (Findley and Jones 1965).

The TPWD lists the species as "threatened" because of its restricted distribution and the scarcity of records of occurrence. The USFWS has not taken any action on its possible status as threatened or endangered. This bat is highly sought after by bat biologists because of its rarity and beauty. This is a monotypic species with no recognized subspecies.

Townsend's Big-eared Bat
Corynorhinus townsendii (Cooper, 1837)

This is a medium-sized bat (forearm 39–48 mm; wingspan 290–320 mm) whose most distinctive features are its extremely large ears, which typically measure over 25 mm in length, and the presence of one large lump on either side of its snout

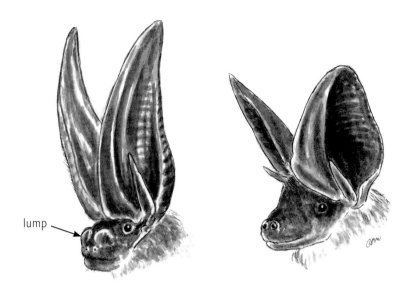

FIG. 39. Facial view of two big-eared bats: left, *Corynorhinus townsendii*; right, *Antrozous pallidus*

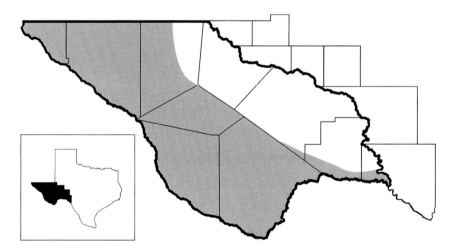

MAP 37. Distribution of Townsend's big-eared bat, *Corynorhinus townsendii*

(fig. 39). The "lump" is a facial gland thought to secrete pheromones important in mating rituals. Pelage is uniformly pale brown to dark brown dorsally and ventrally. *Corynorhinus townsendii* may be easily distinguished from the other big-eared bats (*Antrozous* and *Euderma*) by its nearly uniform color and the lumps on the snout. Average external measurements (in mm) are total length 100; tail 46; hindfoot 11; ear 35; weight is 7–11 g. Females are slightly larger than males.

This bat occupies suitable Chihuahuan Desert habitats across the western and southern parts of the Trans-Pecos but apparently does not occur in the flat, desertlike areas in the eastern part of the region (map 37). It is an uncommon year-round resident that depends more on suitable roosting sites (rocky situations with caves, crevices, or abandoned mine tunnels) than vegetation type, as it has been collected in desert scrub and piñon-juniper forests.

Townsend's big-eared bats are cave dwellers that form small nursery colonies with male bats roosting separately from females. When in the roost, these bats tend to roll or coil their ears backward close to the body, creating what has been termed a "ram's horn" configuration. At BBRSP, they have been captured between March and September at sites near riparian woodlands (Yancey 1997). A single individual was captured in the Whitroy Mine, which is surrounded by desert scrub vegetation (creosote bush habitat). In BBNP captures have occurred over springs and streams in riparian woodlands and in desert scrub and sotol grassland habitats (Higginbotham and Ammerman 2002). In GMNP, they were reported as "not common" at middle to higher elevations (Genoways et al. 1979). Their winter ecology is poorly known. They are nonmigratory and thought to overwinter in the Trans-Pecos.

Available data suggest that females breed in late autumn prior to hibernation, with fertilization delayed until spring. Parturition occurs in May and June, and young bats have been observed flying by mid-July. Females form small nursery colonies with fewer than 200 individuals, and a single young is produced. These bats are relatively late fliers, often absent until long after sunset. They consume a variety of insects including small moths (a preferred food item), flies, lacewings, dung beetles, and sawflies. Their predators are unknown.

Townsend's big-eared bat was formerly placed in the genus *Plecotus*, but it has now been moved back to the genus *Corynorhinus* based on studies using morphological data (Tumlison and Douglas 1992) and both morphological and chromosomal data (Bogdanowicz et al. 1998).

Because of its prominent facial glands, this bat is sometimes referred to as the "lump-nosed" bat. This is one of the rarest bats in the Trans-Pecos, and both federal and state governments classify it as a species of concern. The subspecies that occurs in the Trans-Pecos is *C. t. australis*.

Pallid Bat
Antrozous pallidus (Le Conte, 1856)

This is a large bat (forearm 46–60 mm; wingspan 360–390 mm) with big ears that extend about 20 mm beyond the tip of the nose when laid forward; large

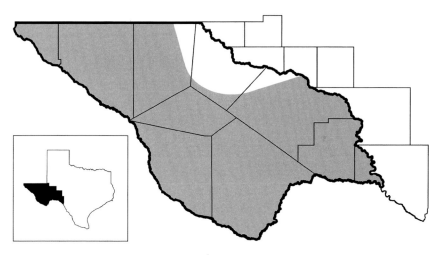

MAP 38. Distribution of the pallid bat, *Antrozous pallidus*

eyes; and broad wings that are dark, thick, and leathery (more so than in other bats). The upperparts are pale yellow washed with brown or gray; underparts are pale cream, almost white. The nostrils are surrounded by a glandular ridge that produces a blunt snout (fig. 39). Average external measurements (in mm) are total length 113; tail 46; hindfoot 12; ear 28; weight is 12–17 g.

Pallid bats occur across the entire Trans-Pecos, except for the extreme northeastern sector, and are most commonly associated with lowland Chihuahuan Desert habitats (Yancey 1997) (map 38). They are among the most abundant bats in many areas. They are cavity roosting bats that prefer caves, abandoned mines, rock crevices, and human-made structures such as bridges and buildings. After feeding, they will often use a separate night roost that is distinct from their day roost.

In BBNP they have been reported from a number of plant communities and habitat types but are more common in lower-elevation riparian (floodplain) communities (Higginbotham and Ammerman 2002); they have also been recorded at higher elevations in mesic montane habitats of oak and pine forests (Easterla 1973). Captures were recorded from mid-March through late November. At BBRSP, pallid bats were most common in desert scrub habitats at lower elevations; they were recorded from mid-March through mid-September, but not in the winter (Yancey 1997). These bats are not known to migrate, and they likely hibernate in the area.

Mating occurs in October after summer colonies are dispersed, and sperm is stored until ova are released in the spring. Maternity colonies are formed in April and early May. Pallid bats typically produce twin pups, although single pups are occasionally reported. In BBNP, pregnant bats have been recorded from late April to late July; lactating females from June and July; and volant young in mid- to late July (Higginbotham and Ammerman 2002).

In addition to preying on flying insects, these bats have a somewhat unusual feeding strategy in that they are able to alight on the ground and forage for larger prey such as arthropods (scorpions and centipedes), small vertebrates (lizards and rodents), and possibly other bats. They have been documented to prey on canyon lizards in CMSNA (Kasper and Yancey 2018a). They hunt, capture, and process these larger prey items by delivering a powerful bite with their formidable teeth. Insects taken include moths (a preferred food item), crickets, ground beetles, ant lions, froghoppers and leafhoppers, and unidentified insects (Easterla 1973). In addition to animal matter, *A. pallidus* will facultatively consume plant nectar, and in the Trans-Pecos it seems to particularly favor that of lechuguilla (Jaquish and Ammerman 2021). Nocturnal snakes, barn owls, sharp-shinned hawks, Brewer's blackbirds, and American kestrels are reported predators (Sparks et al. 2000). The subspecies in the Trans-Pecos is *A. p. pallidus*.

ORDER CARNIVORA (Carnivores)

Carnivores are principally flesh eaters that prey on animals and are adapted for hunting with sharp claws and tearing or cutting teeth. They are characterized by small incisors, enlarged canines, and crushing or shearing cheek teeth. The shearing teeth are known as the carnassial pair, which consists of an enlarged fourth upper premolar and a lower first molar. The carnassial teeth work as shears for meat and bone.

The order Carnivora is subdivided into two suborders—Feliformia (catlike), characterized by a relatively short, rounded snout and retractile claws; and Caniformia (doglike), characterized by a long snout and typically nonretractile claws. Six families are found in the Trans-Pecos—Felidae (cats), Canidae (wolf, coyote, and foxes), Ursidae (bears), Mephitidae (skunks), Mustelidae (badger and weasels), and Procyonidae (raccoon, coati, and ringtail). Twenty-one species have been documented from the Trans-Pecos, making this the most species-rich region in the state for this order of mammals. Of the 21 species, 1 (red fox, *Vulpes vulpes*) has been introduced to the Trans-Pecos, and 5 (ocelot, *Leopardus pardalis*; jaguar, *Panthera onca*; gray wolf, *Canis lupus*; grizzly bear, *Ursus arctos*; and black-footed ferret, *Mustela nigripes*) are now extirpated in the region. Brief summaries for the introduced and extirpated species are provided at the end of this chapter.

1. Total number of teeth 28–30; claws retractile; (Feliformia)
   Felidae (cats) 2
   Total number of teeth 32–42; claws typically nonretractile;
   (Caniformia) 5
2. Tail (100–150 mm) shorter than hindfoot; total number of teeth 28;
   two upper premolars on each side; upperparts gray brown to red
   brown streaked and blotched with black; ears slightly tufted;
   weight 5–9 kg . . . .**Bobcat**, *Lynx rufus*
   Tail (300–1,000 mm) much longer than hindfoot, more than half the
   length of head and body; total number of teeth 30; three upper
   premolars on each side 3
3. Upperparts uniformly tawny colored, not spotted in adults (juveniles
   spotted); total length > 1,400 mm; weight > 45 kg . . . .
   . . . . **Mountain Lion**, *Puma concolor*
   Upperparts with dark to black spots, bars, or blotches at all ages 4
4. Body small to medium, total length < 1,200 mm; upperparts with
   black spots, bars, and irregular black-outlined blotches and bars;
   weight 7–15 kg . . . .**Ocelot**, *Leopardus pardalis*
   Body large, total length > 1,500 mm; upperparts with dark to black
   spots, sometimes with light centers; weight > 45 kg . . . .
   . . . . **Jaguar**, *Panthera onca*
5. Hindfoot with four toes; total number of teeth 40 or 42; Canidae 6
   Hindfoot with five toes; total number of teeth 42 or fewer 10
6. Hindfoot usually > 170 mm; condylobasal length of cranium
   >160 mm; total number of teeth usually 40, occasionally
   42; weight > 9 kg; coyote, wolf, dog. 7
   Hindfoot usually < 170 mm; condylobasal length of cranium
   < 160 mm; total number of teeth 42; weight < 9 kg; foxes 8
7. Hindfoot < 200 mm; nose pad < 25 mm wide; condylobasal
   length usually < 190 mm; weight usually < 18 kg . . . .
   . . . .**Coyote**, *Canis latrans*
   Hindfoot > 200 mm; nose pad > 25 mm wide; condylobasal length
   usually > 210 mm; weight usually > 18 kg . . . .
   . . . .**Gray Wolf**, *Canis lupus*
8. Tip of tail white; hindfoot near 160 mm; upperparts reddish,
   blackish, or yellowish; lower part of legs and feet black;
   weight 3–5 kg . . . . **Red Fox**, *Vulpes vulpes*
   Tip of tail black; hindfoot usually < 150 mm 9

9. Tail with black stripe on dorsal edge terminating with a black tip; hindfoot usually > 140 mm; upperparts grizzled gray; legs and collar reddish; temporal ridges large, converging posteriorly in a lyre or U shape; weight 3–5 kg. . . . . **Common Gray Fox**, *Urocyon cinereoargenteus*
    Tail with no black stripe; tip of tail with some black; hindfoot usually < 140 mm; upperparts lightly grizzled to pale yellowish gray; temporal ridges present but indistinct and converging posteriorly in a V shape; weight 2–3 kg. . . . . **Kit Fox**, *Vulpes macrotis*
10. Tail distinctly shorter than hindfoot; total number of teeth 42; length of adults > 1,200 mm; weight of adults usually > 100 kg; Ursidae (bears) 11
    Tail longer than hindfoot; total number of teeth 40 or fewer; length of adults < 1,100 mm; weight of adults usually < 20 kg 12
11. Claws on front feet 70–120 mm long; head large, with profile of face distinctly concave; upperparts brownish or yellowish brown mixed with white-tipped hairs; mane (hump) present between shoulders; maxillary tooth row > 110 mm; last upper molar nearly twice as large as preceding molar; weight > 130 kg . . . . **Grizzly (Brown) Bear**, *Ursus arctos*
    Claws on front feet about 75 mm long; profile of face nearly straight; upperparts black or brown; no mane (hump) present; maxillary tooth row < 110 mm; last upper molar about 1.5 times as large as preceding molar; weight > 100 kg. . . . . **American Black Bear**, *Ursus americanus*
12. Total number of teeth 40; tail usually with conspicuous alternating dark and light rings; scent glands not well developed; Procyonidae (raccoon, ringtail, coati) 13
    Total number of teeth 32–34; tail not ringed; scent glands well developed 15
13. Tail as long as or longer than head and body; tail with 14–16 alternating dark and light rings with a black tip; hindfoot < 80 mm; ears relatively long (44–50 mm) and less rounded; weight 1–2 kg . . . . **Ringtail**, *Bassariscus astutus*
    Tail shorter than head and body; tail with 6–7 alternating dark and light rings or rings inconspicuous; hindfoot of adults ≥ 85 mm; ears relatively short and rounded 14

14. Snout extending conspicuously beyond mouth and highly flexible; end of snout white; tail about five times as long as hindfoot; alternating rings of tail fading with age and inconspicuous in adults; variable amounts of brown to black across face and upper snout, eyes bordered by white; weight 4–5 kg . . . . **White-nosed Coati**, *Nasua narica*

    Snout not extending conspicuously beyond mouth; tail two to three times as long as hindfoot; alternating rings of tail conspicuous at all ages; distinctive black facial mask covering eyes; weight 4–13 kg . . . . **Northern Raccoon**, *Procyon lotor*

15. Dorsal coloration black interrupted with prominent white markings; bony palate ending near last molar teeth; total number of teeth 32 or 34; Mephitidae (skunks) 16

    Dorsal coloration brown or grayish brown; bony palate extending well beyond last molar teeth; total number of teeth 34; Mustelidae (badger and weasels) 19

16. Total number of teeth 32; two upper premolars on each side; upperparts with a single broad white stripe that is truncate or wedge shaped on the forehead; nose pad large and flexible; weight 1.1–2.7 kg . . . . **Hog-nosed Skunk**, *Conepatus leuconotus*

    Total number of teeth 34; three upper premolars on each side; upperparts with continuous or broken white stripes and marks; nose pad normal 17

17. Upperparts consisting of four or more white stripes broken near midbody; large white area/spot between eyes; weight 368–565 g . . . . **Desert Spotted Skunk**, *Spilogale leucoparia*

    Upperparts usually consisting of one to three continuous white stripes; no large white area/spot between eyes 18

18. Dorsal white stripe bifurcated; divided white stripe patterns variable posteriorly from continuous to broken; tip of tail generally black; length of tail less than length of head and body; weight 1.4–6.6 kg . . . . **Striped Skunk**, *Mephitis mephitis*

    Dorsal white stripe never bifurcated; three color phases based on one to three white stripes; length of tail usually more than length of head and body; distinct ruff of longer hair on upper neck; weight 400–900 g . . . . **Hooded Skunk**, *Mephitis macroura*

19. Body broad and robust; tail and legs short; long and shaggy dorsal pelage silvery brown; prominent white stripe extending from nose

over head to shoulder area variably expressed to near base of tail; claws of forefeet ≥ 25 mm long; bony palate more than half the length of skull; weight 4–10 kg . . . .
. . . . **American Badger,** *Taxidea taxus*
    Body long and slender; tail more than two times length of hindfoot; no stripe over head; bony palate less than half the length of skull
. . . . 20

20.   Feet and legs black; hindfoot > 50 mm; black mask across face and eyes; tail more than two times length of hindfoot; upperparts pale buff yellow with darker hairs on head and back; underparts buff to cream; weight 450–1,500 g . . . .
. . . . **Black-footed Ferret,** *Mustela nigripes*
    Feet brown or tan; hindfoot usually < 50 mm; tail more than three times length of hindfoot; upperparts generally yellowish brown, underparts lighter; colors differentiated along a sharp lateral line; weight < 500 g . . . . **Long-tailed Weasel,** *Mustela frenata*

Family Canidae
DOGS, FOXES, AND WOLVES
Coyote
Canis latrans Say, 1823

The coyote is a long-legged, long-muzzled, doglike carnivore with a bushy tail and long, pointed ears (fig. 40). The color of the back is a grizzled gray or buff overlaid with black; the muzzle, ears, and outer sides of the legs are yellowish buff; the underparts are a light buff. The coyote typically carries its tail down when running, thereby distinguishing it from dogs and wolves (which usually carry theirs up) and foxes (which usually carry theirs straight out). Average external measurements (in mm) are total length 1,169; tail 339; hindfoot 184; ear 101; weight is 12–18 kg.

    Coyotes have been recorded from all counties in the Trans-Pecos and occur throughout the region (map 39). In all the recent mammal surveys of state and national park areas (BBNP, BBRSP, CMSNA, GMNP), coyotes have been shown to be common and widespread (Genoways et al. 1979; Yancey 1997; Bradley et al. 1999; Yancey et al. 2006; C. Jones and Lockwood 2008; C. Jones et al. 2011; Yancey et al. 2019). They occur in all habitats but are less abundant in montane woodlands. They can survive in almost any area because of their adaptability and behavioral flexibility. Coyotes are accused by farmers

FIG. 40. Coyote, *Canis latrans*

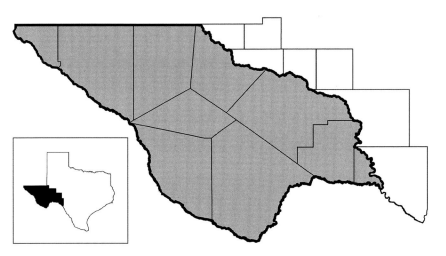

MAP 39. Distribution of the coyote, *Canis latrans*

and ranchers of inflicting serious damage on stocks of sheep and lambs, cattle, and wild game mammals. Consequently, they have been the object of intensive predator control campaigns, which commonly include the use of traps, snares, and M-44s (a type of poison gun triggered by the coyote itself), as well as shooting from the ground and from aircraft.

Their broad diet varies seasonally and includes cottontails, jackrabbits, many kinds of rodents, birds, bird eggs, amphibians, lizards, snakes, insects, carrion, and plant material like prickly pear fruits, mesquite beans, berries, nuts, seeds, and even grasses. When coyotes consume ripe prickly pear fruits, their purple-red scats are obvious. In BBNP, coyotes have been shown to eat primarily insects, birds, reptiles, and lagomorphs (Leopold and Krausman 1986).

Coyotes are primarily nocturnal but may be seen in early and late daylight. They are generally solitary or in bonded pairs; in the fall parents with near-grown young may form packs, and in winter some loose nonfamily members may aggregate. Coyotes are very vocal, with "yappy" barks and high-pitched howls emitted from one to several individuals at a time. They will urinate or defecate to mark certain spots and communicate their presence.

Breeding occurs in January–March following a two-to-three-month courtship involving a bonded pair. Female coyotes can become sexually mature once they experience their first estrus in the first year of life (Kirk 2019). Multiple copulations occur during a two-to-five-day estrous period, which ends in a "copulatory tie" (penis locked in vagina) that lasts from 5 to 25 minutes. A single litter of five to six young is usually produced. Coyote survival in West Texas has been estimated to be 0.659, which means that about 66 percent of the population survives from year to year (Kirk 2019). Predators include hawks, eagles, owls, mountain lions, and other coyotes, but hunting and trapping are major mortality factors.

As apex predators, coyotes can play a keystone role in the population regulation and community structure of microherbivores and mesopredators in certain ecosystems (Henke and Bryant 1999). In a coyote removal study in West Texas, treated plots (coyotes removed) exhibited declines in rodent species richness and diversity but increases in the density and richness of mesocarnivores (badgers, bobcats, and gray foxes), thereby demonstrating how coyote management can influence how other species are managed.

Coyotes may hybridize with domestic dogs and produce offspring called coydogs, which appear more coyote-like than dog-like. The subspecies in the Trans-Pecos is *C. l. texensis*.

Common Gray Fox
Urocyon cinereoargenteus (Schreber, 1775)

This is a medium-sized fox that appears dark gray over most of its body except for the reddish or rusty-brown lateral coloration, the white throat, and the black stripe along the dorsum of the tail (fig. 41). The cranial temporal ridges on top of the skull converge posteriorly in a U shape, and the ventral border of the mandible is characterized by a distinct "step" midway between the tip of

FIG. 41. Common gray fox, *Urocyon cinereoargenteus*

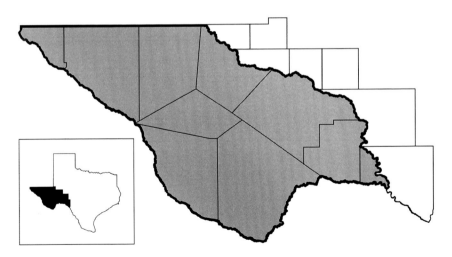

MAP 40. Distribution of the common gray fox, *Urocyon cinereoargenteus*

the angular process and the anterior border of the coronoid process. Average external measurements (in mm) are total length 944; tail 373; hindfoot 130; ear 76; weight is 4–5 kg.

Gray foxes are an adaptive species and can survive in different habitats and situations. They occur throughout the Trans-Pecos, having been recorded from all counties in the region (map 40). Their greatest abundance is in the piñon-juniper forests of montane regions, where they are closely associated with dense brush, trees, cliffs, canyons, rocky gulches, and boulder piles. They are usually not found in open desert country or grassland habitats without rocky outcrops, dense brush, or at least some juniper trees.

These foxes are omnivores that feed on seasonal fruit, berries, seeds, and nuts and prey on insects, birds, bird eggs, snakes, and small mammals such as cottontails, rats, and mice. A study of their food habits (Madsen 1997) in the region revealed that their major food items were insects (37.3 percent), rodents (36 percent), plant items (30.7 percent), lagomorphs (18.7 percent), and birds (13.3 percent). They have also been documented consuming prickly pear fruits and juniper berries (Bailey 1905), stalking a cottontail (Genoways et al. 1979), preying on rattlesnakes (Stangl et al. 1994), and scavenging carrion one to two days old (McKinnerney 1978). Arthropods dominate their diet during summer, prickly pear fruits from fall to early winter, and juniper berries from winter to early spring.

Gray foxes are principally nocturnal and active year-round. Den sites are in dense brushy or rocky habitats and constructed in rock crevices or under boulders, trees, and brush. They are proficient climbers and use trees and large shrubs for escape, resting, and foraging.

Gray foxes are thought to be monogamous, with the pair bond staying intact throughout the year. The male, female, and juveniles form a social family unit that maintains a separate home range from other family units. To communicate with other family units and unpaired individuals in the local population, urine and feces are deposited at scent mark locations. Vocally, they use a short, sharp "yap yap" bark comparable to that of a small dog (Bailey 1931).

Breeding is during January–March followed by a gestation period of about 60 days. A litter of three to six young is born, correlating with six mammae available for nursing (Bailey 1931). Both males and females attain sexual maturity at an early age; females on average breed for the first time at about 9 to 10 months of age (Kirk 2019). Males assist females raising litters, and young remain with parents until fall. The annual survival rate of gray foxes has been estimated at 0.650, which means that about 65 percent of the population survives from year to year (Kirk 2019). Predators include coyotes, bobcats, and golden eagles;

pups may be taken by larger hawks and owls. Rabies, canine distemper, and mange are diseases that may regulate population sizes when at high densities. The subspecies in the Trans-Pecos is *U. c. scottii*.

Kit Fox
Vulpes macrotis Merriam, 1888

The kit fox is a large-eared fox much smaller than the gray fox (fig. 42). The kit fox is a plain buffy gray with clear buff along the sides, midbelly, and underside of the tail. The tip of the tail is black. The cranial temporal ridges on top of the skull are indistinct and in a V shape. Average external measurements (in mm) are total length 790; tail 290; hindfoot 122; ear 85; weight is 2–3 kg.

Although distributional records are scarce, kit foxes have been documented from all counties in the Trans-Pecos except Reeves, and they are projected to range throughout the entire region (map 41). They seem to prefer open desert or grasslands and are absent from rugged, rocky terrain and wooded areas. They occur on mesas and open valley floors and are most abundant in areas where friable soils support the building of earthen dens and burrows and contain sufficient populations of rodents. Their diet includes insects, small birds, rodents, and rabbits, with kangaroo rats (genus *Dipodomys*) appearing to be a preferred food item.

FIG. 42. Kit fox, *Vulpes macrotis*

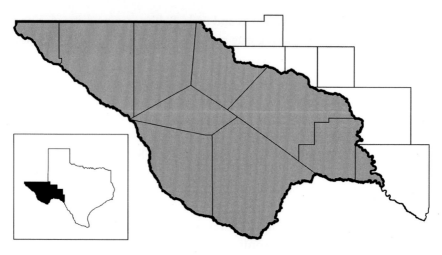

MAP 41. Distribution of the kit fox, *Vulpes macrotis*

A radiotelemetry study of these foxes near Sanderson, Terrell County, produced significant information about their activities, which occur primarily at night (McLaughlin 1979). In this study, four foxes were tagged with radio collars and followed for nearly two years. Two vegetation types were used by the foxes—desert scrubland for denning and vegetated washes for hunting. Each tagged fox lived in a different den. A kit fox spent an average of eight hours outside the den each day, with 60 percent of this time spent foraging in the washes; at least 25 percent of the fox's day was spent hunting. Vegetation on the denning sites was sparse and there were few small animals present, whereas the washes were heavily vegetated and supported more small animals. Active the entire year, kit foxes are primarily nocturnal, emerging after sunset to hunt.

Their underground dens, which may have been vacated by other mammals such as coyotes and badgers, have multiple entrances and are frequently changed during the summer because of a buildup of ectoparasites or depletion of local prey. Urine and feces are used to scent mark their surroundings, and one or more tunnels in the den complex may be used as latrines. Vocalizations include barking to recall straying young or to indicate intrusions.

Prior to the breeding season, females explore possible den sites during September and October. A seasonal monogamous pair bond is established during October–November, and the male is admitted to the den established by the female. Breeding occurs from December to February, followed by a gestation period of 49–55 days. During March and April a single litter of four to five fuzzy young is born with eyes and ear openings closed.

Known predators include humans and coyotes, but bobcats and mountain lions likely prey on them as well. This fox is particularly vulnerable to predator control measures, and extensive poisoning and trapping campaigns carried out against coyotes have likely resulted in serious population declines. An individual who escaped a leghold trap and survived with three legs has been recorded in Culberson County (Stangl et al. 1994). The subspecies in the Trans-Pecos is *V. m. neomexicanus*.

Family Felidae
CATS
Bobcat
Lynx rufus (Schreber, 1777)

Bobcats are short-tailed, long-legged spotted cats, rangy in appearance, with pointed ears that have tufts of black hairs, about one inch in length, at their tips (fig. 43). Their general color is yellowish brown or yellowish gray with varying degrees of black marks, blotches, or spots. Their upperparts have one or two black stripes along the back, with lateral bands over the shoulders, while the underparts are white with black spots, blotches, and bars, especially on the

FIG. 43. Bobcat, *Lynx rufus*

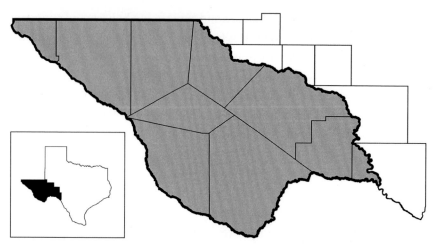

MAP 42. Distribution of the bobcat, *Lynx rufus*

front and inside of the upper forelegs. Melanistic individuals may occasionally be seen. Average external measurements (in mm) are total length 870; tail 146; hindfoot 171; ear 70; weight is 5–9 kg. Tail lengths are variable (132–175 mm).

Bobcats are known from every county in the Trans-Pecos, and they range across the entire region (map 42). They prefer rocky canyons, gulches, cliffs, and dense brush-woodlands. They are habitat generalists, occurring from lowlands along the Rio Grande up to 2,500 m in the Guadalupe Mountains. They are principally carnivorous, with rabbits and rodents constituting most of their diet along with a few birds, lizards, and snakes. In BBNP, bobcats have been shown to feed on a variety of small mammals, primarily lagomorphs (44 percent) and rodents (28 percent); they are also known to take down large prey, such as mule deer and white-tailed deer (Leopold and Krausman 1986; Browning 2014). Occasionally they will consume smaller carnivores, collared peccaries, and feral pigs and scavenge road-killed and hunter-crippled pronghorn and bighorn sheep.

Active throughout the year, bobcats are primarily nocturnal. Fidelity to a home range is strong and advertised to other individuals by scent marking and visual display. Bobcats and coyotes compete for similar prey where their home ranges overlap, with coyotes being dominant, especially around female and subadult bobcats.

Bobcats are polygamous and most females initially breed in their second spring. Breeding in the Trans-Pecos typically begins in February. After a gestation period of 62 days, a litter of one to five young is born in April or May.

Coyotes and mountain lions are known to kill bobcats, and great horned owls may take smaller or younger animals. Interestingly, a case of the plague (*Yersinia pestis*) has been confirmed from a bobcat found dead in Brewster County (S. Tabor and Thomas 1986). The subspecies in the Trans-Pecos is *L. r. fasciatus* (Kitchener et al. 2017; Schmidly et al. 2023).

Mountain Lion
Puma concolor (Linnaeus, 1771)

The mountain lion (also known as puma, cougar, or panther) is a large, long-tailed cat with uniformly light brown or grayish-buff upperparts in adults and no spotting (spotting occurs only in young). The snout and forehead are usually darker and grayer. The underparts are whitish, and the tip of the tail and the backs of the ears are dusky black (fig. 44). Adult males are larger than adult females. A male and a female killed in the Eagle Mountains in Hudspeth County had the following measurements: the male, three years old, had a total length of 213 cm; tail length 88 cm; hindfoot 127 mm; ear 108 mm; height at shoulder 71 cm; weight 50 kg. The female, two years old, had a total length of 178 cm; tail length 70 cm; hindfoot 89 mm; ear 76 mm; height at shoulder 61 cm; weight 36 kg.

The Trans-Pecos is an important location for mountain lions because it lies between two vital sources of these predators, the Rocky Mountains of the western United States and the Sierra Madre of Mexico. Because of their elusive behavior, knowledge about their abundance in the Trans-Pecos has

FIG. 44. Mountain lion, *Puma concolor*

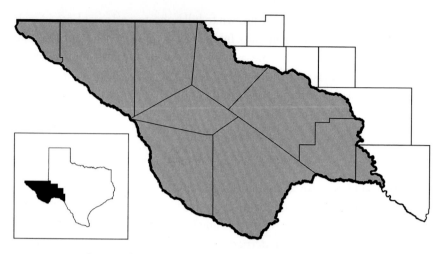

MAP 43. Distribution of the mountain lion, *Puma concolor*

been sketchy. They have been documented from every county in the region (map 43), and historically they were documented as especially abundant in the Santiago and Chisos Mountains and along the canyons of the Rio Grande and the Pecos River (Bailey 1905).

Years of predator control efforts by livestock producers, particularly in the first half of the twentieth century, forced mountain lions into more remote, thinly populated areas (Schmidly et al. 2022). However, with the slowing of predator control efforts since about 1970, it appears that cougars are repopulating portions of their former range. An increasing number have been reported in BBNP since the 1990s (Holtcamp 2008), and the Big Bend area likely supports the largest population in the state. An analysis of their current and potential distribution indicates additional areas that mountain lions could potentially reoccupy given that these habitat patches have adequate connectivity and protection (Hernandez-Santin et al. 2012).

Mountain lions are known to use a variety of habitats, with either dense vegetative cover or severe topography, but canyons, gulches, ledges, and escarpments at higher elevations, and dense brush, such as along the Rio Grande, are preferred. Mountain lions are strict carnivores, consuming meat exclusively. Throughout their distribution their primary prey is deer. Mountain lions are wide ranging and known to move long distances in search of food, water, adequate habitat, and mates. When young mountain lions start to become independent, they will often make long-range or "dispersal" movements away from their natal home ranges, in search of a range to establish as their own.

In their individual movements, mountain lions can occupy large areas of land and use terrain across multiple habitats. Active year-round, mountain lions are primarily nocturnal and terrestrial, although they may climb trees if pursued by dogs. Dens and resting places occur in ledges, crevices, overhangs, canyon walls, open spaces among rocks or boulders, and dense brush. Males are solitary except during breeding; females are associated with young much of their lives. Mountain lions communicate using territorial scrapes (leaves and grasses arranged into piles on which males urinate), scent marking, visual cues, and vocalizations.

Females breed once every two to three years during a two-week estrous period. The female vocalizes and marks nearby objects to attract a mate, and following a short courtship with multiple copulations, the male leaves. Gestation is about 96 days, after which a litter of two to five well-haired, spotted kittens is born.

Other than humans, mountain lions (especially kittens and independent subadults) are killed mainly by other mountain lions. Because they kill livestock, especially goats and sheep, mountain lions are regularly hunted by ranchers. There are reports of a single rancher killing approximately 55 cougars between 1929 and 1939 in the Big Bend area (Borell and Bryant 1942). On the Stockton Plateau one rancher boasted of killing 56 of these carnivores in one year when the country was first being settled (Hermann 1950). Human-caused sources of mortality (e.g., shooting and trapping) in the Trans-Pecos are greater than natural sources of mortality (J. Young et al. 2010). For example, of 21 radio-collared mountain lions in BBRSP, 15 died from predator control trapping and another was shot by a hunter (P. Harveson et al. 2012).

In the Davis Mountains, researchers from Sul Ross State University have used radiotelemetry to track 21 captured mountain lions from 2011 to 2014 (Thompson et al. 2012; Dennison et al. 2016; P. Harveson et al. 2016). One of the most interesting findings is the low survival rate of this population. Annual survival was estimated to be 0.536, which means that only about 54 percent of the population survives from year to year. This is among the lowest survival rates of these elusive animals anywhere in the United States. The average daily movement of the lions was 3.7 km per day. Most males disperse entirely from their natal range, whereas females are more likely to establish residency adjacent to their natal range. Their home range averaged 39,400 ha but varied considerably between individuals. On average an individual lion was shown to be able to move across 25 or more private ranches.

Favored prey items in the Davis Mountains included mule deer, elk, white-tailed deer, javelina, and feral hogs. Interestingly, even though livestock were readily available, no livestock kills were documented. In BBNP, their diet has been demonstrated to vary depending on the availability of prey items (Leopold and Krausman 1986). Based on a scat analysis, lions were found to consume primarily deer during 1972–74 but smaller prey during 1980–81; these changes were partially the result of a significant decline in desert mule deer populations. Overall in the Trans-Pecos, deer and javelina were the predominant prey species, each occurring in 39 percent of stomachs analyzed, followed by nongame wildlife (13 percent) and domestic livestock (9 percent) (Heinen and Canon 1997).

A radiotelemetry study of mountain lions in BBRSP revealed an average density of 0.427 lions/100 km², which is considerably lower than average densities from other geographic areas in the United States (P. Harveson et al. 2012). Seasonal survival rates were lower during the general hunting season (0.750) than during the nonhunting season (0.931). Home ranges averaged 20,639 ha for females and 34,803 for males.

Radiotelemetry studies in BBNP focused on home ranges of lions and the potential for lion-human conflicts (Packard 1991). The results revealed that home ranges of adult males overlapped those of several females, that females in the Chisos Basin had smaller home ranges than females in the desert (perhaps because of greater numbers of prey), and that young females dispersed to areas more distant than did young males (Packard 1991).

In GMNP a decreasing trend in mountain lion sign was observed from fall 1987 to fall 1991, and an increasing trend in sign was observed from spring 1992 to spring 1996 (L. Harveson et al. 1999). A genetic study in GMNP indicated a high number of transient animals and, perhaps, an unstable population in the park as a result of intense hunting pressure outside the protected area (Gilad et al. 2011).

Predators in national and state parks can present risks to domestic animals and untrained people. Records of reported encounters between people and lions in BBNP indicate that interactions have occurred for over 60 years. Since 2004, there have been 21 mountain lion encounters, ranging from lions approaching people in a nonthreatening manner to encounters of greater severity, including actual attacks (P. Harveson et al. 2014).

In Texas, mountain lions are classified as nongame animals with no restrictions on harvest (L. Harveson et al. 1997; Russ 1997), which presents a serious conservation challenge given their low survival rate in the Trans-Pecos. The most important change that would benefit their conservation would be to classify the species as a game animal, as it is in New Mexico, with a limited harvesting season. Molecular genetic studies have revealed that mountain lions in Texas have less genetic variation than previously reported for the species in other parts of its range (Walker et al. 2000), and that there are genetic differences between populations in West Texas (BBRSP) and South Texas, suggesting that mountain lions within the state should be partitioned into two management units (Holbrook et al. 2012).

Two subspecies of the mountain lion were formerly considered to occur in the Trans-Pecos, *P. c. azteca* in El Paso County and the western edge of Hudspeth County, and *P. c. stanleyana* elsewhere in the region. However, a molecular phylogeographic analysis of the entire species has resulted in all North American mountains lions, including Trans-Pecos populations, being assigned to the subspecies *P. c. couguar* (Culver et al. 2000; Kitchener et al. 2017).

Family Mephitidae
SKUNKS
Desert Spotted Skunk
Spilogale leucoparia Merriam, 1890

This small, relatively slender skunk has a long, bushy tail that is white at the tip. Its pelage is black except for four white stripes on the back and two on the sides, which are broken near the middle of the back into large "spots" (fig. 45).

FIG. 45. Desert spotted skunk, *Spilogale leucoparia*

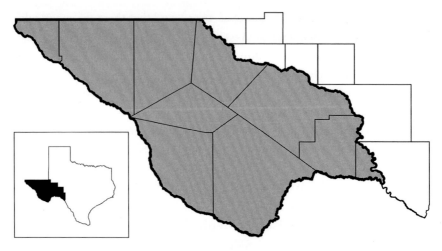

MAP 44. Distribution of the desert spotted skunk, *Spilogale leucoparia*

A large white spot is present between the eyes. Two large scent glands that produce the characteristic skunk odor are located on either side of the anus. They produce a musk unlike that of striped skunks. Average external measurements (in mm) for males are total length 398; tail 138; hindfoot 44; ear 27; weight is about 565 g; for females, respectively, 370; 138; 41; 25; and 368 g. Males are larger than females.

This skunk has been recorded from all counties of the Trans-Pecos except Hudspeth, and, based on the biogeography of the species and the presence of suitable habitat, it is projected to occur throughout the region (map 44). Despite their wide range in the Trans-Pecos, they do not appear to be abundant anywhere in the region. They are most often found in association with rocky canyons, cliffs, or brushy gulches where thickets or rock piles afford protection. They have been captured in rocky situations in piñon-juniper and riparian woodlands in the Guadalupe (Davis and Robertson 1944; Genoways et al. 1979) and Davis Mountains (Bailey 1905; Blair 1940), and in the grasslands around Balmorhea State Park (Patton 1974). A recent study of habitat use has shown that they have a decided preference for grassland areas with stands of honey mesquite. These types of areas are now often brush controlled for livestock management, and there are concerns that this could threaten their habitat (Neiswenter and Dowler 2007).

Spotted skunks are quick, acrobatic, and very adept at climbing trees or scampering over boulders. Their dens may be constructed in rock crevices, abandoned houses, barns, or private residences (Patton 1974). They feed primarily on insects and small vertebrates. Individuals held in captivity readily eat insects, kangaroo rats, pocket mice, fresh fish, and chicken eggs (Patton 1974).

Their reproductive cycle differs from that of other skunks. Copulation occurs in September, at which time both juvenile and adult females breed (Mead 1968). After fertilization, growth of the blastocyst is suspended and the embryos float freely in the uterus for approximately 180 to 200 days. Implantation does not occur until April. Most litters are born in May after a gestation period of 210 to 230 days. The average litter size is about four.

Spotted skunks are strictly nocturnal and extremely secretive. They are active year-round, although less so during extreme cold weather. They are usually solitary except during the fall breeding season. Unlike striped skunks, spotted skunks are quick and more weasel-like and are agile climbers of trees, bushes, and rocky ledges.

Their main predator is the great horned owl (Bailey 1905). When threatened or stressed, they will stamp their forefeet and stand to display their dorsal black-and-white coloration as a warning. If pressed, they will stand on their forefeet with their anal scent glands facing the threat and eject a strong musk usually aimed at the eyes of the aggressor.

Previously, spotted skunks in the Trans-Pecos were regarded as *Spilogale gracilis leucoparia*, a subspecies of the western spotted skunk. However, a recent molecular genetics study using DNA sequence data has resulted in the elevation of *leucoparia* to specific status, and thus spotted skunks in the region are now referred to as *S. leucoparia* (McDonough et al. 2022). This species is monotypic with no subspecies recognized.

Striped Skunk
Mephitis mephitis (Schreber, 1776)

This medium-sized, stout-bodied, bushy-tailed skunk has a narrow white stripe starting on its nose, continuing over its forehead, and dividing into a V behind its head (fig. 46). The snout has a vertical white line. There is considerable variation in stripe pattern in this species, and three different patterns have been observed in Trans-Pecos specimens—broad striped, narrow striped, and short striped (Patton 1974). Of these, the narrow striped is the most common and the short striped the least common (see fig. 47). Average external measurements (in mm) for males are total length 686; tail 325; hindfoot 69; ear 28; for females, respectively, 663; 325; 66; 27; weight is 1.4–6.6 kg, depending on age and sex. In general, males are larger than females.

FIG. 46. Striped skunk, *Mephitis mephitis*

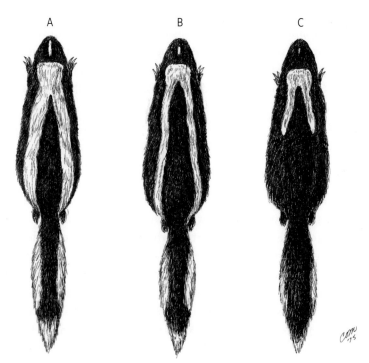

FIG. 47. Three stripe patterns in Trans-Pecos striped skunks: *A*, broad striped; *B*, narrow striped; *C*, short striped

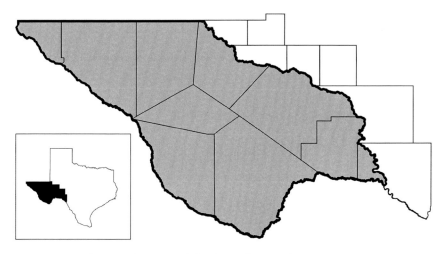

MAP 45. Distribution of the striped skunk, *Mephitis mephitis*

Striped skunks are the most common skunk in the Trans-Pecos, having been recorded from all counties in the region (map 45). They occupy a wide variety of habitats, including desert, grassland, and montane areas ranging from lower elevations to 2,100 m or higher, but they are not common in rough, rocky terrain. They can be especially abundant in agricultural areas and around human constructions. They are highly susceptible to being struck by vehicles, and road-killed animals may commonly be seen along highways throughout the region.

Striped skunks feed primarily on insects (about 78 percent of the diet) and other arthropods, although eggs, nestling birds, mature birds, mice, young rabbits, reptiles, and some plants are consumed (Patton 1974). They have been observed feeding along highways on carrion, in alfalfa fields, in wheat fields, under streetlights where large concentrations of insects are available, along stream beds, in city dumps, and in garbage cans of roadside parks.

Striped skunks are mostly nocturnal but can be active during early and late daylight. They use a variety of places for their dens, including rock crevices, houses, barns, and commercial buildings. Rarely do they construct their own dens, preferring instead to use those made and abandoned by other animals such as badgers. Females are in estrus from late February through March. The young are born in April and May after a gestation period of about 60 days. The average litter size is five.

Known predators include coyotes, foxes, bobcats, mountain lions, badgers, eagles, and great horned owls. Most mammalian predators likely avoid striped skunks except when near starvation. In response to a threat, striped skunks will first face the threat, elevate the back and tail, and stomp the ground with the forefeet. If the threat is not deterred, they will bend into a position where their eyes and anus face the threat and discharge a wide spray of musk from a pair of anal scent glands. They will also spray indiscriminately while being chased. The subspecies in the Trans-Pecos is *M. m. varians*.

Hooded Skunk
Mephitis macroura Lichtenstein, 1832

The hooded skunk is similar in appearance to the striped skunk and is some-times mistaken for the latter in areas where the species are sympatric. Hooded skunks differ in that the dorsal white never bifurcates (splits); the pelage is lon-ger and softer and the hair forms a cape or hood on the back of the head and neck (fig. 48); and the tympanic bullae are larger. Three color phases occur (fig. 49): a white-backed phase, with a single broad dorsal white band from the ears to tail tip; a black-backed phase, with two dorsolateral white stripes from behind the ears to the rump and sometimes the tail; and a combination phase, with a white back and two dorsolateral white stripes (Patton 1974). The first two are common, but the third phase is extremely rare. All phases have a vertical white line on the snout. Average external measurements (in mm) for males are total length 687; tail 357; hindfoot 68; ear 31; weight is 800–900 g; for females, respectively, 662; 365; 62; 28; 400–700 g. Males are larger than females.

The earliest records of the hooded skunk in the Trans-Pecos are from Pecos County in 1925 (Patton 1974) and Jeff Davis County in 1940 (Blair 1940). They were reported to be relatively common near Balmorhea in Reeves County in the early 1970s (Patton 1974). The last specimen-vouchered record was from Davis Mountains State Park in Jeff Davis County in 1999 (Yancey et al. 2017). Since then, camera trap studies have documented five photographic records of hooded skunks from BBNP in Brewster County (Jefferson et al. 2022). Overall, records are available from Reeves, Pecos, Jeff Davis, Presidio, and Brewster Counties, and the species is considered to occur only in the central portion of the Trans-Pecos (map 46). They seem to prefer rocky canyons or heavily vege-tated streams and valleys at lower elevations (Patton 1974). Many mammalo-gists are of the opinion that these skunks have declined dramatically in the last 50 years for unknown reasons (Schmidly et al. 2022). Clearly, they are rare today.

Hooded skunks are omnivores that feed on insects and other arthropods as well as small vertebrates, bird eggs, and some plants. They have been recorded consuming a desert shrew, deermouse, and prickly pear cactus fruit (Patton

FIG. 48. Hooded skunk, *Mephitis macroura*

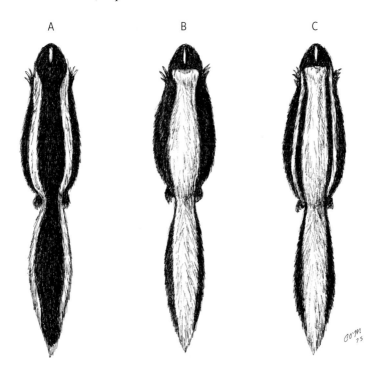

FIG. 49. Three color phases in Trans-Pecos hooded skunks: *A*, black-backed phase; *B*, white-backed phase; *C*, combination black and white backed phase

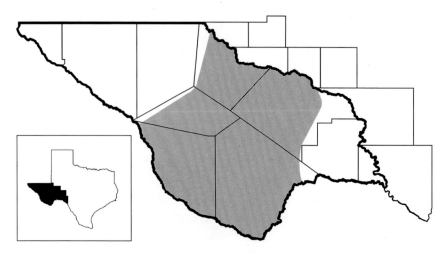

MAP 46. Distribution of the hooded skunk, *Mephitis macroura*

1974). When foraging, they move steadily over the ground and vegetation with their noses down and quickly pounce when a small animal is flushed. These skunks are strictly nocturnal and generally solitary, although multiple individuals may gather at a food source (Patton 1974). They are active throughout the year, and dens have been found in rocky crevices, earthen burrows, and within brush (Bailey 1931).

Mating occurs from mid-February to the end of March, when litters of three to eight young are born (Patton 1974). Females nurse their young until mid-August. Specific predators are unknown but are likely the same as for the striped skunk. Their defensive behaviors are similar to those of striped skunks. Feline distemper has been identified in a female and her litter from the vicinity of Balmorhea in Reeves County (Patton 1974). The subspecies in the Trans-Pecos is *M. m. milleri*.

Hog-nosed Skunk
Conepatus leuconotus (Lichtenstein, 1832)

This large skunk is characterized by a long, naked nose; very long and robust digging claws on the forefeet; a brownish-black body; and a solid white stripe that begins on top of the head and extends the entire length of the body (fig. 50). The tail is mostly white throughout, with a few scattered black hairs beneath, and is shorter in proportion to the body length than in other Trans-Pecos skunks. Average external measurements (in mm) for males are total length 602; tail 233; hindfoot 68; ear 27; for females, respectively, 551; 215; 65; 24; weight is 1.1–2.7 kg.

FIG. 50. Hog-nosed skunk, *Conepatus leuconotus*

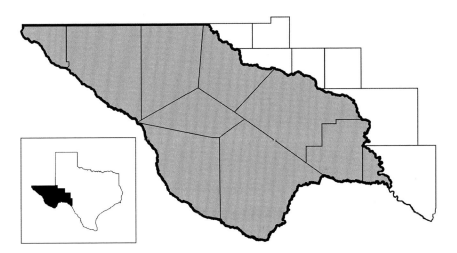

MAP 47. Distribution of the hog-nosed skunk, *Conepatus leuconotus*

Hog-nosed skunks have been documented from all but Hudspeth County in the Trans-Pecos, and based on the biogeography of the species they are projected to span the entire region (map 47). They occur in woodlands, grasslands, deserts, and rugged canyon terrain ranging from the Rio Grande Valley (564 m) to the Guadalupe Mountains (2,195 m). They are the most common skunk of the Guadalupe, Davis, and Chisos Mountains (Bailey 1905; Borell and Bryant 1942; Genoways et al. 1979).

These skunks feed extensively on insects, especially larval forms. They use their piglike nose and long foreclaws to find and dig out insects and larvae, and their diggings resemble those of feral pigs (Patton 1974). They are also known to consume small rodents, reptiles, and nestling rabbits. They are principally nocturnal and active year-round except during cold winter periods. They are solitary and den alone except for females with young. Dens have been found in cavities under boulders and within rock piles, crevices, cutouts, and woodrat nests.

Reproductive information is sparse and incidental. Breeding occurs during February–March, with a gestation of about 60 days. Young are born in late April to early May. Litter size ranges from two to four (average three) and is limited because each female has only three pairs of mammae (Patton 1974). Most litters are weaned by August, at which time young of the year begin to disperse (Patton 1974).

Predators include coyotes, gray foxes, bobcats, mountain lions, badgers, golden eagles, and great horned owls. When threatened, hog-nosed skunks will first attempt to flee into cover. If unsuccessful and further stressed, they will stand upright on their hind legs to appear larger and step toward the threat, stomping the ground with their forefeet and producing a loud hiss. As a last resort, a stressed individual will lay its tail flat against its back, expose its teeth, and bite or spray the threat.

Because they unearth soil for insects and larvae with their naked nose pad, hog-nosed skunks have been called "rooter skunks" (Stangl et al. 1994). In the previous edition of this book, this taxon was referred to as *Conepatus mesoleucus mearnsi*. All Texas hog-nosed skunks are now recognized as *Conepatus leuconotus*, and the subspecies that occurs in the Trans-Pecos is now *C. l. leuconotus* (Dragoo et al. 2003).

Family Mustelidae
WEASELS, OTTERS, AND BADGERS
Long-tailed Weasel
Mustela frenata Lichtenstein, 1831

Long-tailed weasels are slender, elongated little carnivores with a long, black-tipped tail; short legs; and small, rounded ears. The pelage is yellowish brown above and buffy orange below, with a dark brown face and ears and white markings between the eyes and across the snout (fig. 51). Average external measurements (in mm) for males are total length 488; tail 192; hindfoot 51; for females, respectively, 438; 187; 42; weight, in general, is 250–500 g. Males are larger than females.

The long-tailed weasel is rare in the Trans-Pecos. It has been recorded from El Paso, Culberson, Brewster, Pecos, Terrell, and Val Verde Counties (map 48)

FIG. 51. Long-tailed weasel, *Mustela frenata*

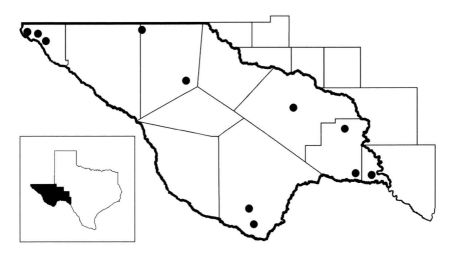

MAP 48. Distribution of the long-tailed weasel, *Mustela frenata*

but likely occurs throughout the entire region. The few specimens that have been collected were associated with permanent water such as near springs and along the Rio Grande. In El Paso County it occurs close to permanent water in the Rio Grande Valley and in the southern portion of the Franklin Mountains where several springs are found (Ederhoff 1971). A long-tailed weasel was shot along a rocky ridge near Kent in Culberson County while it was eating a woodrat (Davis and Robertson 1944). A weasel was sighted, but not collected, just outside the boundary of GMNP on the Hudspeth-Culberson county line (Genoways et al. 1979). In the Rosillos Mountains of BBNP (Harte Ranch), a partial skeleton, with jaw, was salvaged from a dry stock tank in creosote bush–catclaw desert scrub habitat at 1,000 m elevation (Yancey et al. 2006).

Long-tailed weasels are strictly carnivorous, with a diet predominantly of small mammals along with small birds, amphibians, reptiles, and insects. Among small mammals, known prey include shrews, ground squirrels, pocket gophers, woodrats, cotton rats, deermice, harvest mice, and cottontails. Prey are hunted by climbing trees and bushes, ground chasing, and searching burrow openings. The weasels subdue victims by biting the back of the neck while clasping them with their body and legs.

These weasels are active year-round and hunt during both day and night, although primarily at night. Because of their slender bodies, dens may be found in almost any cavity including burrows and under rocks, logs, and roots.

Long-tailed weasels are solitary except during the breeding season and when they are with their young. They are polygamous and breed in July–August. Delayed implantation occurs after an average of 252 days in about March–April. Postimplantation gestation is usually less than 27 days. About four to five young are born during April–May. Their primary predators are foxes and raptors as well as rattlesnakes, coyotes, and bobcats.

Recent work by Patterson et al. (2021) suggests that *Mustela frenata* should now be recognized as *Neogale frenata*, but these researchers also indicate the need for further revision of the taxon to support this generic change. We have taken a conservative approach and continue to regard *Mustela frenata* as the scientific name for the long-tailed weasel pending further research. The subspecies in the Trans-Pecos is *M. f. neomexicana*.

American Badger
Taxidea taxus (Schreber, 1777)

The American badger is a short-legged, squat, robust animal with a short, bushy tail and long front claws ($\geq$ 25 mm) (fig. 52). The upperparts are silvery gray with a tinge of brown; the underparts are yellowish white. A white stripe extends from the nose over the head to the middorsum or sometimes along the center of the back to the rump. The face has "badges"—patches of black surrounded by white pelage. The lower legs and feet are blackish. Average external measurements (in mm) for males are total length 788; tail

FIG. 52. American badger, *Taxidea taxus*

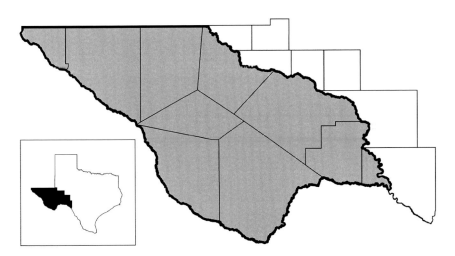

MAP 49. Distribution of the American badger, *Taxidea taxus*

133; hindfoot 120; for females, respectively, 730; 150; 114; weight is 4–10 kg. Males are larger than females.

American badgers have been recorded from all but Hudspeth and Terrell Counties in the Trans-Pecos, and, based on biogeographic patterns of the species, they are thought to occur throughout the entire region (map 49). They are distributed over the Trans-Pecos without much regard for climate or physiographic features, but nowhere do they appear to be common. They occur in both grasslands and deserts where loose soils allow for suitable burrowing. Their occurrence seems to be correlated with the presence of burrowing rodents.

Badgers feed on black-tailed prairie dogs, ground squirrels, and kangaroo rats; other foods include cottontails, birds, eggs, lizards, snakes, insects, scorpions, and carrion. They are substantially fossorial, are active throughout the year, and may be seen day or night. Except during the breeding season, they are mostly solitary. Badgers have been seen traveling with coyotes and teaming with them to catch rodents, especially prairie dogs.

Badgers breed in late summer to early fall. After fertilization, each blastula is held suspended in the uterus until delayed implantation occurs during December–February. Litters of one to five young are born in March–April. Predation on adult badgers is uncommon, but younger individuals have been taken by coyotes and golden eagles. Anal scent glands may be used in defense, although the scent is not disturbing to humans. The subspecies in the Trans-Pecos is *T. t. berlandieri*.

Family Procyonidae
RACCOONS, RINGTAILS, AND COATIS
Ringtail
Bassariscus astutus (Lichtenstein, 1830)

The ringtail is a cat-sized carnivore with the face of a fox, prominent white rings around its eyes, and a long, bushy tail that has about 14 to 16 alternating black and white rings and a black tip (fig. 53). The tail is approximately the same length as the head and body, and the black and white rings give it a resemblance to the tail of a raccoon or coati. The ringtail differs from the raccoon in lacking a black facial mask. It may be distinguished from the coati by its smaller size and more distinctly ringed and bushy tail. Average external measurements (in mm) for males are total length 793; tail 387; hindfoot 67; ear 44–50; weight is 1–1.75 kg; for females, respectively, 756; 383; 67; 44–50; 1–1.5 kg. Males are significantly larger than females in cranial measurements (Stangl et al. 2014).

Ringtails are widely distributed across the Trans-Pecos except in the northeastern sector of the region (map 50). They commonly occur in mountainous, rocky, or eroded environments and are generally associated with rocky habitat and rugged terrain (Ackerson and Harveson 2006). Preferred areas are canyons, rocky gulches, cliffs, and boulder piles; they are not found in flat, open terrain. A study conducted at EMWMA revealed that they prefer

FIG. 53. Ringtail, *Bassariscus astutus*

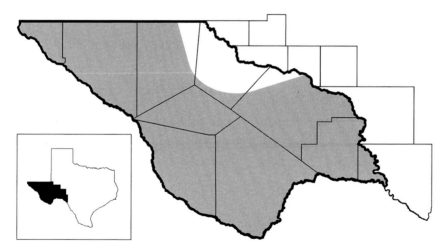

MAP 50. Distribution of the ringtail, *Bassariscus astutus*

red catclaw–persimmon–oak bottom and catclaw–goldeneye–sideoats slope plant communities. Their summer range size averaged 0.28 km², and in the winter range size increased to 0.63 km². The minimum population estimate at this site was 59 ringtails (Ackerson and Harveson 2006).

Ringtails are omnivores that consume predominantly small mammals and seasonal fruit, berries, seeds, and nuts. To a lesser extent, insects, other invertebrates, birds, mammals, and reptiles are taken. At BBRSP, two ring-tails were seen foraging in a small cave high on a sheer cliff that housed a large ghost-faced bat colony. In BBNP, a ringtail was observed climbing a century plant and lapping nectar from open flowers (Kuban and Schwartz 1985). At EMWMA, plant and animal material were found in 75 and 87 percent of their scats, respectively. Typically, they consume more plant material in the non-breeding season and more animal material in the breeding season (Ackerson and Harveson 2006).

Ringtails are active year-round and are almost completely nocturnal. Except for breeding, they are solitary and usually den alone in dense brush piles, hollows in trees, or rocky crevices. At EMWMA they used rock dens exclusively, 80 per-cent of which occurred on slopes (Ackerson and Harveson 2006). Equipped with semiretractile claws, they are quick, dexterous climbers and climb down-ward headfirst by rotating their hindfeet 180°. Ringtails are vocal and use several different calls, including a barking alarm-stress call. They use urine to scent mark and accumulate feces at latrine sites to mark their home ranges.

Little is known of their reproductive habits in the Trans-Pecos. An old female caught near Boquillas on May 27 contained three well-developed fetuses (Bailey 1905). Baby ringtails not more than six inches in total length have been taken on the Stockton Plateau in late June (Hermann 1950). Litter size averages about three, with extremes of two and four. Predators include great horned owls and to a lesser extent bobcats, coyotes, and raccoons (Ackerson and Harveson 2006). The subspecies in the Trans-Pecos is *B. a. flavus*.

White-nosed Coati
Nasua narica (Linnaeus, 1766)

White-nosed coatis are about the size of raccoons. They have long, slender snouts; long, hairy tails often carried erect; flat-footed feet; short ears; and coarse, stiff hair (fig. 54). Their general color is a rusty brown, with indistinct light markings on the nose, face, and shoulders and usually about six or seven indistinct light bands on the tail. They have a long, decidedly flexible snout that extends substantially beyond the lower lip. External measurements (in mm) of an adult male are total length 1,100; tail 500; hindfoot 91; ear 30; weight is 4–5 kg.

The coati is thought to be the largest native animal to make its first appear-ance in the US borderlands in the last 135 years (Gehlbach 1981). They are rare in the Trans-Pecos, having first been documented in the Dead Horse Mountains in Brewster County (F. Tabor 1940). Subsequently, several sightings (none con-firmed by specimens) of solitary individuals have been made in the region within

FIG. 54. White-nosed coati, *Nasua narica*

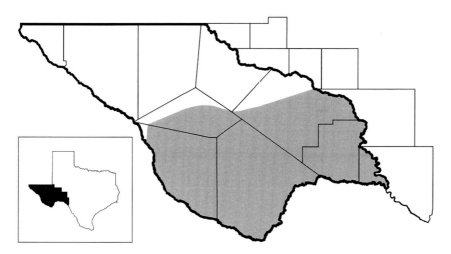

MAP 51. Distribution of the white-nosed coati, *Nasua narica*

about 160 km of the Mexican border (Schmidly et al. 2016a). The most numerous sightings have been from BBNP, but others have been made in Presidio County; the lower canyons of the Big Bend (Brewster, Terrell, and Val Verde Counties); the Pecos County/Crockett County border (near the Interstate 10 bridge over the Pecos River); Terrell County; the Davis Mountains in Jeff Davis County; and Val Verde County (Schmidly et al. 2016a). Based on these accounts, white-nosed coatis are considered to be restricted to the southeastern portion of the Trans-Pecos (map 51). In BBNP, they have been sighted in juniper-oak-pine forests as well as in desert and riparian habitats near the Rio Grande.

Coatis often are indiscriminately called coatimundis, though the latter term properly refers only to solitary adult males. In contrast to most procyonids, they are mainly diurnal in activity and maintain a matriarchal social system with the females and young males in bands and the adult males usually solitary. They spend nights in trees where available and may use crevices and rocky undercuts in areas with few trees. Most of their time is spent on the ground although they are proficient climbers. Bands of females with young have not been sighted in the Trans-Pecos, and there is no evidence that a breeding population has become established in the region (Schmidly et al. 2016a).

Coatis feed predominantly on invertebrates and fruit, although they will consume lizards, snakes, rodents, carrion, nuts, and prickly pear cactus. Predation of young coatis can be high, with coyotes, bobcats, mountain lions, and large raptors the most likely culprits. There is concern about their conservation status in the United States and suggestions that the species likely warrants complete legal protection. It is listed as threatened by the TPWD but is not listed by the USFWS because of the relatively stable populations in southeastern Arizona. The subspecies in the Trans-Pecos is *N. n. molaris*.

Northern Raccoon
Procyon lotor (Linnaeus, 1758)

Northern raccoons are rather stocky, short-legged, grayish to blackish carnivores with a pointed nose, a black facial mask, a bushy tail encircled by five or six black rings, and large, flat-footed feet with naked soles (fig. 55). Average external measurements (in mm) are total length 944; tail 373; hindfoot 130; ear 76; weight is 4–5 kg.

Northern raccoons are one of the most ubiquitous mammals in the Trans-Pecos. They have been documented from every county but Hudspeth, and the species undoubtedly occurs throughout the entire region (map 52). They are most often associated with natural and human-made water sources and are most abundant along the valleys of the Pecos River and Rio Grande. In BBNP they were commonly trapped among the canes, reeds, and salt cedar associated with the Rio Grande (Schmidly 1977b). In El Paso County they live among the mesquites and salt cedars surrounding the few oxbow lakes along the Rio Grande (Ederhoff 1971). In GMNP they are relatively abundant in the riparian communities in the canyons along the eastern and northern slopes of

FIG. 55. Northern raccoon, *Procyon lotor*

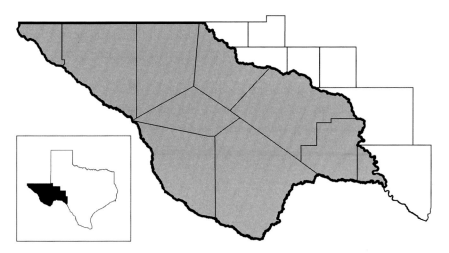

MAP 52. Distribution of the northern raccoon, *Procyon lotor*

the mountains (Genoways et al. 1979). They also occur in similar situations in the Davis and Chinati Mountains, and in BBRSP.

Raccoons are omnivores with a diverse diet that includes seasonal fruits, berries, seeds, nuts, invertebrates, and small vertebrates, especially those in and around water sources. Fruits of prickly pear, persimmon, lotebush, juniper, and oak are also readily consumed. They are known to gorge on seasonal resources and build enormous body fat reserves for the winter months. They are selective when food is abundant but will eat anything available when resources are scarce.

Raccoons are primarily nocturnal and active year-round. They do not hibernate but may remain in a den for long periods during cold winter months. Adults are usually solitary except during the breeding season. Their dens are often associated with trees but where trees are scarce or scattered, they will build dens in rocky ledges, crevices, cutouts, and caves.

Breeding in the Trans-Pecos occurs during February–March. A litter of two to four young is born after a gestation period of about 63 days. Predators include coyotes, bobcats, mountain lions, and great horned owls. Diseases such as rabies, canine distemper, leptospirosis, tularemia, and Chagas disease are known to infect raccoons.

Historically, two subspecies of the northern raccoon, *P. l. mexicanus* in the western two-thirds and *P. l. fuscipes* in the eastern third, have been documented in the Trans-Pecos (Goldman 1950). The difference in the two subspecies was based on coloration, with *P. l. fuscipes* having a darker pelage than *P. l. mexicanus*. However, a recent examination of cranial measurements in specimens from Mexico and Central America concluded that *mexicanus* was not a valid subspecies (Helgen and Wilson 2005). Consequently, modern taxonomic compilers placed *mexicanus* in synonymy with *P. l. hernandezii* (a Mexican subspecies), with the implication that Trans-Pecos populations should be referred to the single subspecies *P. l. fuscipes* (Wilson and Reeder 2005). Further study is needed to determine the exact subspecific status of raccoons in the Trans-Pecos.

Family Ursidae
BEARS
American Black Bear
Ursus americanus Pallas, 1780

Because of its large size and unique appearance, the American black bear cannot be mistaken for any other mammal that occurs in the Trans-Pecos. It is a large, bulky, flat-footed mammal with black or brown hair, an exceedingly short and inconspicuous tail, a rather blunt facial profile, small eyes, and a broad nose pad with large nostrils (fig. 56). It has five toes with short, curved claws on all four feet. Average external measurements (in mm) are total length 1,500; tail 125; hindfoot 175; height at shoulder 625. Typical weight is 100–150 kg, with some individuals weighing as much as 225 kg.

FIG. 56. American black bear, *Ursus americanus*

Prior to the middle of the last century, the American black bear was common throughout the forested mountains and even some rugged lowlands of the Trans-Pecos. Bailey (1905) reported it as abundant in the Chisos, Davis, and Guadalupe Mountains in the early 1900s. Over the 30 years following Bailey's survey, however, the black bear population of the Trans-Pecos became greatly reduced. In the 1930s they remained fairly common in the higher elevations of the Chisos Mountains (Borell and Bryant 1942), but by the end of the decade they were only rarely encountered in the Davis Mountains (Blair 1940), and the population in the Guadalupe Mountains was thought to be no more than 25 individuals (Davis 1940b).

During the latter part of the twentieth century the species continued into a steep decline as a result of hunting pressures and overgrazing by livestock, and by the mid-1970s there was no evidence that resident populations remained anywhere in the region (Schmidly 1977b). From 1974 to 1989 the TPWD documented only three sightings, and those individuals were thought to be transients that wandered into the Trans-Pecos from Mexico or New Mexico. During the 1990s there was a significant increase in reported bear sightings in the Trans-Pecos, but verifiable accounts remained scarce. Then from 2000 to the present a drastic increase in verified accounts of bears in the region led to the confirmation of occurrence in all counties except El Paso and Hudspeth (Light et al. 2021). In 2017 the species was documented in the Chinati Mountains for the

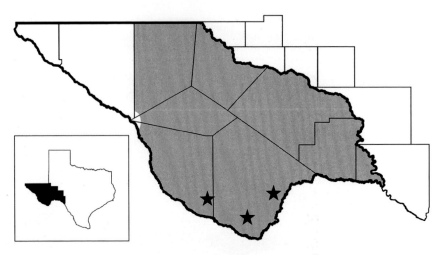

MAP 53. Distribution of the American black bear, *Ursus americanus*. Stars represent currently known established breeding populations; from west to east, BBRSP (Presidio County), BBNP (Brewster County), and BGWMA (Brewster County).

first time (Yancey and Lockwood 2018). During this same period, females with cubs were sighted in Pecos, Jeff Davis, Presidio, Brewster, and Terrell Counties. Of particular significance is the documentation of persistent, naturally reestablished breeding populations (sows with cubs for two or more consecutive years) in BBNP, BBRSP, and BGWMA (Onorato et al. 2004; Yancey and Manning 2018) (map 53).

Natural recolonization of former range by large terrestrial carnivores, such as is happening with black bears in the Trans-Pecos, is rare (Schmidly 2002; Onorato et al. 2004). Recolonization in the region seems to have commenced shortly after a moratorium on black bear hunting in Mexico was implemented in 1986, and it is speculated that an increase in bear numbers as a result of the hunting ban facilitated widespread dispersal into the Trans-Pecos from south of the border. Indeed, genetic studies indicate that the sources of at least some of the immigrant bears are a few isolated mountain ranges in northern Coahuila, Mexico (Onorato et al. 2004). Other possible sources of immigrants include northern Chihuahua, Mexico, and eastern New Mexico.

Black bears are most often found in montane forests and foothill woodlands, especially those dominated by oak and oak-pine associations such as in the Chisos Mountains of BBNP. However, they also occur in desert lowland habitats, and they have been observed in desert scrub habitat dominated by creosote bush, sotol, mesquite, and prickly pear cactus in both BBNP and BBRSP (Onorato et al. 2003; Yancey and Manning 2018).

Throughout most of their range, black bears usually enter a period of dormancy during winter but do not go into a deep state of torpor or hibernation characterized by a marked decrease in body temperature and heart rate. As a result, denning bears may awaken and become active periodically during winter. In the Trans-Pecos, it seems that most bears enter this type of winter inactivity, especially gravid females who give birth to young while denned up for winter. Alternatively, some bears may remain active throughout winter, particularly females with yearlings (Mitchell et al. 2005).

Black bears are generally crepuscular; however, recent camera trap images from BBRSP (Yancey and Manning 2018) and CMSNA (Yancey and Lockwood 2018) indicate they may be active day or night. Although they may occasionally feed on invertebrates and vertebrates, these bears typically consume mostly plant material (McClinton et al. 1992; Hellgren 1993; McKinney and Pittman 2000). Bears from montane woodlands rely heavily on acorns, pine seeds, and juniper berries, whereas lowland desert bears subsist mostly on yucca, sotol, mesquite beans, prickly pear fruits, and acorns from low-elevation oaks (Onorato et al. 2003). Black bears may opportunistically supplement their diet with carrion and livestock kills.

Across their range, American black bears usually breed in June or July, followed by a gestation period of about 210–220 days. Birth of one to four (rarely five, usually two) cubs occurs in January or February while the mother is dormant in her winter den. Humans are the only predators of black bears in the Trans-Pecos. Excessive hunting, poaching, and depredation have had profound effects on populations in the past. Consequently, the species is listed as threatened by the TPWD. American black bear populations in the Trans-Pecos have historically been assigned to the subspecies *U. a. amblyceps*. However, because of the apparent extirpation and subsequent recolonization of the region, potentially from multiple locations with different subspecies (e.g., *U. a. amblyceps* from New Mexico and *U. a. eremicus* from Coahuila), the subspecific status of American black bears in the Trans-Pecos is ambiguous. Therefore, we refrain from referring Trans-Pecos populations to a particular subspecies pending further research on the subject.

ORDER ARTIODACTYLA (Even-Toed Ungulates)

The artiodactyls, or "even-toed" ungulates, include many iconic species (deer, elk, sheep, pronghorn, and bison) that are native to the Trans-Pecos, where they are important for sport hunting and ecotourism. The principal distinguishing feature of the order is the structure of the feet, which have an even number of digits (except for the peccaries, which have three toes on the hind-feet) that are covered by terminal hooves. Some representatives have a complex, four-chambered ruminant stomach that permits them to chew food, regurgitate it (the cud), and rechew it to enhance digestion.

Peccaries (family Tayassuidae) have tusk-like canines, a nonruminant stomach, four toes on the front feet, and three toes on the hindfeet. Deer and elk (family Cervidae) are characterized by the presence of deciduous antlers (found only on males), which are shed and replaced annually. The pronghorn (family Antilocapridae) has unique horns (on both males and females) with bony cores covered by a horny keratin sheath; the sheath but not the core is shed and replaced annually. Bighorn sheep (family Bovidae) have horns (present on both sexes) that are formed over and supported by bony cores; the horns are not shed but grow continuously. Cervids, antilocaprids, and bovids all have ruminant stomachs. Because the feral hog (*Sus scrofa*) has so successfully invaded the Trans-Pecos, it has been included in the key with the native species of artiodactyls.

1. Medium size; body form stocky and barrel-like; snout with terminal nasal disk (piglike in shape); upper incisors present; canines elongated 2
    Large size; body form slender or cowlike; snout never piglike in shape; upper incisors absent; canines absent or, if present, not elongated 3
2. Sparsely covered with coarse bristly hair; some individuals with a scantily haired dorsal mane; tail approximately 300 mm in length; each foot with four toes (the middle two flattened and with hooves; the lateral toes higher up on the limb and normally not touching the ground); incisors 3/3 and cheek teeth 7/7; adults weighing up to 350 kg . . . . **Feral Hog**, *Sus scrofa*
    Pelage thick and bristly; well-developed dorsal mane of long, stiff hairs extending along back from crown to rump; tail length 15–55 mm; front feet with four toes, hindfeet with three toes; incisors 2/3 and cheek teeth 6/6; adults weighing 14–30 kg . . . . **Collared Peccary**, *Pecari tajacu*
3. Males with antlers; females without antlers or horns 4
    Males and females with horns (permanent bony core covered with horny sheath) 6
4. Upper canines present; prominent cream-colored rump patch present in adults . . . . **Elk or Wapiti**, *Cervus canadensis*
    Upper canines absent; no prominent rump patch 5

5. Tail brown above, white below, and fringed with white laterally;
    antlers with one main beam from which tines rise vertically;
    ears about one-half length of head . . . .
    . . . . **White-tailed Deer**, *Odocoileus virginianus*
    Tail tipped with black, and white or black above; antlers branched
    dichotomously, with anterior and posterior beams nearly equal;
    ears about two-thirds length of head . . . .
    . . . . **Mule Deer**, *Odocoileus hemionus*
6. Two toes on each foot; horns forked . . . .
    . . . . **Pronghorn**, *Antilocapra americana*
    Four toes on each foot; horns not forked 7
7. Large, cowlike; height at shoulder more than 1 m; horns smooth
    and directed anteriorly . . . . **Bison**, *Bos bison*
    Small, sheeplike; height at shoulder less than 1 m; horns with annular
    ridges and posteriorly curved . . . .
    . . . . **Mountain or Bighorn Sheep**, *Ovis canadensis*

Family Antilocapridae
PRONGHORN
Pronghorn
Antilocapra americana (Ord, 1815)

The pronghorn is a medium-sized, deerlike artiodactyl with forked horns with outer sheaths that are shed and replaced annually. The horns reach beyond the tip of the ears in males; in females they are shorter and seldom pronged. Conspicuous white patches are present on the sides and rump, and two white bands extend across the throat (fig. 57). Their general color is tan, with a white rump and underparts and black markings on the neck and face. A buck and doe (estimated to be approximately seven years old) taken in Presidio County had the following measurements (in mm): total length of both 1,325; height of buck at shoulder 908; height of doe at shoulder 862; tail of both 62; hindfoot of both 388; ear of buck 150; ear of doe 148.

The general range of pronghorn in the Trans-Pecos includes most of the region except for the southern areas adjacent to the Rio Grande (map 54), although their occurrence throughout this range is sporadic. Before 1880 they were extremely abundant, but because of overhunting they almost disappeared from the area by 1900 (Nelson 1925). Estimates indicate that by 1924 only 692 were left. In 1930 the population began to steadily increase, reaching a maximum of over 12,000 individuals in 1961 (Hailey et al. 1964). Then a drought

FIG. 57. Pronghorn, *Antilocapra americana*

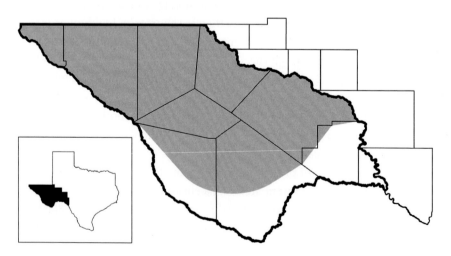

MAP 54. Distribution of the pronghorn, *Antilocapra americana*

began in 1961, and their numbers decreased to about 4,900 (Hailey et al. 1964). From 1964 to 1965, drought conditions severely depleted the range and 60 percent of the population inhabiting the Marfa Plateau in Presidio County died, mostly because of malnutrition coupled with blackbush toxicity (Hailey et al. 1964). With improved conditions, the population increased to an estimated 17,226 animals by 1987 but then sharply declined again to approximately 2,751 individuals by 2010. Current estimates indicate a slight recovery to approximately 6,500 individuals, largely thanks to the translocation of animals from the population in the Texas Panhandle (Woodward 2020).

Initially, the decline of pronghorn in the Trans-Pecos was associated with the encroachment of the sheep industry (Buechner 1950), but recent studies have shown that the primary factors influencing pronghorn numbers are drought, overgrazing, barriers to movement, degradation or loss of suitable habitat, and predation (Sullins 2002). Drought and overgrazing have major impacts on nutrition, fawning cover, and quality of habitat (Sides et al. 2006; Simpson et al. 2007). Barriers (e.g., fencing) can have a critical impact on nutrition, impede access to water and cover, and increase susceptibility to predation. Long-term changes in vegetation composition may degrade suitable habitat and increase vulnerability to predation (Sides et al. 2006). Predation, particularly by coyotes, can reduce local populations where predators are abundant. Worm infestations (barber pole worms and blood worms) are known to cause mortality in malnourished animals (Lightfoot 2010).

The combined cumulative effect of these factors (and possibly others) has lowered the capacity of pronghorn range in the Trans-Pecos to support high numbers of animals under present conditions, and climate change has been predicted to make the conditions "increasingly inhospitable to pronghorn persistence" (Gedir et al. 2015). As populations have declined, the TPWD has reduced the number of hunting permits in an effort to help their recovery.

Today, pronghorn are most numerous near Alpine and Marfa in Presidio, Brewster, and Jeff Davis Counties, and on the Diablo Plateau in Hudspeth County. Optimal habitat is open-country grassland or shrubland communities, with varying successional stages caused mainly by fire, which provides adequate visibility and mobility; pronghorn do not do well in woodlands or agricultural areas (Richardson 2006; Lightfoot 2010; Duncan et al. 2016). South-facing slopes with more food, protection from cold north winds, and more sunshine are particularly important to them. In wet seasons, draws and limestone hills are selected more often, whereas in drier seasons clay flats seem to be preferred (French et al. 2019).

Buechner (1950) documented the life history and ecology of pronghorn in the Trans-Pecos, and in the last two decades additional studies have been undertaken to better understand their natural history. The breeding season extends from the last week in August to the first week in October. Gestation is from seven to seven and a half months, and fawns are born from the first week in April to the latter part of May. The usual number of fawns is two; single fawns are not uncommon and occasionally three are born. Survivorship of fawns is influenced by the quality of their bedding sites (Tucker and Garner 1983).

Pronghorn feed primarily on forbs (broad-leaved weeds, herbs, and wild-flowers). They will significantly browse cholla and prickly pear cactus during drought and will consume moderate amounts of juniper where available (Nelle 2006). Grasses are the least important forage class, but immature grasses may be seasonally important in early spring in some areas (Richardson 2006). Winter energy availability can be important to their survival (French 2015). Important year-round foods include bitterweed, cut-leaf daisy, sideoats grama, blue grama, dalea, eriogonum, deer vetch, paperflower, coneflower, and woolly senecio. Plants toxic to livestock, such as paperflower, groundsel, broom snake-weed, and goathead, are also consumed.

Radiotelemetry studies in Hudspeth County (Canon 1993; Canon and Bryant 1997a, 1997b, 2006) have demonstrated that home ranges average 83.82 km² (Hoffman 2015) and that males have significantly smaller home ranges than females. Females are thought to require larger home ranges because of their greater nutritional demands. Home range size and move-ment rates are significantly higher in dry seasons compared to wetter seasons (Garrison 2015).

About 81 percent of fawn mortality can be attributed to predation (Canon and Bryant 1997b, 2006). Fawn predation comes primarily from coy-otes (81 percent), mountain lions (7 percent), and golden eagles (5 percent). Adequate cover for concealment (bunchgrasses, shrubs, or cacti) is critical to fawn survival (Richardson 2006).

Genetically, populations of pronghorn from the Marathon Basin in Presidio, Jeff Davis, and extreme southern Culberson Counties are different from those in Hudspeth and northern Culberson Counties (T. Lee et al. 1989, 1994; Keleher 2010). This genetic subdivision is thought to result from a combi-nation of isolation caused by Interstate 10 and mountain ranges that bisect the populations, and a restocking legacy from the Texas Pandhandle (subspecies *A. a. americana*) into Hudspeth and northern Culberson Counties that affected the genetic structure of populations in those areas. The subspecies in the non-restocked areas of the Trans-Pecos is *A. a. mexicana*.

Family Bovidae
CATTLE, BISON, ANTELOPE, SHEEP, AND GOATS
Bighorn Sheep
Ovis canadensis Shaw, 1804

The bighorn is a large, dark brown sheep. Males have massive, curled horns and weigh from 75 to 155 kg (fig. 58). Females have slender horns and are slightly smaller. Average external measurements (in mm) of males are total length 1,750; tail 100; hindfoot 450. Females average about 10 percent smaller.

The bighorn formerly inhabited most of the desert mountain ranges in the Trans-Pecos. Approximately 1,500 bighorns were estimated to inhabit the area in the mid-1880s, but then they experienced a steady decline caused by unregulated hunting, disease, interspecific competition with domestic sheep, and predation (Cook 1994). Bailey (1905) estimated that their numbers had declined to 500 individuals by the beginning of the twentieth century. He attributed the primary cause to introductions of large numbers of domestic

FIG. 58. Bighorn sheep, *Ovis canadensis*

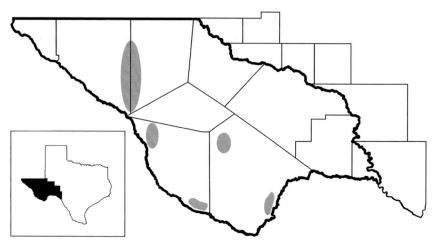

MAP 55. Distribution of the bighorn sheep, *Ovis canadensis*

sheep and goats. As late as 1937 they were still known from at least 15 different mountain ranges in the region (Davis and Taylor 1939). But by 1941 there were no more than 150 wild sheep remaining, and by 1945 there were fewer than 100 in the Beach and Baylor Mountains and in the Sierra Diablo north of Van Horn (Davis 1961). The last known sighting of the native desert bighorn (subspecies *O. c. mexicanus*) occurred in 1960, when two ewes were observed by TPWD employees in the Sierra Diablo (Hailey 1975).

Restoration efforts by the TPWD began in the mid-1950s at BGWMA (Cook 1994), and although early efforts had limited success because of disease and predation, recent efforts have been more successful and bighorn sheep now occur in several mountainous areas of the region (Locke et al. 2005). The reintroduced animals, however, were not the native desert bighorn but came primarily from populations of two other subspecies (*O. c. canadensis* and *O. c. nelsoni*). Since the 1980s, small, wild populations of these two subspecies have been released in the Sierra Diablo–Baylor–Beach Mountains complex, the Sierra Vieja, the Van Horn Mountains, EMWMA, BBRSP, and BBNP (map 55). In 2004, desert bighorn sheep in the Trans-Pecos were estimated at 500 individuals in seven mountain ranges (Brewer and Harveson 2007). Today, an estimated 1,500 individuals exist in 10 mountain ranges, and apparently the number of individuals and their distribution are expanding (Schmidly and Bradley 2016).

Bighorn sheep prefer rough, rocky, mountainous terrain with sparse vegetation. Essential habitat components include topography, food, water, and space. They select habitat based on factors such as proximity of steep-sloped escape terrain, forage availability, and horizontal visibility. They typically avoid areas with dense vegetation, which hampers horizontal visibility and leaves them vulnerable to ambush predators (O'Brien et al. 2014). The amount of habitat available is ultimately determined by the amount of escape terrain close to open landscapes (Brewer 2002).

They are highly opportunistic and adaptable in their food habits. Season, availability, and habitat type are the determining factors in their diet, but in the Trans-Pecos their diet is dominated by shrubs. At EMWMA, desert bighorn diets consisted of 50 percent browse, 35 percent forbs, 11 percent grasses, and 4 percent succulents, with a total of 94 plant genera consumed (Brewer and Harveson 2007). Diets of rams and ewes did not differ annually, but there were seasonal differences, with forbs prevailing during winter (Brewer and Harveson 2007). Another study conducted in the Sierra Diablo and Beach and Baylor Mountains revealed that forbs instead of browse plants made up the majority of the bighorn diet (Fulbright et al. 2001).

Water, which they obtain from freestanding or running sources, condensation, or forage, may be the most crucial factor in their distribution and survival. Most biologists consider freestanding water to be an essential component of desert bighorn habitat (Brewer 2002).

Desert bighorns are social animals that remain in small groups much of the year (Brewer 2002). Ewes and rams reach sexual maturity at about two and three years of age, respectively. During the nonbreeding season, mature males live apart from females. Rams form bachelor groups and inhabit less suitable habitat than ewes and juveniles. Groups are largest during the breeding season or rut when males and females aggregate. The length and season of the rut vary. Carson (1945) reported a rutting period of November–December in the original native bighorn populations. However, July–September seems to a be a more appropriate description of the breeding season among current bighorn populations (Brewer 2002).

The reproductive potential for desert bighorn sheep is considered low (Brewer 2002). Birth occurs after a gestation period of 173–185 days. Ewes typically produce a single lamb, although twinning occurs infrequently. In Trans-Pecos bighorn populations, newborn lambs have been observed as early as November and as late as June (Brewer 2002). Desert bighorns that survive their first year can be expected to reach 10–12 years of age. Their primary predator is the mountain lion.

Success in maintaining large, self-sustaining desert bighorn herds will depend on large blocks of available habitat with sufficient escape terrain. Limiting contact with domestic livestock and exotic sheep species (e.g., aoudad) to prevent disease transmission and interspecific competition is absolutely essential. Predator management to allow populations time to become established is critical, as is the availability of freestanding water to sustain the animals (Locke et al. 2005; Dolan 2006; Drew et al. 2014; O'Brien et al. 2014). A recent incident captured on a camera trap documenting an adult male bighorn sheep within a group of aoudad at BBRSP (Kasper and Yancey 2020) demonstrates how easily diseases could be transmitted from aoudad to bighorn sheep.

Family Cervidae
DEER AND ALLIES
Elk or Wapiti
Cervus canadensis Erxleben, 1777

The elk is the largest of the Trans-Pecos deer. The color of its upperparts is a light grayish brown to dark brown, with a characteristic buffy or whitish patch on the rump. The antlers (grown only by males) consist of long, round beams sweeping up and back from the skull, with each beam usually bearing six points or tines without palmation; they are shed annually (fig. 59). External measurements (in mm) of males are total length 2,032–2,972; tail 80–213; hindfoot 464–660. Males average about 300 kg but may weigh up to 500 kg. Females are smaller than males.

Elk were apparently native in the southern part of the Guadalupe Mountains at one time (Bailey 1905). Those elk belonged to the species *Cervus merriami*, which resembled the American elk but was paler and more reddish, with a more massive skull and more erect antlers. It once roamed in the Guadalupes and perhaps other forested areas in the Trans-Pecos, but the natural distribution of this species decreased substantially in the early twentieth century, presumably resulting in the extirpation of populations in Texas. Shortly afterward, reintroductions of elk from other places began.

The first reintroductions took place in the Guadalupe Mountains in 1928 when 44 individuals of the subspecies *C. c. canadensis* were brought in from the Black Hills of South Dakota and released in McKittrick Canyon (C. Carpenter 1961). They multiplied rapidly and extended their range to nearly all parts of the mountains, although they seemed partial to the higher parts of the Guadalupes despite the scarcity of water at those elevations (Davis 1940b). The estimated population size in 1938 was 400, and by the mid-1960s their numbers had declined to approximately 350 animals. A severe reduction in population size commenced thereafter, with only 108 individuals reported in a 1976-78 census (Moody and Simpson 1979) and a further decrease to 58 by 1983 (McAlpine 1990). Probable factors for this decline include human intervention, decline

FIG. 59. Elk, *Cervus canadensis*

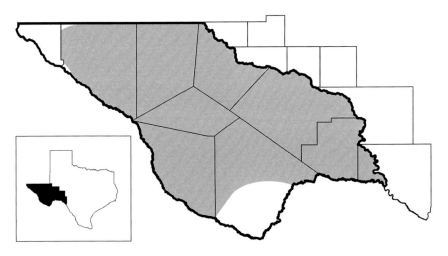

MAP 56. Distribution of the elk or wapiti, *Cervus canadensis*

in forage quality, predation, water loss, and changing climatic conditions (J. Carpenter 1993; Guevara et al. 2008). Today only a few elk remain in the park.

In the past 40 years, other herds of free-ranging elk have become established in the Trans-Pecos (Dunn et al. 2017), primarily the result of reintroductions by landowners and state agencies, and the species can now be found throught most of the region (map 56). Places where major reintroductions have taken place include the Glass Mountains, Wylie Mountains, Davis Mountains, and Eagle Mountains. These introductions were thought to be Rocky Mountain elk (subspecies *C. c. nelsoni*). Aerial surveys conducted by the TPWD in 2010 confirmed free-ranging herds of elk in Brewster, Jeff Davis, Pecos, and Terrell Counties, and landowner surveys indicate that elk now occur in all counties in the Trans-Pecos except El Paso County (Guevara 2009; Pohler et al. 2014). The free-ranging elk in the Glass Mountains (Brewster County) and the Davis Mountains (Jeff Davis County) are thought to represent some of the higher densities in the Trans-Pecos.

In 2009 an extensive study of the elk herd in the Glass Mountains was initiated, involving both aerial and ground surveys as well as radiotelemetry tracking (Pohler et al. 2014). In this region the elk seemed to prefer oak woodland and riparian habitats. The density estimate of 0.39 elk/km² for this herd was considerably higher than the estimate for the entire Trans-Pecos region (0.14 elk/km²), and the annual survival rates of the Glass Mountain herd were astonishingly high (97.1 percent for mature male elk and 94.4 percent for females).

Another study was conducted in the Wylie Mountains in Jeff Davis County, to determine the movements of elk transplanted there in 1988. Ninety-nine elk were released in 1988 and nine cows were radio-collared to monitor their movements. Based on telemetry, it was estimated that the elk dispersed less than 10 km from the release site (Coykendall 1990). The mortality rate of the elk released was 13 percent, with automobile collisions and mountain lion kills accounting for most of the deaths.

Elk diets have received some attention and seem to exhibit seasonal variation. In GMNP their spring diet was estimated to consist of 50 percent browse, 12 percent forbs, and 37 percent grasses; in the summer, these same percentages, respectively, were 49, 12, and 39; in fall they were 47, 28, and 25; in winter, 48, 29, and 24 (Krysl 1979). Preferred browse plants were oak, mountain mahogany, and the flower stalks of *Agave* sp. during the spring and summer. Botanical composition of elk and mule deer diets was also compared in the park (Krysl and Bryant 2001). Annual diets of elk consisted of 48 percent browse, 32 percent grasses, and 20 percent forbs, with oaks, desert ceanothus, curlyleaf muhly, blue grama, and common horehound the major forages. Annual mule deer diets consisted of 77 percent browse, 21 percent forbs, and 2 percent grasses, with oaks, desert ceanothus, mountain mahogany, and bladderpods the primary forages consumed. Overall, annual dietary overlap was moderately high (58 percent). Overlap was greatest in the browse component and was highest during spring (91 percent) and summer (65 percent).

In arid environments such as the Trans-Pecos, limiting factors in elk population density include hunting, habitat quality, and particularly spring precipitation. The amount of water available is especially thought to impact the number of elk the land can support. Throughout their distribution in the Trans-Pecos, elk utilize water provided by private landowners through water troughs intended for cattle operations. Without these permanent water sources, it is likely that elk would not be able to sustain current population numbers (Pohler et al. 2014). Possible mortality factors in the Trans-Pecos include winter starvation, late birth, poor nutrition during the summer, disease, and predation. Primary predators include mountain lions, coyotes, and bobcats (Grace 1983; Coykendall 1990).

Based on the interpretation that the present subspecies in the Trans-Pecos is nonnative, elk were designated as exotic livestock in 1997 by the 75th Texas legislature. The exotic status means that elk in Texas do not have any harvest restrictions. The best thing for their conservation would be to reclassify them as a game species so that populations and harvest limits could be managed more effectively. To justify this requires an answer to the question "Are elk native to Texas?" To date, most wildlife biologists have taken the position that elk were native only in the Guadalupe Mountains, and therefore, elk currently in the state represent exotic imports rather than a native species and subspecies. A recent review of historical and archaeological evidence has concluded that elk are indeed native to Texas, including at least some of the current populations in the Trans-Pecos (Gill et al. 2016). This study also refutes that there ever was a separate species or subspecies called *C. merriami*, and it concludes based on morphological, statistical, and genetic evidence that today's free-ranging elk in the Davis and Glass Mountains are the result of natural immigration from the Lincoln National Forest of New Mexico to areas of their former native range in the Trans-Pecos (Gill et al. 2016).

A separate genetic analysis, using both mitochondrial and microsatellite loci as well as simulations of population genetic parameters, seems to support the conclusion that the origin of contemporary elk in Texas is likely natural emigrants from New Mexico, or descendants of previously introduced individuals from New Mexico, as well as reintroduced animals from outside the state (Dunn et al. 2017). From these observations it seems best to conclude that the subspecies of elk in the Trans-Pecos is *C. c. nelsoni*.

Mule Deer
Odocoileus hemionus (Rafinesque, 1817)

The mule deer is a moderate-sized deer with large ears that are especially conspicuous in antlerless animals. The antlers are dichotomously branched and are restricted to males (fig. 60). The tail is white or tan dorsally and conspicuously tipped with black. The general color is grayish buff (suffused with black) in winter and more reddish in summer. External measurements (in mm) for males are total length 1,370–1,830; tail 106–230; hindfoot 330–585; ear from crown (dry) 118–250; for females, 1,160–1,800; 115–200; 325–475; and 118–243, respectively. Especially in areas where their ranges overlap, mule deer and white-tailed deer can be confused by the untrained eye. Table 7 presents the more important characters for distinguishing them.

The mule deer is the most important game animal in the Trans-Pecos. It has been recorded from every county and is considered to occur throughout the entire region (map 57). It occupies almost all types of habitat in the

FIG. 60. Mule deer, *Odocoileus hemionus*

TABLE 7. Morphological features for distinguishing between white-tailed and mule deer.

Character	White-tailed deer	Mule deer
Facial features	Brown with white circles around eyes and ears	Large patches of white on face
Coat color	Reddish-brown fur	Grayish fur
Ears	Smaller in proportion to head; about 50% or less of head length	Larger in proportion to head; about 55% or more of head length
Tail	Broad and dark with completely white underside	Thin and white above with a black tip
Rump patches	Barely visible when tail is down	Distinctly white and highly visible
Antlers	Have a main shaft with tines pointing upward	Fork out repeatedly
Metatarsal glands	Oval, about 25 mm (1 inch) long; surrounded by whitish or light colored fur	Elongated, about 80–150 mm (3–6 inches) long; heavily surrounded by light brown fur

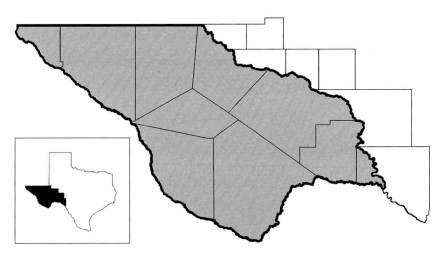

MAP 57. Distribution of the mule deer, *Odocoileus hemionus*

Trans-Pecos to some extent, from elevations of about 750 m in the Big Bend to 2,400 m in the Guadalupe Mountains. Mule deer populations seem to vary from one decade to the next. From 1978 through 2001 the Trans-Pecos mule deer population fluctuated from lows of fewer than 100,000 to a high of 221,220 (Sumner 2002). Subsequently, the numbers have continued to fluctuate but the overall trend has been downward. As landowners are producing less revenue from livestock and more revenue from wildlife, more importance is being placed on understanding and improving mule deer habitat and numbers.

The major factors affecting their numbers are drought, habitat degradation, reproduction, water distribution, and predator density (Sumner 2002). Of these the most significant is drought. The all-time low population count of 99,790 in the Trans-Pecos occurred in 2000 following seven years of drought. Lack of grass and forb cover for fawns during droughts makes them more susceptible to predators and lowers herd nutrition. Mismanagement of the range, overgrazing, and subsequent brush encroachment are major forms of habitat degradation and are amplified during a drought. Brush encroachment makes mule deer more susceptible to predators, as they depend on their eyesight and their ability to quickly move away from danger. Poor fawn survival reduces recruitment and causes deer populations to dwindle. Overgrazing in desert environments removes most of the herbaceous cover and is detrimental to mule deer populations. They generally avoid areas occupied by large numbers of cattle and are more abundant in ungrazed areas. Water availability from springs and seeps as well as from water troughs and windmills, when reduced, is another factor in mule deer survival. Finally, the increased density of mountain lions and coyotes resulting from a decline in predator control efforts can be particularly detrimental to mule deer survival.

Locally, mule deer are most abundant on the mesas and slopes of barren foothills and outlying desert ranges between 900 and 1,500 m. Desert grassland and mixed desert scrub are key resources. In Brewster County the upper part of their range overlaps the lower range of the white-tailed deer. The whitetail usually selects rough, brushy habitats, whereas the mule deer favors more open areas, such as the heads of draws that support good stands of sotol and grass (Borell and Bryant 1942). In BBNP, the distribution and productivity of desert mule deer are significantly influenced by plant density and rainfall (Leopold and Krausman 1991). In the park, deer abundance is correlated with total plant, forage, and succulent densities as well as perennial water abundance and bed site availability (Leopold 1984). Fawn production and survival are related to spring rainfall.

There have been estimates of mule deer densities as high as 1.4 individuals/km² in BBRSP, indicating a relatively healthy population in the park (Yancey

1997). The mule deer was formerly common on the Stockton Plateau at places such as Langtry in Val Verde County, where it occupied the small, rough canyons covered with abundant growth of cactus and sotol (Bailey 1905).

A major assessment of mule deer survival in the Trans-Pecos was conducted from 1991 to 1993 at EMWMA (Lawrence et al. 2004). The annual survival rate over this period was 0.76 (suggesting that about 76 percent of the population survives from year to year), but there was considerable variation in survivorship among age classes within the population. Young deer (8–20 months) exhibited the lowest survival (0.64), followed by adult bucks (0.70) and adult does (0.79); subadult deer (21–33 months) had the highest survival rate (0.92). Predation and natural stress-related maladies (disease, old age, malnutrition) were the most common causes of death in all three years of this study. Predation was an important cause of mortality, with mountain lions responsible for at least 83 percent of all predation-related deaths. The most stressful periods for Trans-Pecos desert mule deer are when there is increased activity associated with searching for mates, reduced nutrient reserves associated with the period following the rut (typically January–March), and seasonal lows in forage quality and quantity (Lawrence et al. 2004).

The annual diet of mule deer comprises primarily browse (brush species) and forbs (broad-leaved weeds), with grasses and cacti making up less of the diet. Generally, browse use decreases from spring to winter with a corresponding increase in forb use. Use of succulents is greatest during dry seasons. Deer in xeric conditions consume evergreen browse species during periods of low rainfall and rely less on forbs. Deer in the more mesic foothills also demonstrate lower forb consumption during drought periods (Leopold and Krausman 1987). In the deserts of southwestern Arizona, jojoba and ironwood make up most of their browse diet in all seasons, but forbs (buckwheat, globemallow, lupine, filaree, desert vine, vetch, and spurge) account for much of the winter-spring diet (Heffelfinger 2008). In the Sierra Vieja of the Trans-Pecos (Presidio County), Emory oak, yucca, scrub sumac, lechuguilla, and grama grass were important dietary items (Anderson 1949).

Mule deer follow a daily rhythm or activity pattern controlled by temperature (Anderson 1949). As the warm hours of the morning approach, they move from mesas and flats toward canyons. As the temperature rises, they move into the shade of larger trees, which afford them protection from the sun. They rest through the day in the shade and leave only when the temperature drops enough to permit comfortable feeding. They are seldom active in the middle of the day.

Mule deer are gregarious, occurring in herds of up to 23 individuals (Anderson 1949). Groups of as few as two, three, or four deer sometimes feed together.

If frightened they form a tight group and run or lope in a single line. *Odocoileus hemionus* is polygynous, with a tending-bond type of breeding system rather than a harem breeding system (Yancey 1997).

The breeding season (rut) occurs in the late fall (November–December), with single or twin fawns born in summer (late June through mid-August) after a gestation period of about 200 days. Typically, litters consist of two fawns, but one is not uncommon; triplets are rare. A specimen collected in the Guadalupe Mountains on June 27 contained a nearly full-term fetus (Davis 1940b). At BBRSP, a gravid female, freshly killed by a mountain lion, was examined on May 13 and had a single fetus with a crown-rump length of 1,300 mm (Yancey 1997).

Males grow antlers covered with a skin called "velvet" throughout the summer. In the fall the velvet dries and is rubbed off to reveal the bony, sharp antlers in time for the rut. The antlers are shed in spring and regrown annually, with antler size increasing each year until peaking when the buck reaches six to eight years of age.

Home ranges of radio-collared mule deer in the Apache Mountains were quite large, averaging 6,230 ha (range, 1,745–12,252 ha), with mature bucks having substantially larger home ranges than females (Sumner and Harveson 2008). Deer home ranges are usually larger at lower elevations where precipitation is less. In desert areas, mule deer does reduce their home ranges and move to mountainous areas to have their fawns (Heffelfinger 2008).

Vernon Bailey's 1890 mammal report for the Davis Mountains includes a comment that white-tailed deer and mule deer may be producing hybrids in this area. It is now common knowledge that these two species do in fact interbreed and produce hybrid offspring (Carr et al. 1986; Stubblefield et al. 1986; Bradley et al. 2003), although the production of first-generation hybrids appears to be rare (Ballinger et al. 1992). A relatively high proportion of backcross individuals has been documented (Cathey et al. 1998). Surveys of whitetails and mule deer in a five-county area of West Texas in the early 1980s revealed that on average 5.6 percent of the deer tested had evidence of hybridization (Stubblefield et al. 1986; Heffelfinger 2008). The subspecies of mule deer in the Trans-Pecos was previously considered to be *O. h. crooki*. However, it was determined that the type specimen used to describe this taxon is a mule deer × white-tailed deer hybrid, thus invalidating *crooki* as a subspecies. Consequently, the subspecies of mule deer in the Trans-Pecos is now considered to be *O. h. eremicus* (Heffelfinger 2000). A record 27-point mule deer, named "Hank," was recently shot by a hunter in Culberson County, and this animal shattered the Boone and Crockett score for Texas (M. Williams 2020).

White-tailed Deer
Odocoileus virginianus (Zimmermann, 1780)

The white-tailed deer is similar to the mule deer but is smaller and has relatively shorter ears; a larger tail that dorsally is the color of the back edged with white and ventrally is white and lacks a distinct black tip; and smaller antlers (present in males only) in which all major points come off the main beam (fig. 61; table 7). External measurements (in mm) of a male and female, respectively, are total length 1,512 and 1,386; tail 214 and 182; hindfoot 403 and 365.

White-tailed deer range throughout the eastern two-thirds of the Trans-Pecos (map 58). They generally occur in two distinct habitats in the Trans-Pecos—the mesquite–creosote bush thickets typical of the plains and plateaus of Culberson, Reeves, Pecos, Terrell, and Val Verde Counties, and the higher elevations of the mountains in Brewster, Presidio, and Jeff Davis Counties where oaks and conifers grow. These deer are plentiful throughout the Chisos Mountains above 1,300 m. There, they favor the canyons and steep slopes of the

FIG. 61   White-tailed deer, *Odocoileus virginianus*

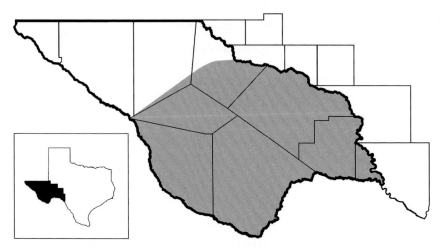

MAP 58. Distribution of the white-tailed deer, *Odocoileus virginianus*

mountains where junipers, piñons, several species of oaks, and brush afford
dense shelter (Borell and Bryant 1942). They occur as low as the oak belt, where
their range overlaps that of the mule deer. White-tailed deer are also relatively
abundant at the higher elevations of the Chinati Mountains of Presidio County
in and around oak woodlands (C. Jones et al. 2011). These deer are rare in eastern Culberson County, where they reach the western limits of their range in
the Trans-Pecos (Davis and Robertson 1944). At one time they were present
in the Guadalupe Mountains, but they no longer occur there (Genoways et al.
1979). Bailey (1905) reported a specimen from San Elizario in El Paso County,
but according to Ederhoff (1971), it was probably brought in by hunters or other
persons traveling through the area.

Peak reproductive activity in the Big Bend area is from mid-December
through mid-January. After a gestation period of about 201 days, fawns are
born in July and August (Krausman and Ables 1981). Most litters consist of
twins, but litter sizes range from one to three (Yancey 1997). Although primarily crepuscular, they may be active at any time of day or night (Krausman and
Ables 1981).

Their diet varies by location and season. In the Big Bend they feed on
browse, forbs, succulents, and grasses (Krausman and Ables 1981). A comparison of the diets of white-tailed deer with mule deer in BBNP has revealed
that browse constitutes 35 percent, succulents 28 percent, forbs 14 percent, and
grasses 4 percent of the whitetail's diet. Mule deer ate more succulents than
browse (38 vs. 27 percent), whereas forbs and grasses made up 19 and 3 percent,
respectively, of their diet. From this information it appears that factors other

than forage are responsible for habitat separation of the Big Bend's two deer species (Krausman 1978).

The specimens Bailey described from the Chisos Mountains are now recognized as *O. v. carminis*. This subspecies is represented by a small, isolated population in the Chisos Mountains, where it is well protected within the borders of BBNP, and by populations from the Rosillos, Christmas, Chinati, and Davis Mountains (Krausman et al. 1978). These small mountain whitetails differ markedly in size from the Stockton Plateau whitetails (*O. v. texanus*), which are larger. White-tailed deer from the Davis Mountains, although assignable to the subspecies *O. v. carminis*, approach *O. v. texanus* in size, suggesting that the two subspecies show evidence of intergradation in this area (Krausman et al. 1978).

Family Tayassuidae
PECCARIES
Collared Peccary
Pecari tajacu (Linnaeus, 1758)

This small, wild pig–like mammal has straight, daggerlike canines, a short tail, and a light stripe or collar encircling its shoulders (fig. 62). It has four hoofed toes on the front feet, but only three on the hindfeet. The pelage is harsh and grizzled black or grayish; a distinct black "mane" extends from the crown to the rump. External measurements (in mm) of an adult female specimen are total length 940; tail 55; hindfoot 180; ear 100; weight is 13–25 kg.

Interestingly, Vernon Bailey and the federal agents who surveyed the Trans-Pecos at the end of the nineteenth century did not find peccaries in the Big Bend or the Davis Mountains, places where they are common today. Apparently, their range expanded to the west during the last century (Schmidly et al. 2022). Based on surveys and information from biologists on the ground, the TPWD has determined that they occur in all counties in the Trans-Pecos (Schmidly and Bradley 2016) (map 59).

Collared peccaries are numerous on the slopes and flats surrounding the northern end of the Chisos Mountains in BBNP, where they seem to prefer lechuguilla-creosote-cactus and sotol-grassland associations with a good growth of catclaw, mesquite, sotol, creosote bush, persimmon, and prickly pear (Borell and Bryant 1942; Bissonette 1982). In BBRSP, they are common in brushy habitats where prickly pear is abundant; they have also been observed in grassland and riparian habitat as well as near human dwellings (Yancey 1997). Highly preferred habitat in the Davis Mountains included Emory oak–plains bristlegrass, gray oak–hackberry woodland, and juniper–catclaw–mimosa–sideoats grama

FIG. 62. Collared peccary, *Pecari tajacu*

shrubland associations (Green et al. 2001). On the Stockton Plateau they have been observed in the heads of canyons among cedar-oak and cedar-ocotillo plant associations (Hermann 1950). At one time peccaries were probably more numerous than they are today. Heavy grazing by domestic livestock has greatly reduced forage and shelter and has probably been an important factor in reducing their numbers.

The increasing presence of feral hogs in the Trans-Pecos is another factor that could reduce peccary populations. There have been no studies of competitive interactions between the two in the Trans-Pecos, but in the Pantanal of South America, a recent study has demonstrated that feral pigs may influence the foraging period of collared peccaries (Galetti et al. 2015). In the absence of feral pigs, collared peccary activity peaked in the early evening, whereas in the presence of feral pigs, collared peccaries foraged mostly in early morning.

An extensive study of the ecology and social behavior of collared peccaries has been made in BBNP (Bissonette 1978, 1982). Peccaries travel in small bands of a half dozen or so and seem to concentrate their activities near surface

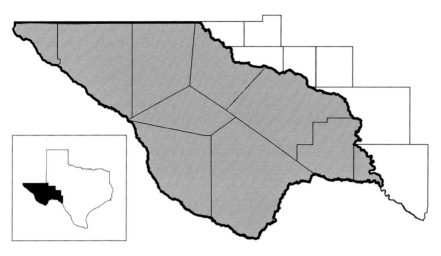

MAP 59. Distribution of the collared peccary, *Pecari tajacu*

water or places where prickly pear and other succulent plants form a conspicuous part of the flora. Group size is positively correlated with the percentage of
vegetative cover and the percentage composition of prickly pear, lechuguilla,
and forbs, and negatively correlated with the percentage of woody cover.
Variability in herd size results from herds subdividing into smaller groups,
and solitary individuals may be regularly observed that are members of a herd
(Green et al. 2001).

Feeding activities account for most of their active time. Depending on
temperature, peccaries generally bed down from 0800 to about 1200. In the
hot summer period, their daily activity is usually initiated in late afternoon or
early evening, around 1700 to 1800. During winter, they leave their beds earlier in the afternoon, around 1500 (Bissonette 1978). Individuals communicate via vocalizations that vary in pitch, intensity, and continuity to express an
aggressive, submissive, or alert notification. After "alert-alarm patterns" peccaries may either disperse or stay tightly grouped. Social interactions are mostly
nonaggressive.

Herds establish definite territories and there is little to no overlap with
adjacent groups. Peccaries are thought to mark their territories with secre-
tions from a gland on the hind part of the back, placed on vegetation and
rocks, and by placing scat piles "at a boundary line" (Ellison and Harwell 1969;
Bissonette 1982). Herds exhibit a clear dominance hierarchy with dominant
males first, followed by submissive males and females in lower positions determined mostly by size, and young three months and older at the lowest rank.

The young are defended by their mother. Dispersal apparently occurs over relatively short distances among neighboring herds, dictated primarily by males (Cooper et al. 2010).

The breeding season extends from December into January, with male success influenced by dominance position; the most dominant male in a group does most of the breeding (Bissonette 1982). Multiple males may sire offspring within a herd, and some litters have multiple sires (Cooper et al. 2011). The number of young produced is usually two, but litters range in size from one to five. The gestation period is 142–149 days. Parturition is seasonal and correlates with the annual onset of heavy rainfall. The sex ratio is essentially 1:1. At birth, the young are reddish or yellowish and weigh about 500 g. Mortality is high (50 to 100 percent), especially for young of the year.

Collared peccaries show a decided preference for prickly pear in their diet throughout the year, especially when the fruits are ripe. Lechuguilla is most heavily consumed during spring and fall, but only the core leaves and basal parts of the modified leaves and the root are eaten. There are camera images of peccaries feeding on desert gourds, and they are potential dispersal agents for these plants (Platt et al. 2014).

The mountain lion is the primary predator of collared peccaries. Two incidents of observed predation by mountain lions have been recorded in BBNP (Van Pelt 1977). In one instance an attempted predation of a large male javelina in Ward Canyon ended with the javelina defending himself and the cat subsequently departing after 15 seconds. In the other instance, just below Panther Junction, a mountain lion was observed chasing and bringing down a "female" javelina and partially consuming the carcass before carrying it off to a more protected area.

Interestingly, in BBNP four incidents of collared peccaries bedding in close association with mule deer have been observed; one was at a distance of only 5 m. From these observations it has been suggested that deer and peccaries apparently share the shade in a passive manner, and that the association of the two may increase opportunities to detect predators (Krausman et al. 1999).

Collared peccaries occupy the same habitat as domestic livestock and could serve as possible reservoirs for parasites and infectious diseases. Serological examination of 55 individuals from the Trans-Pecos revealed evidence of Lyme disease (*Borrelia burgdorferi*) and plague (*Yersinia pestis*), suggesting they could potentially be reservoirs for these diseases and expose both livestock and humans (Gruver and Guthrie 1996). Genetic analysis indicates there is little variation among populations of peccaries in Texas (Cooper et al. 2010). The subspecies in the Trans-Pecos is *P. t. angulatus*.

ORDER RODENTIA (Rodents)

Rodents are remarkably uniform in structural characters. They have two ever-growing incisors in each jaw; canines and anterior premolars are lacking, leaving a space (diastema) between the incisors and the cheek teeth.

Rodents have adapted to many different habitats. Some, like the beaver and muskrat, are semiaquatic; others, such as kangaroo rats and pocket mice, are terrestrial and live in arid desert country; still others, such as pocket gophers, are fossorial and live underground; and yet some, such as many squirrels, are arboreal. Although there are exceptions, rodents are basically vegetarians, living on grass, herbaceous plants, bark, leaves, seeds, berries, roots, buds, and agricultural crops.

There are 46 species, representing eight families, of rodents in the Trans-Pecos, making this the largest major group of mammals in the area. Of these species, nine occur in Texas only in the Trans-Pecos. Two of the eight families that occur in the region, namely the family Echimyidae (represented by the nutria, *Myocastor coypus*) and the family Muridae (represented by the house mouse, *Mus musculus*; the Norway rat, *Rattus norvegicus*; and the black rat, *R. rattus*), are not native to North America and thus the Trans-Pecos. Because of the size and diversity of this order, we present a key to the families of Trans-Pecos rodents, including a brief introduction and, when applicable, a separate key to the species for each family.

Key to the Families of Rodents of the Trans-Pecos

1. Dorsal hairs in part modified to form sharp quills or spines . . . .
       FAMILY ERETHIZONTIDAE (**Porcupine**, *Erethizon dorsatum*)
       Dorsal hairs not modified to form sharp quills or spines 2
2. Tail paddle shaped (dorsoventrally flattened), broad, scaly, and naked
       . . . . FAMILY CASTORIDAE (**Beaver**, *Castor canadensis*)
       Tail not paddle shaped, rounded or nearly so 3
3. Hindfeet fully webbed . . . .
       . . . . FAMILY ECHIMYIDAE (**Nutria**, *Myocastor coypus*)
       Hindfeet not fully webbed 4
4. External, fur-lined cheek pouches present 5
       External, fur-lined cheek pouches absent 6
5. Front feet much larger than hindfeet; ears short and inconspicuous;
       tail about one-third length of head and body . . . .
       .FAMILY GEOMYIDAE (pocket gophers)

Front feet much smaller than hindfeet; ears conspicuous;
  tail as long as (or longer than) head and body . . . .
  FAMILY HETEROMYIDAE (pocket mice and kangaroo rats)
6. Lower jaw with four cheek teeth on each side . . . .
  FAMILY SCIURIDAE (squirrels, chipmunks, and prairie dogs)
  Lower jaw with three cheek teeth on each side 7
7. Hair on tail conspicuous; cusps of upper molars in two longitudinal
  rows . . . . . FAMILY CRICETIDAE (New World mice and rats)
  Hair on tail inconspicuous, tail appearing naked and scaly; cusps of
  upper molars in three longitudinal rows . . . .
  . . . . FAMILY MURIDAE (Old World mice and rats)

Family Sciuridae
SQUIRRELS, CHIPMUNKS, AND PRAIRIE DOGS

Members of this family in the Trans-Pecos include chipmunks, ground squirrels, prairie dogs, and tree squirrels. The ground squirrels recently underwent a major taxonomic revision (Helgen et al. 2009); herein we follow the nomenclature of that work and no longer use the generic name *Spermophilus* for Trans-Pecos species.

1. Upperparts striped or distinctly spotted 2
  Upperparts without stripes or distinct spotting 5
2. Upperpart pattern with stripes 3
  Upperpart pattern with scattered spots or rows of spots 4
3. Single dorsal pale stripe on each side of grayish pelage; sides of head
  without stripes; ears small and rounded . . . .
  . . . . **Texas Antelope Squirrel**, *Ammospermophilus interpres*
  Two dorsal pale stripes on each side alternating with three to five
  brownish stripes; each side of head with a light stripe above and
  below eye; ears prominent . . . .
  . . . . **Gray-footed Chipmunk**, *Tamias canipes*
4. Dorsal spots pale, squarish, and usually arranged in 9–10 rows; tail
  about three times as long as hindfoot and moderately bushy
  . . . . **Rio Grande Ground Squirrel**, *Ictidomys parvidens*
  Dorsal spots pale, irregularly shaped, and randomly arranged or
  scattered; tail about two times as long as hindfoot and slightly
  bushy . . . . **Spotted Ground Squirrel**,
  Xerospermophilus spilosoma

5. Tail not bushy, short and black tipped, about one and a half times
    length of hindfoot; overall color yellowish brown. . . . .
    . . . . **Black-tailed Prairie Dog**, *Cynomys ludovicianus*
    Tail bushy and long, more than three times length of hindfoot, half
    of total length; overall color brown, gray, or blackish 6
6. Upperparts generally mottled brownish gray with variegated black,
    white, and buff appearing as wavy markings crossing the back;
    sometimes head, shoulders, and anterior back completely black;
    belly grayish; hindfoot ≤ 70 mm. . . . .
    . . . . **Rock Squirrel**, *Otospermophilus variegatus*
    Upperparts grizzled brownish cinnamon, rusty, grayish, or blackish;
    belly reddish to rusty; hindfoot ≥ 70 mm. . . . .
    . . . . **Eastern Fox Squirrel**, *Sciurus niger*

Gray-footed Chipmunk
Tamias canipes (V. Bailey, 1902)

Gray-footed chipmunks are small, grayish squirrels with alternating light and
dark stripes (four whitish and three to five dark brownish) that extend from the
shoulders to the rump, a pair of whitish eye stripes on each side of the head,
and prominent ears (fig. 63). Average external measurements (in mm) are total
length 225; tail 102; hindfoot 35; ear 15; weight is about 70 g.

FIG. 63. Gray-footed chipmunk, *Tamias canipes*

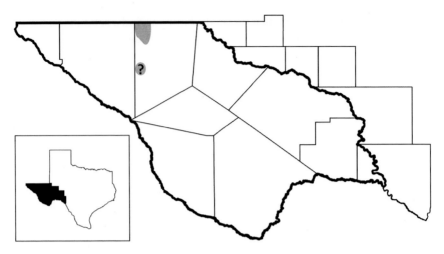

MAP 60. Distribution of the gray-footed chipmunk, *Tamias canipes*. Question mark indicates population in the Sierra Diablo that is likely extirpated.

Gray-footed chipmunks have been recorded in the Trans-Pecos only from the Guadalupe Mountains and Sierra Diablo of Culberson County, but the latter population is apparently extirpated today (Schmidly et al. 2022) (map 60). They remain relatively common in the Guadalupe Mountains, where they occupy brushy hillsides with logs, rocky crevices, and rocky outcrops in oak and higher coniferous forests at elevations of 1,800–2,500 m. They are less common in lower piñon-juniper-oak associations, although during late summer to fall they may shift to these areas to exploit acorns (Bailey 1905).

Gray-footed chipmunks are diurnal, with most activity and feeding occurring soon after sunrise (Davis 1940b). They are skillful climbers of rocks, trees, and shrubs and will forage aboveground when resources dictate. They are shy and cautious, and their rapid "chipper" sounds are sometimes the only evidence of their presence. When alarmed they become silent and retreat to burrows, rocks, crevices, or brush. Dens are found in variable locations including under logs and stumps or in rock piles and crevices (Davis 1940b). Their diet includes seeds of Douglas fir and sunflowers, acorns, mushrooms, currants, green vegetation, and insects. Food items are gathered and carried to dens using cheek pouches; acorns constitute a seasonal resource that is cached.

A single litter of four young is produced each year from mid-May to August (Bailey 1931). A female from the Guadalupe Mountains had four embryos in August (Genoways et al. 1979). No specific predators have been documented, although diurnal raptors, snakes, and mammalian carnivores occur in the same habitat and likely prey on them.

In the first edition of this book, this chipmunk was included as *Eutamias canipes*, but a subsequent taxonomic revision resulted in its placement into the genus *Tamias* (Nadler et al. 1977; Levenson et al. 1985). The current status of this peripheral species appears to be stable within the protected confines of GMNP (Schmidly et al. 2022). The subspecies in the Trans-Pecos is *T. c. canipes*.

Texas Antelope Squirrel
Ammospermophilus interpres (Merriam, 1890)

The Texas antelope squirrel is a chipmunk-like ground squirrel with a conspicuous white stripe (not bordered with black) on each side of the back from shoulder to rump (fig. 64). The underside of the moderately bushy tail is conspicuously white and the upperparts are grayish brown. The ears are short, each appearing as hardly more than a rim above the head. Crescent-shaped white stripes are located over and under the eye. Average external measurements (in mm) are total length 226; tail 74; hindfoot 38; ear 10; weight is 85–120 g.

The Texas antelope squirrel ranges across most of the Trans-Pecos and has been recorded from all counties in the region except Reeves (map 61). It inhabits rocky foothills in desert, grassland, and lower woodland environments at elevations between 550 and 1,850 m. It is strongly associated with canyons, cliffs,

FIG. 64. Texas antelope squirrel, *Ammospermophilus interpres*

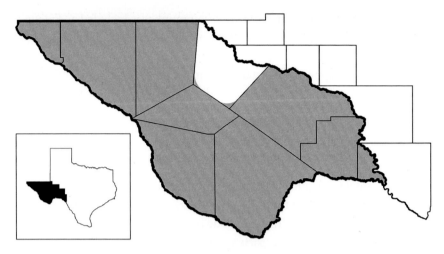

MAP 61. Distribution of the Texas antelope squirrel, *Ammospermophilus interpres*

boulders, rock piles, gravelly washes, and rocky slopes and is effectively absent from level ground, sandy terrain, and flat mesas.

Texas antelope squirrels are diurnal, with peak periods of daily activity varying with the season. They are active from earliest morning to early evening but seem to prefer the hottest part of the day (Hermann 1950). Their dens are usually located under large boulders or rock ledges, and in crevices. A shallow burrow system is utilized, which may meander around rocky intrusions and include several entrances and a nest chamber (Stangl et al. 1994). One squirrel was observed using five separate burrows (Hermann 1950), all probably entrances to the same system. Burrows are inconspicuous because no mounds of soil are evident.

Texas antelope squirrels are nervous and cautious, and when alarmed they sprint long distances to get out of sight instead of opting for nearby refuges (Stangl et al. 1994). They make a distinctive mellow, rolling trill call, identifying them audibly. As proficient climbers, they may be seen in tops of bushes, yuccas, prickly pear cactus, and junipers foraging for fruit and will often sit on tops of large boulders. They apparently do not hibernate as they are active throughout the year, although fat accumulations suggest that long periods underground are possible during cold episodes (Bailey 1931).

Early in the season green vegetation is typically consumed, followed by a variety of seeds, berries, fruits, and insects as the season progresses. Diet includes fruits of cholla, prickly pear cactus, Mexican walnut, Mexican buckeye, mountain laurel, and seeds of mesquite, sotol, and creosote bush. Cheek pouches are used to transport gathered foods back to their dens.

They are early breeders, generally producing a litter of 5–14 young (based on embryo counts) during February–March. Mammae are in five pairs (Bailey 1931), indicating their capacity to handle large litters. A second litter may be produced by some females. Females lactating or with embryos have been found in March in the Trans-Pecos (Stangl et al. 1994). Specific predators are unknown, although diurnal raptors and snakes likely capture and eat these squirrels.

Texas antelope squirrels are named for their escape behavior of running away and flashing white in their tails, such that they resemble a fleeing pronghorn (Stangl et al. 1994). The species is monotypic, and subspecies are not recognized.

Rio Grande Ground Squirrel
Ictidomys parvidens (Mearns, 1896)

This ground squirrel is distinctive in possessing a brownish-buff dorsum with 9 to 10 rows of pale squarish dorsal spots (fig. 65). The tail is moderately bushy and about two-fifths of the total length. The ears are short and rounded. Average external measurements (in mm) are total length 302; tail 126; hindfoot 40; ear 9; weight is 140–330 g. Males are typically larger than females (Yancey et al. 1993).

Rio Grande ground squirrels have been recorded from throughout the Trans-Pecos, except for Hudspeth and El Paso Counties and the southern portions of Presidio and Brewster Counties (map 62). In the eastern part of the region, they are common in flat lowland grassland and open brushy areas associated with mesquite, creosote bush, and prickly pear cactus; they avoid rocky areas, preferring instead sandy or gravelly soils in which they construct burrows.

Rio Grande ground squirrels are diurnal, but seasonal activity may be curtailed by cold weather. Burrows are central to daily activity, refuge, and nesting. There is generally a main burrow with one or more openings, as well as shallower secondary burrows used for refuge. Although they are not known to be social except during the breeding season, they can be locally colonial. When alarmed they will stand upright for a visual survey and emit a high-pitched, trilling alarm call.

These squirrels are omnivores and prefer seeds, grasses, forbs, and flowers, and they sometimes climb into low bushes to forage. They will also eat insects

FIG. 65. Two ground squirrels of the Trans-Pecos: above, *Xerospermophilus spilosoma*; below, *Ictidomys parvidens*

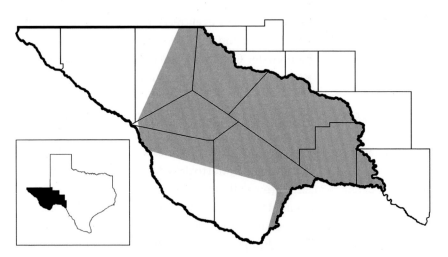

MAP 62. Distribution of the Rio Grande ground squirrel, *Ictidomys parvidens*

as well as other arthropods and their larvae. In spring green vegetation dominates their diet, whereas during summer about 50 percent of the diet includes insects. They occasionally prey on small vertebrates such as snakes and lizards (Kasper 2014). Well-developed cheek pouches are used to gather and carry food back to burrows where they hoard seeds and nuts.

Breeding occurs for about two to three weeks in late March or early April (R. Edwards 1946). The gestation period is about 30 days followed by birth of about five young. Predators include diurnal raptors and snakes, as well as open-land mammalian predators such as badgers, bobcats, and coyotes.

In the first edition of this book, this ground squirrel was included as *Spermophilus mexicanus parvidens* (Mexican ground squirrel), but a recent taxonomic revision (Helgen et al. 2009) split *Spermophilus* into multiple genera, including three in the Trans-Pecos: *Ictidomys, Otospermophilus,* and *Xerospermophilus.* Consequently, *Spermophilus mexicanus parvidens* was changed to *Ictidomys parvidens.* The species is monotypic, and subspecies are not recognized.

Spotted Ground Squirrel
Xerospermophilus spilosoma (Bennett, 1833)

The spotted ground squirrel is distinctive among Trans-Pecos ground squirrels in having small, irregular pale spots mottling its reddish-brown back and sides (fig. 65). During winter months, it molts into a grayish pelage (Stangl and Goetze 1991). The tail is not especially bushy and is about one-third of the squirrel's total length. The ear pinnae are small and rounded. Average external measurements (in mm) are total length 229; tail 72; hindfoot 33; ear 11; weight is 100–125 g.

Spotted ground squirrels occur throughout the Trans-Pecos except for the southeastern portion of the region (map 63). They are generally found in desert scrub, arid grasslands, and sand dunes at low to moderate elevations. They have been found in areas associated with mesquite, buckthorn, creosote bush, sagebrush, and blackbush (Davis and Robertson 1944). These squirrels may occur in a variety of soil conditions but prefer deep sandy soils that facilitate burrow construction. In areas of loose sandy soil, burrow entrances are usually adjacent to structures such as shrubs or rocks.

Spotted ground squirrels are diurnal, with peak periods of daily activity varying with the season. During hot days they are active in the morning and late afternoon. As with other ground squirrels, seasonal activity is curtailed by late-season cold weather or depletion of resources, which can initiate torpor or

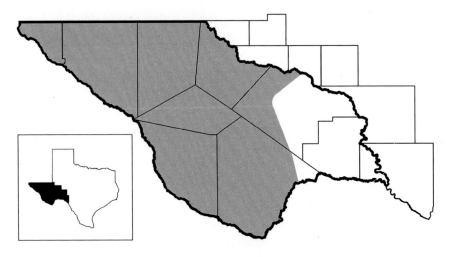

MAP 63. Distribution of the spotted ground squirrel, *Xerospermophilus spilosoma*

hibernation. The species has been collected in all months except December and January (Stangl and Goetze 1991).

The spotted ground squirrel is a capable burrower but is known to occasionally invade and take over the burrows of kangaroo rats, sometimes evicting the inhabitant (Stangl et al. 1994). Individuals are wary and, except during breeding periods, generally stay near the burrow. They are vocal and use a high-pitched trilling whistle to warn nearby conspecifics.

Their diet is primarily green vegetation, prickly pear cactus, and the seeds of mesquite beans, saltbush, nightshade, and sunflower. Grasshoppers are heavily utilized in late summer, and lizards and Ord's kangaroo rat have been noted as prey.

Breeding occurs in late April to May. Gestation is about 28 days and litters of five to eight young are born around June. Predators include snakes and diurnal raptors. Gopher snakes may be a major predator, as they have been seen methodically checking burrow openings.

In the first edition of this book, this ground squirrel was included as *Spermophilus spilosoma*, but a recent taxonomic revision placed it in the genus *Xerospermophilus* (Helgen et al. 2009). For an explanation, see the account of *Ictidomys parvidens*. Two subspecies occur in the Trans-Pecos: *X. s. marginatus* in the eastern two-thirds of the region and *X. s. canescens* in El Paso and Hudspeth Counties; the former, compared to the latter, has larger hindfeet and fewer and less distinct dorsal spots (Howell 1938).

Rock Squirrel
Otospermophilus variegatus (Erxleben, 1777)

The rock squirrel is a large ground squirrel with grayish-brown upperparts resulting from a variegated pattern of black, white, and buff that appears as wavy markings crossing the back (fig. 66). The upperparts of some individuals are completely black. The tail is bushy and the same color as the back. Individuals may have black patches on their head and rump. Average external measurements (in mm) are total length 475; tail 194; hindfoot 56; ear 27; weight is 450–875 g.

Rock squirrels are common throughout the Trans-Pecos except in flat, boulderless areas. They have been recorded in all counties except Reeves and are projected to occupy suitable habitat across the entire region (map 64). They prefer rough, rocky areas such as cliffs, canyon walls, talus slopes, and boulder piles where they seek refuge and make dens. They occur from about 550 to 2,400 m elevation and can be found high into juniper–piñon pine habitat. On the lower elevations of the Stockton Plateau in Terrell County, they are restricted to persimmon-oak associations (Hermann 1950).

FIG. 66. Rock squirrel, *Otospermophilus variegatus*

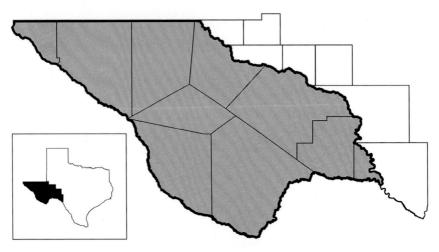

MAP 64. Distribution of the rock squirrel, *Otospermophilus variegatus*

Rock squirrels are diurnal, with peak periods of daily activity varying with the season. They are generally inconspicuous because of their vigilance to nearby intrusion. In the Big Bend region, obese rock squirrels were observed to enter hibernation, whereas individuals of normal weight remained active throughout the winter (Layton 1973). Local populations form maternal colonies that contain a group of breeding females and a dominant male, with subordinate males relegated to the periphery (Layton 1973; Johnson 1981).

Their diet is predominantly plant based, including acorns, pine nuts, juniper berries, Mexican walnuts, mesquite beans, agave, century plants, and prickly pear cactus, but they will also opportunistically consume invertebrates and small vertebrates. They are known to climb trees to forage for nuts, seeds, and berries, and high-stalked agaves to feed on flowers and buds.

Breeding begins after emergence from hibernation, usually during March–April, but may continue into June. Unlike other female ground squirrels, female rock squirrels reach their estrous condition after males are ready to breed. Usually one litter of one to seven (average four) young is born (Layton 1973), but a second litter may be produced depending on elevation, seasonal rain, and food resources. Two females collected in the Guadalupe Mountains in June contained four and five embryos (Genoways et al. 1979). Predators include many diurnal raptors, bobcats, ringtails, raccoons, coyotes, gray foxes, badgers, and rattlesnakes.

The history of the status and distribution of subspecies of *O. variegatus* in the Trans-Pecos is enigmatic and confusing. Bailey (1905) recognized two subspecies in the region, *O. v. grammurus* from the Franklin and Guadalupe

Mountains, and *O. v. couchii* from the Chisos and Davis Mountains and along the Pecos River and Rio Grande (as well as along the Devils River east of the Pecos). Then Howell (1938), in his revision of the North American ground squirrels, restricted the range of *couchii* in the Trans-Pecos to a small area in what is now BBNP that extends from the Rio Grande north to the Chisos Mountains, while assigning all other populations in the region to *grammurus*. Over the next half century, this configuration was recognized and accepted as valid in virtually all major publications that addressed *O. variegatus* taxonomy (e.g., Miller and Kellogg 1955; Hall and Kelson 1959; Hall 1981; Oaks et al. 1987). Then, from 1988 to the present, major publications that focused on the species' taxonomy and distribution specifically in Texas seem to have, without explanation, dropped recognition of *couchii* occurring in the Trans-Pecos, and thus considered *grammurus* as the single subspecies in the region (see J. Jones et al. 1988; J. Jones and Jones 1992; Schmidly 2002; Manning et al. 2008; Schmidly and Bradley 2016; Schmidly et al. 2022). Schmidt (1999), in an overview of subspecies distributions in North America, reported *couchii* to occur only in Mexico, also without explaining why it was no longer deemed present in Trans-Pecos Texas. During this time several major taxonomic works continued to recognize the validity of *couchii* as a subspecies but made no comment on its distribution (i.e., whether it was considered to occur in the Trans-Pecos) (see Thorington and Hoffmann 2005; Helgen et al. 2009; Thorington et al. 2012). Because we can find no systematic work that rendered *couchii* in the Trans-Pecos invalid, we tentatively recognize two subspecies of *O. variegatus* in the region: *O. v. couchii* in BBNP from the Chisos Mountains south to the Rio Grande, and *O. v. grammurus* from throughout the remainder of the Trans-Pecos. Clearly, there is serious need for additional research to clarify this issue.

Black-tailed Prairie Dog
Cynomys ludovicianus (Ord, 1815)

Black-tailed prairie dogs are large, thick-bodied, short-tailed, ground-dwelling sciurids with small, rounded ears (fig. 67). They are light reddish brown, with the tip and underside of the tail black. The underparts are buffy white. Average external measurements (in mm) are total length 365; tail 82; hindfoot 60; ear 12; weight is 1–2 kg.

The general range of the black-tailed prairie dog includes most of the Trans-Pecos except for the southern portions of those counties adjacent to the Rio Grande in the Big Bend region (map 65). Over the last century their numbers have plummeted drastically, to the point that only a few isolated colonies

FIG. 67. Black-tailed prairie dog, *Cynomys ludovicianus*

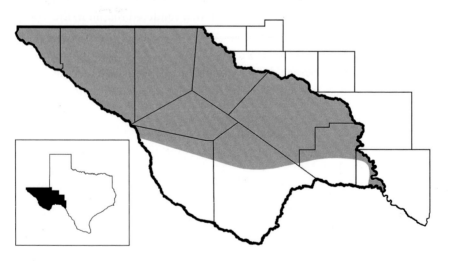

MAP 65. Distribution of the black-tailed prairie dog, *Cynomys ludovicianus*

exist sporadically across this range. This steep decline was the result of purposeful control efforts by humans, episodes of sylvatic plague, and habitat loss to agriculture (Truett et al. 2014). Today it is estimated that only 5,162 ha of prairie dog colonies in the Trans-Pecos remain (Truett et al. 2014), with most of the large colonies occurring in arid ranch country that lacks roads and therefore has limited human access (as in Hudspeth County). Several projects are currently underway to restore black-tailed prairie dog numbers in the Trans-Pecos.

Their preferred habitat is shortgrass prairie on alluvial deposits, flatlands, and in low valleys (Truett et al. 2014). They have been located at elevations between 1,770 and 2,100 m in mesquite–creosote bush, grama-needlegrass, and burrograss-cholla associations that lack dense brush or tall grass (Bailey 1905). They survive poorly where vegetation grows tall and dense because they depend on unobstructed views of the landscape to avoid predators.

These are highly social animals that live in colonies ranging from a fraction of a hectare to many square kilometers. Their populations are structured as "coteries," or harems, of two to eight females defended by one or several dominant males. Activity and breeding occur within coterie territory, and members cooperatively defend the coterie. Several coteries together form "wards," which are separated by unoccupied or unsuitable habitat. Prairie dogs have a behavioral inclination to crop tall plants within a ward to maintain an unobstructed view, and they use alarm calls described as repetitive barks and chuckles.

Burrows consist of a steep 2–5 m tunnel that levels to a horizontal tunnel with a nest chamber and blind tunnels. Multiple entrances are marked by two types of mounds—relatively high, conical "crater" mounds; and lower, broader "dome" mounds. Differences in mound height and shape help provide underground ventilation and prevent flooding of burrows.

Prairie dogs are diurnal and are most active during morning and evening hours. They do not hibernate and are active aboveground on mild winter days; sometimes they will go into torpor for long periods and expend summer fat reserves. Their diet shifts seasonally. In early spring, green vegetation, especially grasses (e.g., burrograss and purple needlegrass), is mainly consumed, but later in the summer seeds become an important resource. After most plants become dormant in the fall, stems, roots, and winter herbs are frequently eaten. Insects may be consumed when available.

Females in estrus are receptive for one day a year, and almost all copulation events occur underground. Breeding occurs in March and April, with a single litter producing four or five blind and hairless young. To avoid inbreeding, juvenile males disperse to another coterie, and dominant males move about every two years to preclude mating with their two-year-old female offspring.

Colonies of prairie dogs attract an array of other species that prey on them, use their burrows for shelter, or both (Truett et al. 2014). One species, the endangered black-footed ferret, historically coexisted with *C. ludovicianus* and apparently cannot survive in the wild without prairie dogs. Other species, such as the burrowing owl, ferruginous hawk, and kit fox, also benefit greatly from their presence but are not prairie dog obligates. Predators include the afore-mentioned species as well as badgers, coyotes, bobcats, and snakes. In response to predators, several coterie members will perch on their mounds to act as sentinels and produce bark-like alarm calls. The subspecies in the Trans-Pecos is *C. l. arizonensis*.

Eastern Fox Squirrel
Sciurus niger Linnaeus, 1758

The fox squirrel is a large tree squirrel with buffy to orangish underparts and feet; the upperparts are grizzled brownish or grayish (fig. 68). The tail is usually less than half the total length and is cinnamon mixed with black. The ears are prominent. Average external measurements (in mm) are total length 550; tail 240; hindfoot 70; ear 25; weight is 600–1,300 g.

Fox squirrels are widespread throughout eastern Texas but barely enter the Trans-Pecos in Pecos and Terrell Counties. In addition to the native populations that occur in these eastern counties, a far disjunct human-introduced population has been reported in the western Trans-Pecos in the Lower El Paso Valley of the Rio Grande, in El Paso County (map 66). The native populations seem to be limited to riparian areas along the Pecos River and its tributaries. One of us (DJS) obtained several specimens in pecan and oak trees along Independence Creek, south of Sheffield, in Pecos County. They have been recorded from walnut-willow, hackberry, live oak, and mesquite-creosote associations in Terrell County (Hermann 1950). The introduced squirrels of El Paso County have been labeled a severe threat to the pecan industry in the area, and some biologists have recommended the complete eradication of this species from its nonnative range in the Trans-Pecos (Frey et al. 2013).

These squirrels are diurnal, do not hibernate, and are active throughout the year. They prefer more open woodlands and are comfortable traveling between trees or on the ground. Preferred den sites are in cavities or hollows in trees if available, and nests are made from masses of twigs and leaves attached to higher branches. Little is known about their diet in the Trans-Pecos, but elsewhere it includes green vegetation, flowers, fruits, seeds, insects, eggs, and baby birds as well as acorns and other nuts.

FIG. 68. Eastern fox squirrel, *Sciurus niger*

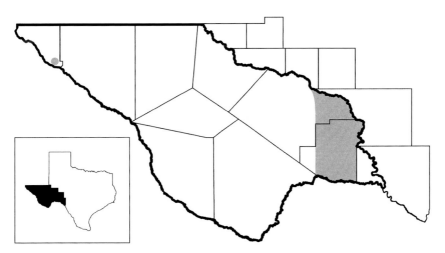

MAP 66. Distribution of the eastern fox squirrel, *Sciurus niger*. The shaded area in the far eastern part of the map represents the range of native individuals; the shaded dot in the far western part of the map represents a human-introduced population.

Breeding occurs during the spring. Males aggregate near a female until she begins her one-day estrus; she may mate with several males. After a gestation period of around 45 days, an average of four blind, nearly hairless young are born. Predators include opossums, raccoons, foxes, coyotes, bobcats, large hawks, and snakes. Eastern fox squirrels get their name from their tail, which resembles that of a fox. The subspecies in the Trans-Pecos is *S. n. limitis*.

Family Geomyidae
POCKET GOPHERS

Pocket gophers are medium-sized burrowing rodents that spend most of their life belowground. They have strong incisors, long claws on the front feet for digging, external fur-lined cheek pouches for transporting food, small eyes, and a short, nearly naked tail. Their presence may be detected by their characteristic gopher mounds—piles of dirt that they push to the surface from their underground burrows and deposit in cone-shaped mounds. One gopher may produce many such mounds. Three species in three genera (*Cratogeomys, Geomys,* and *Thomomys*) live in the Trans-Pecos.

1. Upper incisors smooth, without longitudinal grooves on anterior surface (fig. 69); claws of forefeet small and slender . . . .
    . . . . **Bailey's Pocket Gopher**, *Thomomys baileyi*
    Upper incisors with one or two longitudinal grooves on anterior surface (fig. 69); claws of forefeet large and long 2
2. Upper incisors with one deep, nearly centered, longitudinal groove on anterior surface (fig. 69); feet blackish . . . .
    . . . . **Yellow-faced Pocket Gopher**, *Cratogeomys castanops*
    Upper incisors with two longitudinal grooves on anterior surface, one that is deep and nearly centered and one that is less robust along inner (medial) edge (fig. 69); feet whitish . . . .
    . . . .**Desert Pocket Gopher**, *Geomys arenarius*

Bailey's Pocket Gopher
Thomomys baileyi Merriam, 1901

Bailey's pocket gopher is the smallest of the three species of pocket gophers that occur in the Trans-Pecos. Its dorsal coloration varies from pale gray to russet brown, and the tops of the forefeet are white. The anterior surface of the upper incisors is smooth and without grooves, and claws on the front feet are

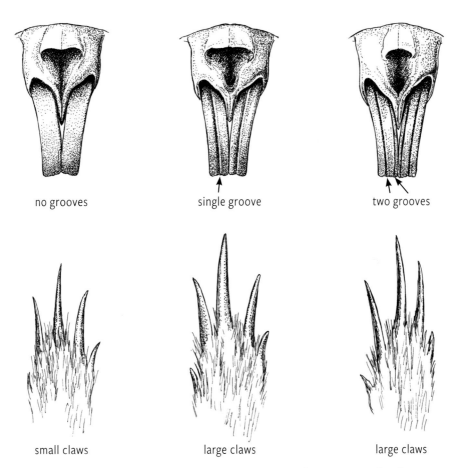

no grooves   single groove   two grooves

small claws   large claws   large claws

FIG. 69. Frontal views of the upper incisors and front feet of three genera of pocket gophers: left, *Thomomys*; center, *Cratogeomys*; right, *Geomys*

relatively small (< 10 mm). Average external measurements (in mm) are total length 205; tail 58; hindfoot 29; ear 7; average weight is 106 g. Males average larger than females. *Thomomys baileyi* can be differentiated from other pocket gophers in the Trans-Pecos as shown in fig. 69.

This gopher is broadly distributed and common across the Trans-Pecos in suitable habitat. It has been recorded from all counties except Reeves (map 67). It occupies a wide variety of substrata including loose sands, caliche, silts, and tight clays, and in vegetative zones grading from dry deserts to montane meadows. It can be found at elevations of approximately 500 to 2,300 m. In desert habitats it occurs in close association with lechuguilla,

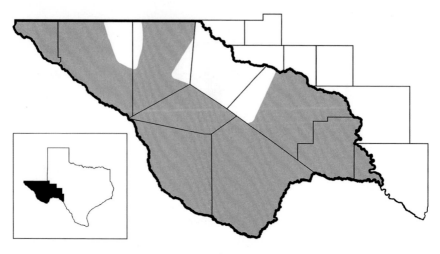

MAP 67. Distribution of Bailey's pocket gopher, *Thomomys baileyi*

yucca, sotol, catclaw, and several cacti. In the Guadalupe, Apache, Delaware, and Chisos Mountains and the Sierra Vieja it is restricted largely to the poorer, thinner soils on dry, rocky flats and mountain slopes and is mostly absent from the deeper soils at mountain bases (Davis 1940b; Borell and Bryant 1942; Blair and Miller 1949; Stangl et al. 1994). It has been taken in desert grassland with large stands of lechuguilla and along a grassy stream-side with scattered mesquite in BBRSP (Yancey 1997), and along rimrocks just below the tops of limestone mesas in the cedar-ocotillo plant associa-tion on the Stockton Plateau (Hermann 1950). It is especially common in the ash-cottonwood association of floodplain areas bordering Limpia Creek in the Davis Mountains (Blair 1940).

Bailey's pocket gopher is active year-round, and individuals are known to store food for periods of drought and food shortage. It feeds on a wide variety of plant species, but lechuguilla seems to be a particular favorite (Stangl et al. 1994). Long (up to > 30 m), extensive burrow systems allow these gophers to forage over large areas without exposure to aboveground predators. They can pull entire plants from the roots into the burrow for leisurely consumption. Individuals are known to occasionally leave the burrow opening, clip all vege-tation in reach, and then retreat back into the burrow. Two of us (FDY, RWM) observed a burrow in BBNP that had recently been plugged with lechuguilla stems, each with the sharp tip facing outward, seemingly in an effort to fortify the burrow entrance with these spear-like structures.

Breeding occurs throughout the year, with periods of increased fertility in spring, summer, and early winter. Females produce an average of two litters a year, typically with five young each. Predators include rattlesnakes, gopher snakes, red-tailed hawks, barn owls, great horned owls, bobcats, coyotes, long-tailed weasels, and badgers.

A recent molecular genetics analysis has revealed that populations of this taxon from Texas, formerly called *T. bottae*, represent a separate species, for which the name *T. baileyi* is now used (Bradley et al. 2023). These gophers show extensive morphological variation, and historically 10 subspecies have been described from the Trans-Pecos. A recent morphological study by Beauchamp-Martin et al. (2019) reduced that number to six as follows: *T. b. texensis* from the northern Front Range of Hudspeth and Culberson Counties southward to the Davis Mountains in Jeff Davis County; *T. b. lachuguilla* from El Paso County to the Big Bend in Brewster County; *T. b. limpiae* from the lower Limpia Canyon in Jeff Davis County; *T. b. spatiosis* from Alpine, Brewster County; *T. b. baileyi* from Sierra Blanca in Hudspeth County; and *T. b. robertbakeri* from the Stockton Plateau in Terrell County. However, the molecular genetics data do not support this interpretation and the recognition of subspecies of *T. baileyi* remains unresolved (Bradley et al. 2023). Despite numerous attempts, *T. b. baileyi* has not been collected in several decades and may now be extinct (Schmidly et al. 2022).

Desert Pocket Gopher
Geomys arenarius Merriam, 1895

This dull, pale brown pocket gopher is readily differentiated from the other pocket gophers in the Trans-Pecos by its medium size and the presence of two longitudinal grooves on the anterior surface of each upper incisor (see fig. 69). The feet and underparts are white. Average external measurements (in mm) for males and females, respectively, are total length 260, 243; tail 85, 74; hindfoot 33, 32; weight is 198–254 g for males and 165–207 g for females.

This pocket gopher occurs in Texas only in the Trans-Pecos, where it is known from several localities in the cottonwood-willow association along the Rio Grande in El Paso and Hudspeth Counties (map 68). Although its distribution is restricted, this gopher may be locally common within its range. It prefers loose soils of disturbed terrain and sandy areas along riverbanks and cannot tolerate clay or gravelly soils. Individuals are commonly found along edges of bodies of open water, such as rivers, ponds, and irrigation canals.

The desert pocket gopher is active year-round, and specimens have been taken during every month in the Trans-Pecos. It is active day or night and will occasionally leave its burrow to forage nearby. Diet is not well known but

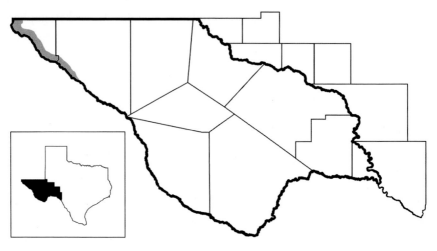

MAP 68. Distribution of the desert pocket gopher, *Geomys arenarius*

presumably consists mostly of roots and stems of a variety of plants. Its burrow systems can be extensive, with up to 20 to 30 mounds covering an area as large as 25 m².

This gopher seems to have a prolonged breeding season during the summer months that allows for more than one litter per year. Pregnant females with four to six embryos have been captured in June and August, and juveniles have been taken from late June through August (J. Jones and Lee 1962). There are no reports of specific predators of *G. arenarius*, although they likely include a variety of hawks, owls, and small mammalian carnivores.

Because of its patchy and restricted distribution, *G. arenarius* is listed as near threatened by the IUCN (IUCN 2022), but it is not included on any federal or state list of threatened and endangered species. The subspecies in the Trans-Pecos is *G. a. arenarius*.

Yellow-faced Pocket Gopher
Cratogeomys castanops (Baird, 1852)

This pocket gopher is moderately large and can be readily distinguished from other gophers by the presence of a single groove on each upper incisor (see fig. 69). The color of the upperparts is dull yellowish brown; the feet are blackish (whitish in other Trans-Pecos pocket gophers; fig. 70). Average external measurements (in mm) are total length 290; tail 78; hindfoot 36; ear 6; weight is 220–410 g. Males are larger than females.

FIG. 70. Yellow-faced pocket gopher, *Cratogeomys castanops*

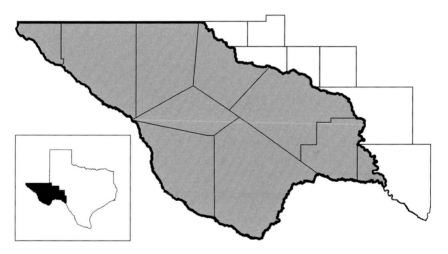

MAP 69. Distribution of the yellow-faced pocket gopher, *Cratogeomys castanops*

The yellow-faced pocket gopher ranges across the entire Trans-Pecos, having been recorded from every county in the region (map 69). It is partial to low to midelevation (about 550–1,750 m) areas with deep soils (17–20 cm of topsoil) that are mostly free of rocks, although on occasion it may be found in habitats with shallow, rocky soils (Russell 1968; Hollander 1990; Yancey 1997; Yancey et al. 2006; C. Jones et al. 2011). The species is common in the sandy loam soils of the Rio Grande in BBNP (Borell and Bryant 1942). It is less common in the deep sandy and shallow rocky desert lowlands of BBNP and BBRSP, where it is often found in association with mesquite or creosote bush (Yancey 1997; Yancey et al. 2006). On the Stockton Plateau it is restricted largely to the mesquite–creosote bush association and floodplain areas where the soil has a greater depth and moisture content (Hermann 1950).

Over the last half century, the yellow-faced pocket gopher has replaced Bailey's pocket gopher (*Thomomys baileyi*) in many places in the Trans-Pecos (Reichman and Baker 1972). Bailey's pocket gopher was previously the predominant pocket gopher in fluvial soils across the region, but an increase in xeric conditions seems to have favored *C. castanops* at the expense of *T. baileyi*.

This pocket gopher is active throughout the year and has been captured during every month. A variety of plant material including roots, tubers, stems, woody parts, and green vegetation serves as its primary food source. Lechuguilla, in particular, seems to be a favorite in parts of the Trans-Pecos (Hermann 1950). Most foraging occurs from within the gopher's burrow system as plants are pulled into the burrow by the roots. Green vegetation may also be gathered from near the openings of burrows. Plants are severed at the base and pulled into the burrow, where they are cut into pieces suitable for transport in the gopher's cheek pouches (Bailey 1931). Yellow-faced pocket gophers breed throughout the year and produce litters of one to four (average two) up to three times annually. Large hawks and owls (particularly great horned owls) and small carnivorous mammals are their primary predators.

The taxonomic history of *C. castanops* in the Trans-Pecos is complex. In the first edition of this book (Schmidly 1977b), this species was included as *Pappogeomys castanops*, but it is now recognized as *Cratogeomys castanops*. Three subspecies were recognized within the Trans-Pecos by Russell in 1968 and then again by Hollander in 1990: *parviceps*, *lacrimalis*, and *clarkii*. Then, in their 2008 revision of the species using DNA sequence data, Hafner and colleagues concluded that all Texas, and thus Trans-Pecos, populations should be lumped into a single subspecies, *C. c. castanops*. However, because their sample size was small and little attention was given to morphological characters, Schmidly and Bradley (2016) and Schmidly et al. (2022) did not accept the conclusion of Hafner et al. (2008) and instead continued to recognize the three subspecies proposed by Russell (1968) and Hollander (1990). Then, to add more confusion to the issue, Schmidly et al. (2023) discovered a misinterpretation of the type locality of *C. c. clarkii* that resulted in the inappropriate and thus invalid use of *clarkii* for populations from the Trans-Pecos. They proposed, based on priority considerations, that individuals from the Trans-Pecos formerly considered *C. c. clarkii* be referred to as *C. c. pratensis*. In addition, a recent molecular genetics study indicates that specimens of this taxon from independent regions in the northern Trans-Pecos are similar to those of *C. c. lacrimalis* and *C. c. parviceps* (Bradley et al. 2023). Therefore, we tentatively recognize three subspecies of *C. castonops* in the Trans-Pecos: *C. c. parviceps* from the extreme northwestern part of the region, *C. c. lacrimalis* from the northeastern part of the region, and *C. c. pratensis* from across the southern two-thirds of the region.

Family Heteromyidae
POCKET MICE AND KANGAROO RATS

Members of this family are among the most characteristic desert rodents in North America. Their general form indicates that jumping is their primary means of locomotion. They usually have large heads, compact bodies, long hind limbs, and long, frequently tufted tails. They also have external fur-lined cheek pouches like those of pocket gophers. Perhaps their most remarkable adaptation is their ability to survive for long periods on a diet of dry seeds with no free water. Ten species representing three genera (*Chaetodipus*, *Dipodomys*, and *Perognathus*) occur in the Trans-Pecos.

1. Hindfeet more than twice as long as forefeet; auditory bullae of skull greatly enlarged 2

   Hindfeet less than twice as long as forefeet; auditory bullae of skull not greatly enlarged 4

2. Hindfeet with five toes, inner toe minute . . . .
   . . . . **Ord's Kangaroo Rat**, *Dipodomys ordii*

   Hindfeet with four toes 3

3. Prominent white tip on end of tail; hindfeet > 45 mm; tail > 170 mm
   . . . . **Banner-tailed Kangaroo Rat**, *Dipodomys spectabilis*

   No white tip on end of tail; hindfeet < 45 mm; tail < 170 mm
   . . . . **Merriam's Kangaroo Rat**, *Dipodomys merriami*

4. Size small, total length < 140 mm; pelage soft, smooth, and silky 5

   Size large, total length > 140 mm; pelage harsh, often bristly, never smooth and silky 6

5. Total length usually > 120 mm; tail length usually ≥ 60 mm; length of skull usually > 21 mm; postauricular patch inconspicuous
   . . . . **Plains Pocket Mouse**, *Perognathus flavescens*

   Total length usually < 120 mm; tail length usually < 60 mm; length of skull usually < 21 mm; postauricular patch conspicuous
   . . . . **Silky Pocket Mouse**, *Perognathus flavus* or

Merriam's Pocket Mouse, *Perognathus merriami* (these two cryptic species cannot be reliably differentiated without chromosomal and DNA sequence data comparisons)

6. Length of tail less than length of head and body (when tail laid forward over back, tip does not reach snout); tail without terminal tuft . . . .**Hispid Pocket Mouse**, *Chaetodipus hispidus*

   Length of tail greater than length of head and body (when tail laid forward over back, tip extends beyond snout); tail with terminal tuft 7

7. Rump pelage with spiny hairs (spines) that project beyond guard hairs 8

   Rump pelage without spines . . . . Chihuahuan Desert Pocket Mouse, *Chaetodipus eremicus*

8. Rump spines numerous, well developed; total length usually > 180 mm . . . . **Highland Coarse-haired Pocket Mouse**, *Chaetodipus collis*

   Rump spines fewer, not well developed, with some spines lighter; total length usually < 180 mm . . . .**Rock Pocket Mouse**, *Chaetodipus intermedius*

Silky Pocket Mouse
Perognathus flavus Baird, 1855
and
Merriam's Pocket Mouse
Perognathus merriami J. A. Allen, 1892

The silky pocket mouse (*Perognathus flavus*) and Merriam's pocket mouse (*Perognathus merriami*) are two closely related species of pocket mice with a long, complex, and confusing taxonomic history. In the first edition of this book, they were lumped together as a single species, *P. flavus*, based on morphological examination (Wilson 1973). However, subsequent chromosomal and biochemical studies (T. Lee and Engstrom 1991), additional morphometric examination (Brant and Lee 2006), and DNA sequence data (Coyner et al. 2010) clearly indicate the two taxa are separate species that both occur in the Trans-Pecos. However, because of their taxonomic complexity, it is difficult to separate much of the natural history and biogeographic information for the two species. Therefore, we have taken a conservative approach and address them in a single account as the *P. flavus/P. merriami* complex.

Both species are small pocket mice with a short, silky, dorsal pelage of various shades of buff washed with black-and-white underparts (fig. 71). A spot

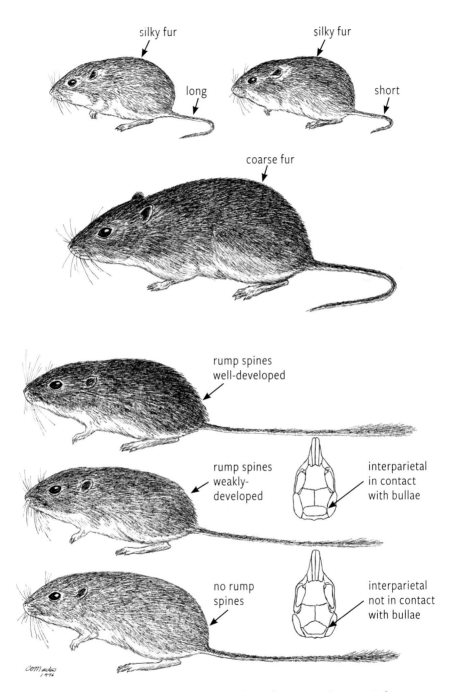

FIG. 71. Pocket mice, genera *Perognathus* and *Chaetodipus*, top to bottom: *P. flavescens* (left); *P. flavus/P. merriami* (right); *C. hispidus*; *C. collis*; *C. intermedius*; *C. eremicus*. See text for a discussion of cranial differences between *C. eremicus* and *C. intermedius*

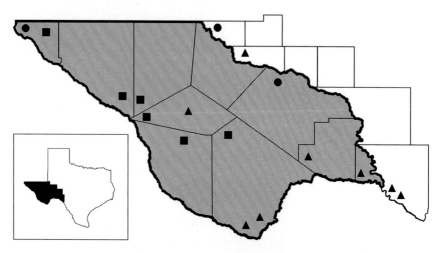

MAP 70. Distribution of silky/Merriam's pocket mice in Trans-Pecos Texas. Localities of populations of *Perognathus flavus* (circles) and *Perognathus merriami* (triangles) and general regions where both species occur (squares) are indicated based on specimens definitively identified by biochemical or molecular analyses.

behind each ear is clear buff and conspicuous. Their tails are relatively long and their ears relatively short. Average external measurements (in mm) for both species are total length 113–116; tail 50–57; hindfoot 16; ear 6; weight is 6–9 g. There are no reliable morphological features that will distinguish all individuals of the two species. However, there are seven measurements (two external and five cranial) in which specimen samples of *flavus* differ significantly on average (in a statistical sense at the $p < 0.05$ level of probability) from those of *merriami* (Brant and Lee 2006). The external measurements are tail length and length of hindfoot; *merriami* averages larger than *flavus* in tail length (58.00 mm vs. 55.96 mm) but smaller in hindfoot length (16.92 mm in *flavus* vs. 16.56 mm in *merriami*). For the cranial measurements of bullar length, mastoid breadth, and interparietal length, *flavus* is statistically larger on average than *merriami*, whereas for interorbital breadth and interparietal length, *merriami* is larger than *flavus* (see Brant and Lee 2006).

Collectively, these two silky pocket mice range across the entire Trans-Pecos (map 70). To examine possible distributional differences, we considered 167 specimens housed in the Angelo State Natural History Collection that had previously been accurately identified to species using either biochemical (allozyme markers) or morphological (discriminant function analysis) data (T. Lee and Engstrom 1991; Brant and Lee 2006). To that sample, we added 10 specimens that had been definitively identified to species using molecular genetic markers (cytochrome b gene) (Coyner et al. 2010) and then overlaid the localities of these specimens on a map of the Trans-Pecos (see map 70).

Based on definitively identified specimens, it appears that *P. flavus* occupies the western, central, and northern regions of the Trans-Pecos, whereas *P. merriami* is particularly widespread across the central, southern, and eastern regions. The two species have been found together at localities in eastern El Paso County (Hueco Basin), southern Hudspeth County (Diablo Plateau), southern Culberson and western Jeff Davis Counties (Valentine Basin), northern Presidio County (Marfa Plateau), and near Alpine and Marathon in Brewster County (Marathon Basin). The distribution map suggests that *P. flavus* is absent from the Stockton Plateau in the southeastern part of the Trans-Pecos (Terrell and Val Verde Counties), where only *merriami* occurs. Clearly, further analyses of specimens from across the entire Trans-Pecos using molecular genetic and biochemical markers are needed to completely refine the distribution of the two species in the region.

These small silky pocket mice occur in a wide variety of habitats, including rocky, gravelly, and sandy substrata, and in a number of plant associations, such as cottonwood-willow, shortgrass-mesquite, grama-bluestem, catclaw-tobosa, creosote bush–mesquite, shortgrass-yucca, juniper-sacahuista, and juniper sotol (Blair 1940; Blair and Miller 1949; Hermann 1950; Tamsitt 1954; Denyes 1956; Ederhoff 1971; Stangl et al. 1994; Yancey 1997). However, because of the uncertain identification of the two species in these studies it is not possible to determine the extent to which habitat differences between them might exist.

Silky and Merriam's pocket mice seem to enter a state of torpor during cold temperatures but apparently do not hibernate. They are mostly nocturnal but may show some aboveground activity during daylight. Seeds are their primary food, with some green vegetation, juniper berries, and insects consumed to a lesser degree.

Their breeding season extends from April to November, and two or more litters may be produced each season. Gestation is probably about 22–26 days, after which one to seven young (average about four) are born (Yancey 1997). Reported predators include numerous snakes, barn and burrowing owls, loggerhead shrikes, coyotes, gray foxes, kit foxes, badgers, striped skunks, ringtails, and grasshopper mice. Ten specimens captured in BBRSP were submitted for hantavirus testing in 2017, and seven proved positive. Although not definitively resolved at this time, the subspecies of *P. flavus* in the Trans-Pecos is probably *P. f. flavus*, and that of *P. merriami* is *P. m. gilvus*.

Plains Pocket Mouse
Perognathus flavescens Merriam, 1889

This is a small pocket mouse with a short, silky, yellow-buff dorsal pelage washed with black, and white underparts. The tail is relatively long and the ears relatively short. It is noticeably larger and has relatively smaller ears and a longer tail than either *P. flavus* or *P. merriami* (fig. 71). Average external measurements (in mm) are total length 132; tail 64; hindfoot 19; ear 7; weight is 8–11 g.

This pocket mouse is known from the Trans-Pecos only from El Paso County in the far western part of the region (map 71), where its preferred habitat is stabilized sand dunes 1–2 m in height that support mesquite, some creosote bush, and scattered yucca and Russian thistle (J. Jones and Lee 1962).

Little life history information is available from the Trans-Pecos. *Perognathus flavescens* is primarily nocturnal. Its diet is almost exclusively granivorous, including a wide variety of grass and weedy plant seeds. Some green herbaceous vegetation may be consumed on occasion. Females are normally pregnant from April through August, with a single annual litter, though in favorable years multiple litters may be produced. Gestation is about 25–26 days, with two to seven young born. Reported predators include barn and great horned owls, although snakes, other raptors, and small mammalian carnivores also likely consume them.

In the first edition of this book, this taxon was regarded as the Apache pocket mouse, *Perognathus apache*. Subsequent examination of morphology

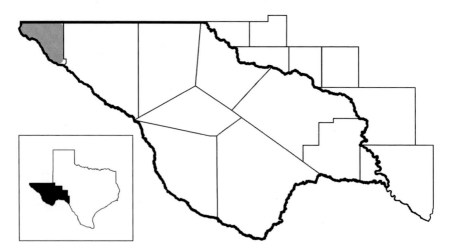

MAP 71. Distribution of the plains pocket mouse, *Perognathus flavescens*

and chromosomes concluded that *P. apache* and *P. flavescens* constituted a single species (D. Williams 1978a, 1978b). As a result, the Trans-Pecos population is now designated *P. flavescens*. The subspecies in the Trans-Pecos is *P. f. melanotis*.

Hispid Pocket Mouse
Chaetodipus hispidus (Baird, 1858)

This is a large, stout pocket mouse with a coarse, buffy blond dorsal pelage mixed with black above and buff on the sides. The underparts are white. The tail is relatively short (< head and body length), bicolored, and scantily haired and lacks a tuft at the end. *Chaetodipus hispidus* is easily distinguished from other species of pocket mice based on its large size and coloration (fig. 71). Average external measurements (in mm) are total length 208; tail 96; hindfoot 24; ear 12; average weight is 37 g.

This is the rarest pocket mouse in the Trans-Pecos, with a distribution confined primarily to the grasslands in the central part of the region (map 72). Apparently, it is absent from much of the harsh desert habitats in the western, southern, and northern parts of the region. It seems to prefer areas of sandy or friable soils covered with scattered to moderate stands of desert grassland vegetation. Although seemingly uncommon, it is known to occupy a number of vegetation associations, including shortgrass-yucca, shortgrass-juniper, juniper-ocotillo, mesquite-cholla, and shortgrass-mesquite (Blair 1940; Denyes 1956; Yancey 1997; Yancey and Jones 2000).

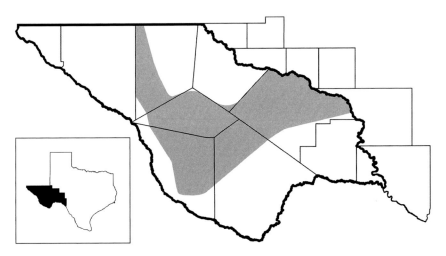

MAP 72. Distribution of the hispid pocket mouse, *Chaetodipus hispidus*

This pocket mouse does not hibernate but may briefly enter torpor during periods of food shortage. It is active at night, foraging mostly for seeds of a variety of plants such as mesquite, cactus, sagebrush, and sunflower. Some green vegetation and insect material may be consumed as well. Most foraging occurs on the ground, but on occasion individuals will climb vegetation in search of food. Seeds collected during the warmer months may be cached and used for subsistence during winter.

Breeding is thought to occur throughout the year during favorable conditions. One to two litters of four to seven offspring may be produced annually. Nothing is known of the gestation period and development of young. Predators include western diamondback rattlesnakes and great horned owls as well as other snakes, raptors, and mammalian carnivores. A high percentage of *C. hispidus* individuals have been found to harbor *Trypanosoma cruzi*, the protozoan responsible for Chagas disease, and thus it may serve as a reservoir for this parasite.

In the first edition of this book, this taxon was regarded as *Perognathus hispidus*. In 1983, based on anatomical, chromosomal, and biochemical differences, four species of Trans-Pecos pocket mice of the genus *Perognathus* were reassigned to the genus *Chaetodipus*, and thus *P. hispidus* became formally known as *Chaetodipus hispidus* (J. Hafner and Hafner 1983). The subspecies in the Trans-Pecos is *C. h. paradoxus*.

Chihuahuan Desert Pocket Mouse
Chaetodipus eremicus (Mearns, 1898)

This medium-sized pocket mouse has a long, tufted tail, short ears, and a rather harsh, grayish-brown pelage that lacks spines on the rump. This species closely resembles *C. intermedius*, but the two may be separated by close examination of the pelage and the skull (fig. 71) or the karyotype. Upon close examination, weak rump spines can be noticed on *C. intermedius*, whereas they are completely absent from *C. eremicus*. The latter has also been shown to have a significantly larger greatest length of skull measurement than *C. intermedius* (Wilkins and Schmidly 1979). Furthermore, the interparietal of *C. eremicus* is not in contact with the mastoid bullae, being separated by narrow projections of the parietals and supraoccipitals. In contrast, the interparietal of *C. intermedius* is in contact, or nearly so, with the mastoid bullae (Wilkins and Schmidly 1979). *Chaetodipus eremicus* and *C. intermedius* are readily distinguished by their karyotypes, with the former having a diploid number (2N) of 46 and a fundamental number (FN) of 56, whereas in the latter the 2N is 46 and the FN is 58. Average external measurements (in mm) are total length 175; tail 96; hindfoot 23; ear 8; weight

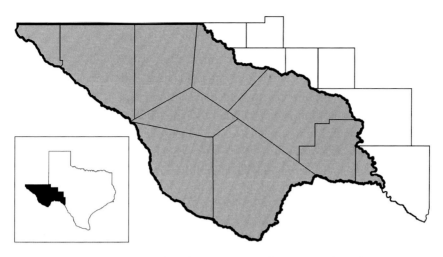

MAP 73. Distribution of the Chihuahuan Desert pocket mouse, *Chaetodipus eremicus*

is about 15 g. Interestingly, males are slightly larger than females in a population from CMSNA (Goetze et al. 2018), whereas no sexual dimorphism occurs in a population from BBNP (Manning et al. 1996).

The Chihuahuan Desert pocket mouse is the most common and widespread species of *Chaetodipus* in the Trans-Pecos. It has been recorded from every county except Terrell and is projected to occur throughout the entire region in suitable habitat (map 73); it has been recorded from both sides of the Pecos River (J. Jones and Manning 1991). It prefers sandy, rock-free, alluvial soils, and the presence of these kinds of substrata may be more of a factor in determining its occurrence than vegetation type. However, the species may occasionally be found on rocky, gravelly substrata (Yancey 1997; Yancey and Jones 2000). While it has been found in many different vegetation associations, it seems to favor desert scrub situations that include such plants as creosote bush, mesquite, catclaw, and tasajillo. It is common in these habitats in CMSNA and BBRSP (Yancey 1997; C. Jones et al. 2011). In BBNP it is especially common in the false willow–mesquite river-bottom habitat along the Rio Grande (Boeer and Schmidly 1977), as well as in desert scrub habitats with sandy soils (Borell and Bryant 1942; Yancey et al. 2006).

Trapping records indicate that it is not as active in winter as in summer (Manning et al. 1996; Yancey 1997; Porter 2011). It may enter torpor and become lethargic and inactive for several days in extremely cold winter periods to facilitate survival when food is not readily available. Like other pocket mice, *C. eremicus* is strictly nocturnal and forages for seeds, preferring those of mesquite, creosote bush, and broomweed. Green vegetation and insect larvae are eaten on occasion.

Peak periods of reproduction occur in spring and early and late summer (Porter 2011), although observations of gravid females in March, November, and December suggest that some reproduction takes place during autumn and winter (Yancey 1997; Yancey et al. 2006). After a gestation of three to four weeks, 2 to 5 (average 3.6) young are born (Porter 2011). Curiously, Manning et al. (1996) reported a female with 15 placental scars, thus suggesting a potentially larger range in litter size. Specific predators have not been reported, although they likely include several species of snakes, raptors, and small carnivorous mammals.

In the first edition of this book, this taxon was included as *Perognathus penicillatus*, but the generic assignment was subsequently changed from *Perognathus* to *Chaetodipus*, which resulted in designating this taxon *Chaetodipus penicillatus* (for an explanation, see the account of *C. hispidus*). Following that change, another study using mitochondrial DNA split *C. penicillatus* into two species, with Trans-Pecos populations assigned to *C. eremicus* (T. Lee et al. 1996). This is a monotypic species, with no subspecies recognized.

Rock Pocket Mouse
Chaetodipus intermedius (Merriam, 1889)

The rock pocket mouse is a medium-sized pocket mouse with a buff-brown dorsal pelage sprinkled with black. The pelage is coarse, with weak, often inconspicuous rump spines. The tail is long and tufted, and the ears are short. *Chaetodipus intermedius* closely resembles the desert pocket mouse, *C. eremicus*, from which it differs in features of the pelage, skull, and karyotype (see the account of *C. eremicus* and fig. 71). Average external measurements (in mm) are total length 173; tail 97; hindfoot 21; ear 8; average weight is 13 g. Males are slightly larger than females.

The rock pocket mouse is known in Texas only from the Trans-Pecos, where it has been commonly recorded from all counties except Pecos, Terrell, and Val Verde (map 74). This pocket mouse seems to prefer low to midelevation zones with boulders, rocks, pebbles, and gravelly substrate. In GMNP, CMSNA, BBRSP, and areas around El Paso, it seems to have an affinity for desert scrub habitats where creosote bush is present (Ederhoff 1971; Genoways et al. 1979; Yancey 1997; C. Jones et al. 2011).

Chaetodipus intermedius is known to enter periods of torpor and become less active during the hottest and coldest times of the year (Stangl et al. 1994).

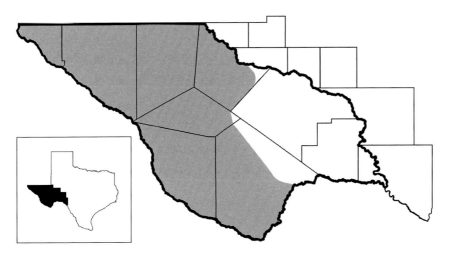

MAP 74. Distribution of the rock pocket mouse, *Chaetodipus intermedius*

It is nocturnal, foraging at night for seeds of small plants. Seeds are often cached for later consumption during times of food shortage. Breeding is thought to begin in February or March and continue for several months. Pregnant females have been captured in May, June, and July (Davis 1974); litter size varies from three to six. Little is known about its predators, although they likely include a variety of snakes, birds of prey, and small carnivorous mammals.

In the first edition of this book, this taxon was regarded as *Perognathus intermedius*, but for reasons explained in the account of *C. hispidus*, the generic assignment was subsequently changed from *Perognathus* to *Chaetodipus*, which resulted in renaming this taxon *Chaetodipus intermedius*. The subspecies that occurs in the Trans-Pecos is *C. i. intermedius*.

Highland Coarse-haired Pocket Mouse
Chaetodipus collis (Blair, 1938)

This tawny-olive to drab gray, medium-sized pocket mouse has a harsh pelage and a tail that is longer than the head and body, bicolored (dark above, light below), and distinctly tufted at the distal end. The ears are short and have a small white patch at the base. It may be distinguished from *C. eremicus* and *C. intermedius* in having numerous elongated, black-tipped, spiny hairs on the rump that overreach the normal guard hairs (fig. 71). Soles of their hindfeet are dark blackish brown and sparsely haired (pale brownish or flesh colored in the other two species). Because these distinctions are subtle, geographic locality and habitat type are often heavily weighted in identification (Stangl et al. 1994). Rock-loving *C. collis* and *C. intermedius* are often distinguished from sand-dwelling *C. eremicus* based on substrate, whereas *C. collis* and *C. intermedius* are frequently identified based on perceived geographic separation. However, varying degrees of sympatry in the Trans-Pecos have been reported for these three pocket mice (Wilkins and Schmidly 1979; Yancey 1996, 1997; Yancey and Jones 2000), and therefore range and habitat should not be the primary criteria for definitive identification of specimens. Average external measurements (in mm) are total length 184; tail 104; hindfoot 22; ear 8; average weight is 15 g. Males are slightly larger than females in CMSNA (Goetze et al. 2018), but there is no significant size difference between the sexes in BBNP (Manning et al. 1996).

This pocket mouse is common in suitable habitat throughout most of the Trans-Pecos, being absent from only the extreme northern and far western parts of the region (map 75). It has been found on both sides of the Pecos River (J. Jones and Manning 1991). It shows a strong preference for rocky places in grassland and desert scrub vegetation and is most abundant on steep, rocky slopes between about 1,200 and 1,700 m (Baccus 1971). As long as rocks or boulders are present, vegetation seems to be of secondary importance to this species. Some of the many plant associations in which it has been trapped include lechuguilla–chino grass (BBNP; Porter 2011); creosote bush–guayacan–acacia, yucca-lechuguilla-ocotillo, mesquite–prickly pear (BBRSP; Yancey and Jones 2000); catclaw-grama (Sierra Vieja; Blair and Miller 1949); catclaw (Davis Mountains; Blair 1940); ocotillo-catclaw (La Mota Ranch area;

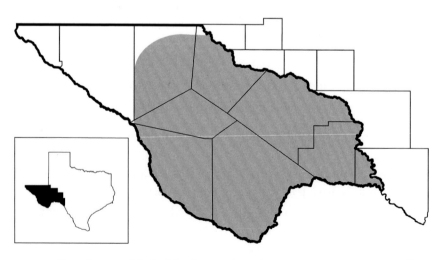

MAP 75. Distribution of the highland coarse-haired pocket mouse, *Chaetodipus collis*

Tamsitt 1954); grama–prickly pear and sotol-lechuguilla (BGWMA; Tamsitt 1954); and persimmon–shin oak (Stockton Plateau; Hermann 1950).

Unlike some other species of pocket mice, *C. collis* is active year-round in the Trans-Pecos, although activity seems to peak from July through September (Manning et al. 1996; Porter 2011). This species is strictly nocturnal and feeds primarily on seeds, especially those of creosote bush, mesquite, prickly pear, and wild buckwheat (Judd 1967), although it will also consume other plant materials and insects.

Seasonal records of gravid females and males in reproductive condition indicate that mating begins in February and continues through July. The percentage of pregnant females reaches a peak in March, declines in April, reaches a secondary peak in May, declines again in June, and increases slightly in July (Porter 2011). Litter size based on embryo counts ranges from 2 to 5 (Manning et al. 1996), with an average of 3.2 (Porter 2011). A female taken in August in BBNP had seven obvious placental scars, suggesting that on occasion litter size may exceed the normal range (Manning et al. 1996). Gestation is about 30 days and development of neonates is rapid. Documented predators include western diamondback rattlesnakes, great horned owls, and barn owls, although other small carnivorous mammals, raptors, and snakes likely prey on it as well.

In the first edition of this book, this taxon was regarded as *Perognathus nelsoni*. Subsequently, pocket mice of the genus *Perognathus* were reassigned to the genus *Chaetodipus*, and thus *P. nelsoni* became formally known as *Chaetodipus nelsoni*. Then in 2019, a study using morphological, chromosomal, and molecular data split *C. nelsoni* into three species, with populations in the Trans-Pecos regarded as *C. collis*. The subspecies in the Trans-Pecos is *C. c. collis* (Neiswenter et al. 2019).

Banner-tailed Kangaroo Rat
Dipodomys spectabilis Merriam, 1890

The long, tufted (well haired at the end) tail, long hind legs and feet, small front feet, external fur-lined cheek pouches, and greatly enlarged auditory bullae (figs. 72 and 73) serve to distinguish kangaroo rats from other rodents. The general yellowish-brown color with pure white underparts, a white stripe over each hip, and a conspicuous white patch behind each ear are more or less similar in all kangaroo rats, but differences in other traits are sufficient for species recognition. *Dipodomys spectabilis* is the largest of the Trans-Pecos kangaroo rats and may be distinguished from the other species by the distinct white tuft at the end of its tail (fig. 72). This species has four toes on each hindfoot.

FIG. 72. Banner-tailed kangaroo rat, *Dipodomys spectabilis*

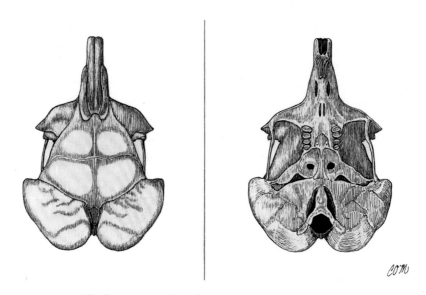

FIG. 73. Dorsal (left) and ventral (right) views of the skull of *Dipodomys spectabilis*

Average external measurements (in mm) are total length 348; tail 214; hind-foot 54; ear 16; weight is 100–150 g. Males are significantly larger than females.

The banner-tailed kangaroo rat ranges throughout the northern two-thirds of the Trans-Pecos, having been recorded from every county except Terrell and Val Verde (map 76). It seems to prefer the dry, gravelly, hard soil of the barren

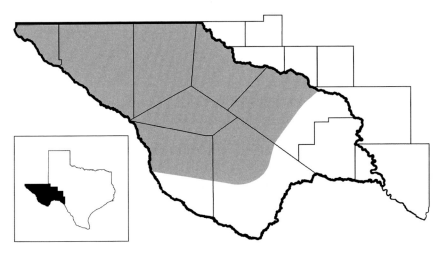

MAP 76. Distribution of the banner-tailed kangaroo rat, *Dipodomys spectabilis*

mesa tops and the foothill slopes of desert ranges where it is uncommon in well-developed grasslands at elevations between about 1,200 and 1,500 m. It has been found on slopes covered with scattered mixed stands of creosote bush and acacia, and in the yucca-sotol association on foothill slopes. Its presence in a particular habitat is indicated by the large (up to over 1 m in height and 3 m in diameter) earthen mounds produced at the opening of its burrows (Stangl et al. 1994).

Dipodomys spectabilis does not hibernate or aestivate, although individuals may remain in their burrows for extended periods when the weather is rainy or cold. It is primarily nocturnal, although there may be some daylight activity during periods of drought. These kangaroo rats are almost entirely vegetarian, consuming mostly seeds but also green vegetation to a lesser degree.

There is evidence that *D. spectabilis* may be coprophagous to some extent (Stangl et al. 1994). Large amounts of foods are stored for consumption during times of shortage. Water is rarely consumed, even if present.

The breeding season begins in January and continues into at least August. Gestation is about 22–27 days, and one to three litters are produced annually. A limited sample size of gravid females from the Trans-Pecos suggests litter sizes of one to three in the region (Stangl et al. 1994). Known predators include diamondback rattlesnakes, barn owls, great horned owls, coyotes, kit foxes, badgers, and long-tailed weasels (Stangl et al. 1994).

The IUCN lists *D. spectabilis* as near threatened (IUCN 2022), but the species does not appear on state or federal lists of threatened and endangered species. However, because it has apparently been declining in numbers in the Trans-Pecos, its status in the region should be carefully monitored. The subspecies in the Trans-Pecos is *D. s. baileyi*.

Ord's Kangaroo Rat
Dipodomys ordii Woodhouse, 1853

Ord's kangaroo rat is the only species of kangaroo rat in the Trans-Pecos with five toes on its hindfoot. It is about the same size as *D. merriami* but considerably smaller than *D. spectabilis*. Dorsal coloration is pale buff washed with gray above, and white on the flanks and underside. Its face is white with dark markings, and the tail is long with a dark tuft at the end (fig. 74). Average external measurements (in mm) are total length 243; tail 137; hindfoot 37; ear 13; weight is 60–70 g. Males are significantly larger than females.

Ord's kangaroo rat occurs throughout the Trans-Pecos in suitable habitat. It has been recorded from all counties except Terrell and Val Verde, and its range is projected to span the entire region (map 77). *Dipodomys ordii* is partial to friable soils, especially sand, and substrate seems more critical in determining its presence than does vegetation. In areas of desert pavement or tough clay soils, it is restricted to pockets of windblown sand or to alluvial soils along arroyos. It reaches its maximum abundance in El Paso County, where it is the most common rodent in the mesquite-covered sand dunes east of El Paso. It is also common in the sand dunes on the western boundary of the Guadalupe and

FIG. 74. Ord's kangaroo rat, *Dipodomys ordii*. Merriam's kangaroo rat, *Dipodomys merriami*, is virtually identical in external appearance but differs in having four instead of five toes on its hindfoot

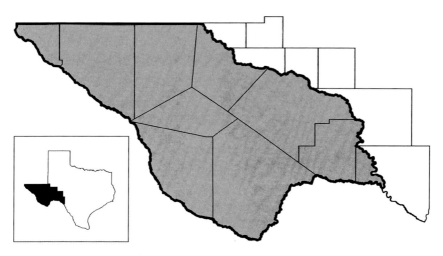

MAP 77. Distribution of Ord's kangaroo rat, *Dipodomys ordii*

northern Apache Mountains. It is uncommon to rare in BBNP, BBRSP, and the Davis Mountains area (Blair 1940; Borell and Bryant 1942; Baccus 1971; Yancey 1997; Yancey et al. 2006; DeBaca 2008). Creosote bush, shortgrass-yucca, and mesquite are plant associations for which *D. ordii* has shown an affinity in the Trans-Pecos (Blair 1940; O'Connell 1979; Stangl et al. 1994).

This kangaroo rat is active throughout the year. It is nocturnal but may occasionally be active outside its burrow during the daytime. It feeds primarily on seeds of shrubs, forbs, and grasses, although green vegetation and insects are consumed periodically (O'Connell 1979). Seeds are collected mostly from the ground, although on occasion individuals may scale vegetation to collect seeds from elevated flowers. Once collected, seeds are placed in cheek pouches for transport back to the burrow. Interestingly, *D. ordii* has recently been shown to be capable of ultraviolet vision (McDonald et al. 2020).

Breeding and parturition occur from August through May. After a gestation period of 28–32 days, a litter of 1 to 6 (average 3.5) is produced up to two times a year. Predators include owls (barn, great horned, long-eared, burrowing), badgers, long-tailed weasels, coyotes, and kit foxes. Several human pathogens infect *D. ordii*, including organisms that cause plague, tularemia, and Rocky Mountain spotted fever.

There are two subspecies of this kangaroo rat in the Trans-Pecos: *D. o. ordii* occurs over most of the region, and *D. o. obscurus* in the Big Bend. The latter differs from *D. o. ordii* in having smaller cranial measurements and darker coloration (Setzer 1949; Baumgardner and Schmidly 1981).

Merriam's Kangaroo Rat

Dipodomys merriami Mearns, 1890

This small, four-toed kangaroo rat is distinguished from *D. spectabilis* by its smaller size and lack of a white tuft at the tip of the tail. *Dipodomys merriami* differs from *D. ordii* in having four instead of five toes on its hindfoot; otherwise the two appear almost identical (see image of *D. ordii* in fig. 74). Average external measurements (in mm) are total length 248; tail 145; hindfoot 38; ear 14; average weight is 40 g. Males are significantly larger than females.

Merriam's kangaroo rat has been documented in every county of the Trans-Pecos, and it is one of the most common and widespread mammals in the region (map 78). As a habitat generalist, it resides in a wide variety of habitats and is equally successful on sandy, clayey, gravelly, or rocky soils. It occurs in many vegetation associations but seems to favor lower-elevation desert scrub habitats, particularly those with creosote bush or mesquite (Davis and Robertson 1944; Blair and Miller 1949; Hermann 1950; Tamsitt 1954; Baccus 1971; Ederhoff 1971; Stangl et al. 1994; C. Jones et al. 2011). In addition to desert scrub, *D. merriami* has been found in desert grassland, juniper roughland, and riparian habitats (Blair 1940; Yancey 1997). Where it occurs with *D. ordii*, *D. merriami* usually occupies the gravelly, desert pavements, whereas *D. ordii* is found in the loose, friable soils of sand dunes and along arroyos.

Merriam's kangaroo rat is not known to enter seasonal torpor and is active during every month (Yancey 1997). It is considered nocturnal but is often observed active during the day (C. Jones et al. 2011). Individuals tend to be

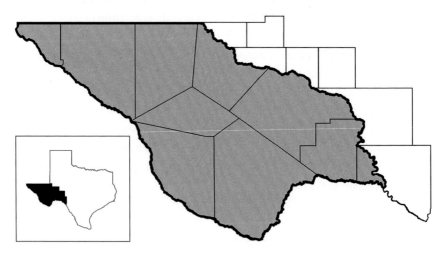

MAP 78. Distribution of Merriam's kangaroo rat, *Dipodomys merriami*

more active on warm nights than cold nights. It feeds mostly on seeds, particularly those of creosote bush and mesquite, as well as forbs, grasses, and succulents. Green vegetation and insect matter are seasonally important supplements to its diet (O'Connell 1979). Most foraging occurs on the ground, but this kangaroo rat may occasionally climb vegetation to gather food.

Seasonal records of pregnant females and males in reproductive condition indicate breeding occurs throughout the year in the Trans-Pecos. After a gestation period of about 29 days, one to five (average three) young are born. Snakes, raptors, and small carnivorous mammals are likely predators. Merriam's kangaroo rat is known to be infected by *Coccidioides immitis*, the fungus responsible for valley fever. The subspecies in the Trans-Pecos is *D. m. ambiguus*.

Family Castoridae
BEAVERS

The beaver is the largest North American rodent. It occurs in favorable river systems from central Alaska and northern Canada south to northern Mexico. Long sought for its fur in North America, the beaver was an important reason for the early explorations of the western United States.

American Beaver
Castor canadensis Kuhl, 1820

The American beaver is a large, robust, semiaquatic rodent with a broad, dorso-ventrally flattened, scaly tail, and webbed hindfeet (fig. 75). Color of upperparts is deep chestnut brown in early autumn, fading to paler brown by spring; the belly is pale buff gray. Average external measurements (in mm) are total length 1,160; tail 400; hindfoot 178; weight is 18–27 kg. Because of its size, unique tail, and webbed hindfeet, the American beaver is easily distinguished from all other mammals that occur in the Trans-Pecos.

Beavers are semiaquatic mammals and therefore require a pond, stream, lake, or river for their existence. Consequently, in the Trans-Pecos they occur only along the Pecos River and Rio Grande (map 79). Specimens have been recorded from the mouth of the Pecos River and from Langtry in Val Verde County, from Independence Creek in Terrell County, from BBNP in Brewster County, and from west of Fort Hancock in Hudspeth County. In addition, there are reports of beaver sightings or sign from along the Rio Grande in El Paso County (Davis 1940a), BBRSP in Presidio County (Yancey 1997), and near Lajitas in Brewster County, as well as from the Pecos River in Terrell County

FIG. 75. American beaver, *Castor canadensis*

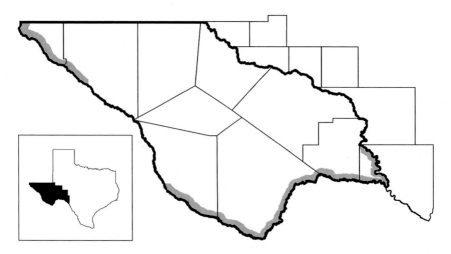

MAP 79. Distribution of the American beaver, *Castor canadensis*

(Hermann 1950). Vegetation in these areas typically consists of salt cedar, giant reed, common reed, willow, and a variety of grasses. Beavers seem to be especially thriving in the deeper pools of the Rio Grande between Terlingua Creek and Boquillas Canyon (BBNP), where a recent camera trap study resulted in an estimate of 185 individuals occupying 98 colonies, an increase of 38 percent in the area since 1981 (Reich 2015).

American beavers are active year-round and do not hibernate. They are crepuscular and nocturnal but may occasionally be active during the day. They are obligate herbivores, typically consuming the leaves, fruits, bark, and woody parts of plants. They are regarded as choosy generalists, meaning that although their diet may include a variety of plant species, only a few make up the majority of foods consumed (Barela and Frey 2016). Along the Rio Grande, beavers seem to favor willow and seepwillow, with common reed, giant reed, cottonwood, tamarisk, sedges, prickly pear, and cocklebur (Reich 2015) serving as less important food species along with desert willow, tree tobacco, and mesquite.

Unlike beavers throughout most of their range, those of the Rio Grande seldom construct dams but instead dig burrows in the bank at the normal level of the river, with about half the opening above water and the remaining half below. When the river water line is low, it is common to see openings above the water level with a mud slide leading from the entrance to the water.

In general, beavers begin breeding in January or February. One of us (DJS) observed young beavers on the Rio Grande that appeared to be only a few weeks old as early as March. Three to four fully furred, precocial young are typically produced once a year. Kits can develop the ability to swim in as few as four days. Predators of beavers include coyotes, mountain lions, and bobcats. Potential threats to the species in the Trans-Pecos include loss or destruction of surface waters and competition with the introduced nutria. Although some authorities do not recognize subspecies of the American beaver because of widespread reintroductions across its range (Helgen 2005), we tentatively consider the subspecies that occurs in the Trans-Pecos to be *C. c. mexicanus* pending future taxonomic review (Schmidly and Bradley 2016).

Family Cricetidae
NEW WORLD MICE, RATS, AND VOLES

This family contains a great variety of mouselike rodents such as deermice, cotton rats, woodrats, harvest mice, grasshopper mice, voles, and muskrats and is the second largest family of mammals worldwide. Most cricetids have a "standard" mouselike form, with a long tail and a generalized limb structure. Voles are somewhat atypical, as they have short tails and a stout, short-legged appearance. In many terrestrial communities, cricetid rodents are the most important small mammals in terms of their effect on the environment and their importance as a staple food item for many predators. There are seven genera and 20 species in the Trans-Pecos.

1. Enamel pattern of molar teeth with two rows of interlocking cusps that may disappear with wear, producing a few relatively formless dentine lakes bordered by enamel; chewing surface usually not flat prior to heavy wear 2

    Enamel pattern of molar teeth with elaborate arrangements of prisms, folds, and angles; chewing surface nearly flat 13

2. Upper incisors each with a deep groove on anterior surface 3

    Upper incisors smooth, without a deep groove on anterior surface 5

3. Tail longer than head and body (fig. 76); color of upperparts usually rich orangish to golden brown ....
    .... **Fulvous Harvest Mouse,** *Reithrodontomys fulvescens*

    Tail about as long as head and body, or shorter; color of upperparts not orangish to golden brown 4

4. Tail shorter than head and body, with a dark, narrow dorsal stripe (fig. 76); braincase breadth < 9.6 mm ....
    .... **Plains Harvest Mouse,** *Reithrodontomys montanus*

    Tail length about length of head and body, without dorsal stripe (fig. 76); braincase breadth > 9.6 mm ....
    .... **Western Harvest Mouse,** *Reithrodontomys megalotis*

5. Tail length < 60% of length of head and body; coronoid process of lower jaw relatively high and strongly recurved 6

    Tail length > 60% of length of head and body; coronoid process of lower jaw relatively low and not strongly recurved 7

6. Tail length usually < 30% of total length, usually less than half length of head and body; crown length of maxillary tooth row ≥ 4.0 mm
    .... **Northern Grasshopper Mouse,** *Onychomys leucogaster*

    Tail length usually > 30% of total length, usually more than half length of head and body; crown length of maxillary tooth row ≤ 3.9 mm
    .... **Mearns's (Chihuahuan) Grasshopper Mouse,**
    Onychomys arenicola

7. Tail shorter than, or as long as, head and body, ≤ 50% of total length... 8

    Tail longer than head and body, > 50% of total length................ 10

8. Ear length more than hindfoot length.........**Piñon Deermouse,**
    Peromyscus truei

    Ear length less than hindfoot length 9

9. Total length of adults usually > 170 mm; tail with broad, not sharply bicolored middorsal stripe; greatest length of skull usually > 26 mm
    .... **White-footed Deermouse,** *Peromyscus leucopus*

Total length of adults usually < 170 mm; tail with narrow, distinctly bicolored middorsal stripe; greatest length of skull usually < 26 mm . . . . **Elliot's Deermouse**, *Peromyscus labecula*

10. Sole of hindfoot naked to end of ankle; nasal bones of skull decidedly exceeded by premaxillae; two principal outer angles of first two upper molars simple, without accessory tubercles and enamel lophs . . . . **Cactus Deermouse**, *Peromyscus eremicus*

    Sole of hindfoot haired on proximal fourth to ankle; nasal bones of skull not exceeded by premaxillae, or only slightly so; two principal outer angles of first two upper molars with well-developed accessory tubercles and enamel lophs 11

11. Tail heavily haired, with tufted tip and sharply bicolored; ear usually > 20 mm; greatest length of skull > 28 mm; molar tooth row length usually > 4.2 mm; mastoid breadth of skull usually > 12 mm . . . . **Northern Rock Deermouse**, *Peromyscus nasutus*

    Tail sparsely haired and not sharply bicolored; ear usually < 20 mm; greatest length of skull < 28 mm; molar tooth row length usually < 4.2 mm; mastoid breadth of skull usually < 12 mm 12

12. Tarsal joints of ankles white, like upper side of hindfeet; tail sparsely haired, without tuft; length of molar tooth row usually < 4 mm . . . . **Lacey's White-ankled Deermouse**, *Peromyscus laceianus*

    Tarsal joints of ankles dusky, color extending from hind leg to or over tarsal joints; tail sparsely haired, with slight tuft; length of molar tooth row usually > 4 mm . . . . **Brush Deermouse**, *Peromyscus boylii*

13. Tail laterally compressed, long, and scaly in appearance; large bodied; toes of hindfeet partly webbed . . . . **Common Muskrat**, *Ondatra zibethicus*

    Tail rounded, sparingly hairy or bushy; mouse to rat sized; hindfeet not webbed. 14

14. Mouse sized, total length usually < 150 mm; tail < 50 mm; ears nearly hidden in fur . . . . **Mogollon Vole**, *Microtus mogollonensis*

    Rat sized, total length of adults > 225 mm; tail > 100 mm; ears conspicuous or only partly hidden in dense fur 15

15. Ears prominent and naked; eyes black, large and bulging in life; color of dorsal to ventral pelage with a sharp bicolored transition; tail in adults usually > 150 mm; pelage rather soft; whiskers long, usually > 50 mm; molars with V-shaped notches (reentrant angles) 16

Ears partly hidden in dense pelage and furred; eyes not obviously large or bulging in life; color of dorsal to ventral pelage with a blended transition; tail in adults about 100–125 mm; pelage rather harsh; whiskers 25–35 mm; molars somewhat rounded, without V-shaped notches 18

16. Tail not sharply bicolored, but darker dorsally; underparts white, but base of hairs usually grayish buff; first upper molar with a well-developed, deep anterointernal reentrant angle extending halfway across crown .. **Mexican Woodrat**, *Neotoma mexicana*

Tail sharply bicolored; underparts white to base of hairs; first upper molar with a poorly developed, shallow anterointernal reentrant angle not extending across crown 17

17. Upperparts bluish steely gray to grayish buff in adults, dispersed with black along back; throat hairs dark at base; hindfoot usually > 40 mm . . . . **Southern Plains Woodrat**, *Neotoma micropus*

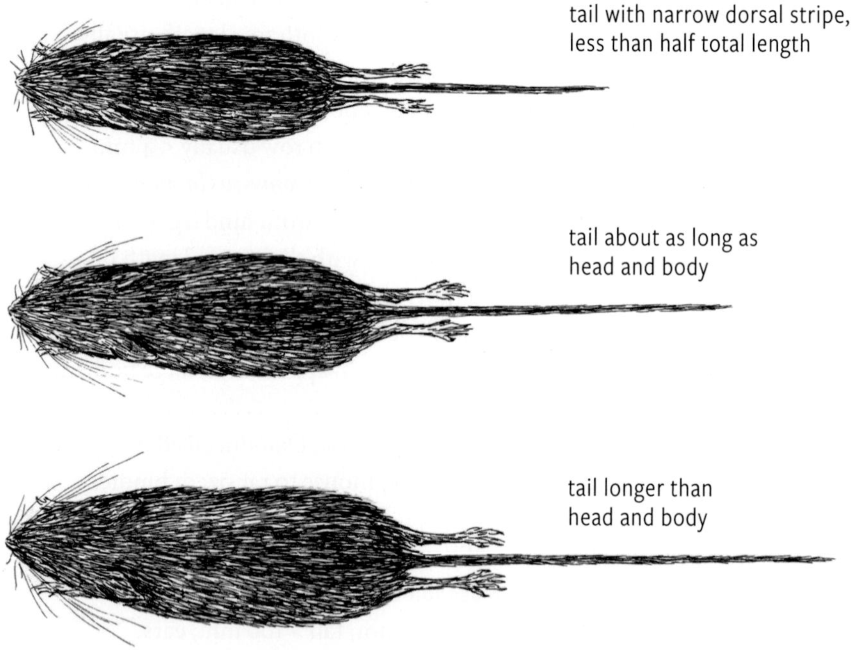

tail with narrow dorsal stripe, less than half total length

tail about as long as head and body

tail longer than head and body

FIG. 76. Three species of harvest mice, genus *Reithrodontomys*: top, *R. montanus*; center, *R. megalotis*; bottom, *R. fulvescens*

Upperparts dull pinkish buff in adults, suffused with nominal black
along back; throat hairs white at base; hindfoot usually < 40 mm
.... **White-toothed Woodrat**, *Neotoma leucodon*
18. Underparts buffy to ochraceous; tops of feet buffy brown; tail
uniformly black...................**Tawny-bellied Cotton Rat**,
Sigmodon fulviventer
Underparts grayish white, not buffy to ochraceous; top of feet
whitish; tail bicolored, dark above grading to light below 19
19. Nose area yellowish to rusty orangish; total length of adults < 260 mm
.... **Yellow-nosed Cotton Rat**, *Sigmodon ochrognathus*
Nose area colored like rest of body, not yellowish to rusty orangish;
total length of adults > 260 mm...........................
......................**Hispid Cotton Rat**, *Sigmodon hispidus*

Fulvous Harvest Mouse
Reithrodontomys fulvescens J. A. Allen, 1894

The fulvous harvest mouse is readily distinguished from other Trans-Pecos
harvest mice by its larger size, longer tail, and bright fulvous (golden buff) coloration on its sides (fig. 77). The upperparts are ochraceous buff mixed with
dark brown; the underparts are white or pale buff. The tail is longer than the
head and body, naked, scaly, and not bicolored (see fig. 76). The ears are short
and have reddish-orange hairs inside. As in all species of *Reithrodontomys*,
the anterior surface of each upper incisor is grooved. Average external measurements (in mm) are total length 159; tail 91; hindfoot 19; ear 14; weight is
8–12 g.

The distribution of the fulvous harvest mouse includes most of the Trans-
Pecos except for the far northern and eastern parts of the region (map 80).
Although it is widespread, populations seem somewhat localized, and
nowhere is the species especially common (Stangl et al. 1994). It is primarily
a grassland species and is well known from rough mixed shrub grassland habitats of the Davis Mountains and BBNP (Blair 1940; Borell and Bryant 1942).
However, at BBRSP and CMSNA it is more common in riparian habitats and
may occasionally be found in desert scrub habitats (Yancey et al. 1995a; Yancey
1997; C. Jones et al. 2011; Yancey and Manning 2018). *Reithrodontomys fulvescens* rarely occurs at the same locality and in the same habitat as its congener
R. megalotis (Yancey et al. 1995a).

FIG. 77. Fulvous harvest mouse, *Reithrodontomys fulvescens*

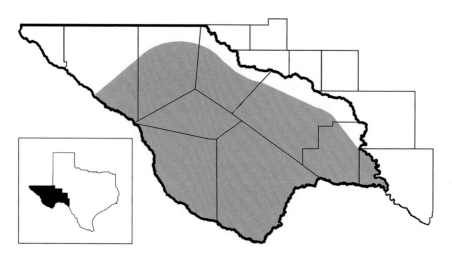

MAP 80. Distribution of the fulvous harvest mouse, *Reithrodontomys fulvescens*

The fulvous harvest mouse does not appear to enter seasonal torpor and has been captured during all seasons in the Trans-Pecos (Stangl et al. 1994; Yancey et al. 1995a; Yancey 1997). It is strictly nocturnal, with activity commencing immediately after sunset and slowing after midnight. Vegetation (primarily seeds) and invertebrates dominate its diet. This mouse is an excellent climber and is known to nest in shrubs well above the ground.

Reproductive data indicate that breeding occurs throughout the year (Stangl et al. 1994; Yancey et al. 1995a; Yancey 1997). After a pregnancy of about 21 days, a litter of two to six (average three to four) is produced two or more times a year. Barn owls and red-tailed hawks are known predators, but numerous species of owls, hawks, snakes, and mammalian carnivores undoubtedly capture them as well. The subspecies in the Trans-Pecos is *R. f. canus*.

Plains Harvest Mouse
Reithrodontomys montanus (Baird, 1855)

This small-bodied, grayish-brown harvest mouse has a narrow, diffuse dark brown band extending down the middle of its back and a tail with a narrow dark dorsal stripe that usually is less than half its total length (see fig. 76). As in other harvest mice, the anterior surface of each upper incisor is grooved. Average external measurements (in mm) are total length 113; tail 52; hindfoot 15; ear 11; weight is 6–10 g.

The plains harvest mouse is the rarest of the harvest mice in the Trans-Pecos and is known only from disparate localities in the western counties (El Paso, Hudspeth, Culberson, Jeff Davis, Presidio, and Brewster) (map 81), and never in large numbers (J. Jones et al. 1993). Because of the paucity of specimens, details of its natural history are not well understood. It is a resident primarily of grassy habitats, especially in areas where vegetation cover is sparse. Specimens have been obtained from shortgrass associations in the Davis Mountains (Blair 1940), grassy areas of creosote bush flats in the Apache Mountains (Stangl et al.

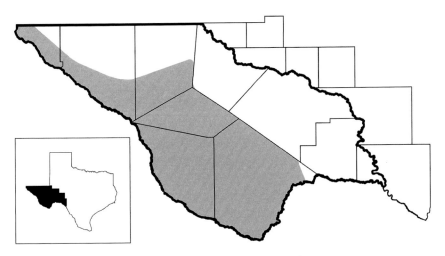

MAP 81. Distribution of the plains harvest mouse, *Reithrodontomys montanus*

1994), and buffalo grass–lechuguilla–ocotillo–creosote bush associations in BBNP (J. Jones et al. 1993).

It does not hibernate and is mostly nocturnal. Seeds of grasses and weedy plants, cactus fruits, and insects are its primary foods. Reproductive data are wanting, but it is thought to breed year-round. Gestation is about 21 days, after which a litter of one to nine (average four) is born. Snakes, raptors, and mammalian carnivores are thought to be major predators. The subspecies in the Trans-Pecos is *R. m. montanus.*

Western Harvest Mouse
Reithrodontomys megalotis (Baird, 1858)

This is a medium-sized harvest mouse with a tail about as long as its head and body, covered with short hairs, and indistinctly bicolored. The ears are short. *Reithrodontomys megalotis* is smaller than *R. fulvescens* but larger than *R. montanus* (see fig. 76). *Reithrodontomys megalotis* may be separated from *R. montanus* only with great difficulty; distinguishing features between these two species are noted in the key and fig. 76. Average external measurements (in mm) are total length 137; tail 71; hindfoot 18; ear 14; weight is 7–10 g.

The western harvest mouse is the most common and widely distributed harvest mouse in the Trans-Pecos. It has been captured in every county except for Terrell and Val Verde in the southeastern part of the region (map 82). It is most often found in grassy habitats and at any elevation associated with diverse vegetation such as yucca, cholla, creosote bush, mesquite, desert willow,

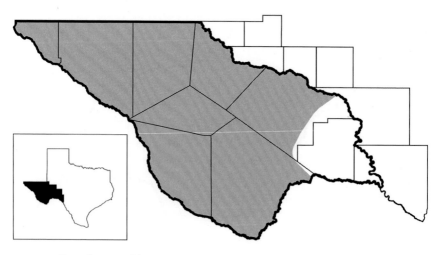

MAP 82. Distribution of the western harvest mouse, *Reithrodontomys megalotis*

sacahuista, sotol, juniper, piñon pine, and ponderosa pine (Blair 1940; Stangl et al. 1994; Yancey et al. 1995a; Yancey 1997). It also has an affinity for riparian situations, particularly in BBRSP and the Davis and Chinati Mountains (Blair 1940; Yancey et al. 1995a; Yancey 1997; C. Jones et al. 2011). On occasion, it may be found in desert scrub habitats (Stangl et al. 1994; Yancey et al. 1995a; Yancey 1997; Yancey et al. 2006).

This mouse does not hibernate, although individuals may become torpid when exposed to cold temperatures or water stress. It is nocturnal and seems to be most active on moonless and rainy nights. *Reithrodontomys megalotis* is primarily a granivore and dependent on grass seeds (Genoways et al. 1979). Other plant material and insects are consumed to a lesser degree.

Reproductively active individuals have been recorded in all seasons, indicating that this mouse breeds throughout the year (Genoways et al. 1979; Stangl et al. 1994; Yancey 1997; Yancey et al. 2006). After 23–25 days of gestation, a litter of one to nine (usually three to seven) is born; multiple litters are produced annually. Predators include snakes, owls, hawks, jays, coyotes, foxes, weasels, and bobcats. The subspecies in the Trans-Pecos is *R. m. megalotis*.

Cactus Deermouse
Peromyscus eremicus (Baird, 1858)

The cactus deermouse is a medium-sized mouse whose tail is so sparsely haired that annulations are evident; it also lacks a prominent terminal tuft of hairs (fig. 78). The soles of the hindfeet, including the heels, are naked (in other Trans-Pecos *Peromyscus* species the heels are haired). The skull can be distinguished by the absence of accessory lophs on the upper molars and by the extension of the premaxillae beyond the nasals. The upperparts are buffy ochraceous, but the head is gray and the venter is white. The tail is longer than the head and body and not distinctly bicolored (but darker above than below). The ears are short to moderate in length (about length of hindfeet). Average external measurements (in mm) are total length 185; tail 97; hindfoot 20; ear 18; average weight is 18 g. Manning et al. (2006) discuss morphological variation in Trans-Pecos populations.

The cactus deermouse occurs throughout the Trans-Pecos, primarily in lowland desert where it occupies rocky substrata in the lowlands and foothills of desert mountain ranges (map 83). It is the most abundant species of *Peromyscus* in the lower and middle elevations of the mountains of GMNP, CMSNA, BBNP, and BBRSP in desert scrub vegetation (especially associated with creosote bush, mesquite, lechuguilla, prickly pear, and other cacti), grasslands, riparian vegetation, and juniper roughlands (Baccus 1971; Ederhoff 1971;

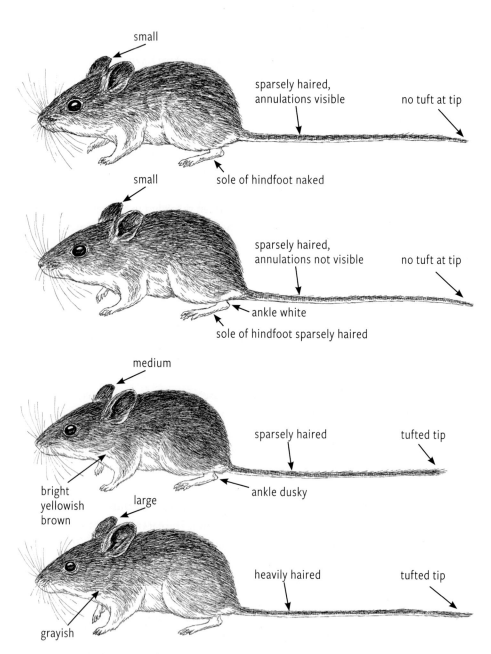

small

sparsely haired, annulations visible

no tuft at tip

sole of hindfoot naked

small

sparsely haired, annulations not visible

no tuft at tip

ankle white

sole of hindfoot sparsely haired

medium

sparsely haired

tufted tip

bright yellowish brown

ankle dusky

large

heavily haired

tufted tip

grayish

FIG. 78. Four species of mice of the genus *Peromyscus* in which the tail is longer than the head and body: top to bottom, *P. eremicus, P. laceianus, P. boylii,* and *P. nasutus*

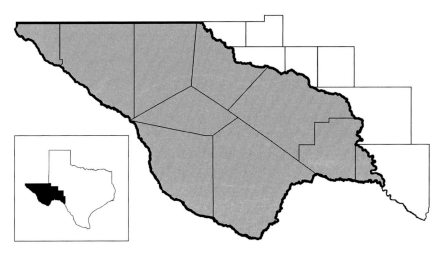

MAP 83. Distribution of the cactus deermouse, *Peromyscus eremicus*

Genoways et al. 1979; Cornely et al. 1981; Yancey 1997; Yancey et al. 2006). It is much less common in oak and piñon-oak woodlands of the Davis and Chinati Mountains and the Sierra Vieja (Blair 1940; Blair and Miller 1949; C. Jones et al. 2011). In addition to its natural habitats, this deermouse frequently occupies human dwellings and structures (Yancey et al. 2006; C. Jones et al. 2011).

The cactus deermouse is reported to aestivate to conserve food and water during the hot summer months. Trapping records reveal that it does not invariably enter prolonged summer torpor in the Trans-Pecos but is active throughout the year, although to a much lesser degree during the summer (Yancey 1997). *Peromyscus eremicus* is nocturnal and, in contrast to most other nocturnal rodents, seems to be most active on moonlit nights, possibly a strategy to reduce competition with other species of *Peromyscus*. It feeds on seeds of desert annuals, green vegetation, fruits and flowers of shrubs, and insects.

Reproductive data suggest that breeding occurs throughout the year (Cornely et al. 1981; Stangl et al. 1994; Yancey 1997; Yancey et al. 2006). After a gestation period of about 21 days, a litter of one to four (average three) young is produced up to four times a year. Female cactus deermice possess only two pairs of inguinal mammae (three pairs in other *Peromyscus*) and thus have reduced litter sizes relative to many of their congeners. Barn owls and king snakes are primary predators, although several other raptors, snakes, and mammalian predators likely hunt this species. The subspecies in the Trans-Pecos is *P. e. eremicus*.

Lacey's White-ankled Deermouse
Peromyscus laceianus V. Bailey, 1906

Lacey's white-ankled deermouse is a small-eared (17–19 mm), long-tailed *Peromyscus* with white fur over the tarsal joint on each hindfoot. Its tail is longer than its head and body, scantily haired, and not sharply bicolored. The upperparts are grayish brown; the underparts and feet are white. Average external measurements (in mm) are total length 192; tail 103; hindfoot 21; ear 18; weight is 13–21 g. *Peromyscus laceianus* may be easily confused with *P. boylii* but differs in that the dark color of its hind legs does not extend over the tarsal joint, thus leaving its ankles completely white (see fig. 78). Male specimens of *laceianus* and *boylii* are easily separated by examining the baculum (a bone in the penis); that of *laceianus* has a long, cartilaginous spine at its terminal end, whereas in *boylii* the spine is short (W. Clark 1953).

Lacey's white-ankled deermouse ranges throughout the Trans-Pecos except for the far western part of the region (map 84). It occurs in a wide variety of habitats, including grasslands, desert scrub, and juniper woodlands (Blair 1940). It has an affinity for rocky situations and is most often observed near cliffs, talus deposits or slopes, and large boulders. In the Trans-Pecos, the species favors elevations below 2,000 m. *Peromyscus laceianus* has commonly been trapped in persimmon–shin oak, cedar-oak (Hermann 1950), catclaw-grama, grama-bluestem, lechuguilla–bear grass (Blair and Miller 1949), sotol-juniper-lechuguilla, shortgrass-juniper (Denyes 1956), and sotol-lechuguilla associations (Stangl

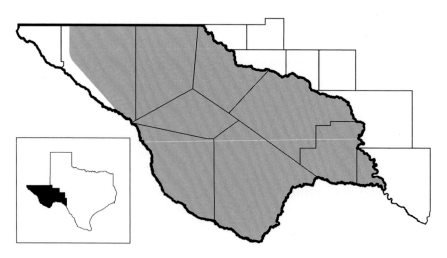

MAP 84. Distribution of Lacey's white-ankled deermouse, *Peromyscus laceianus*

et al. 1994). At BBRSP, it occupies grasslands, desert scrub, and juniper woodlands but seemingly has an atypical affinity for riparian habitat (Yancey 1997). It is less common on the plateaus and mesas of BBNP and is the least abundant species of *Peromyscus* in CMSNA (C. Jones et al. 2011).

Peromyscus laceianus has been captured during all seasons in the Trans-Pecos and is active year-round (Yancey 1997). It is nocturnal, with nighttime activities focused on climbing and foraging among cliffs, trees, and shrubs (Mullican and Baccus 1990). It feeds primarily on seeds, juniper berries, acorns, cactus fruits, and hackberries; other fruits, lichens and fungi, and some insects are also consumed.

Breeding occurs throughout the year based on dates of observed gravid females and males in breeding condition (Yancey 1997). Females may produce multiple litters annually. After a gestation period of about 23 days, two to five (average three) young are born. Rare cases of hair loss associated with post-juvenile molt have been observed in maturing individuals in the Trans-Pecos (Yancey and Jones 1999). Snakes, birds of prey, and mammalian carnivores are common predators.

Peromyscus laceianus was formerly considered a subspecies of *P. pectoralis*, but DNA sequence and morphometric data resulted in its elevation to specific status (Bradley et al. 2015). *Peromyscus laceianus* is a monotypic species, and subspecies are not recognized.

Brush Deermouse
Peromyscus boylii (Baird, 1855)

The brush deermouse is a large species of *Peromyscus* whose tail is usually longer than its head and body. Its ears are large but shorter than its hindfoot, and its tail is sparsely haired, slightly tufted, and indistinctly bicolored. The upperparts are a dull brownish gray; the lower parts and feet are white; the ankles are dusky. Average external measurements (in mm) are total length 185; tail 97; hindfoot 21; ear 20; weight is 22–36 g. *Peromyscus boylii* may be easily confused with *P. laceianus* (for differences, see the account of *P. laceianus*) and *P. nasutus* (see fig. 78). It is especially difficult to distinguish *boylii* from *nasutus*. Generally, the brush deermouse is smaller in external and cranial measurements (see table 8) and more richly colored on the flanks. The flanks of adult *boylii* are bright yellow brown, whereas in adult *nasutus* they are predominantly grayish.

Peromyscus boylii occurs throughout the mountainous areas of the western part of the Trans-Pecos (map 85), where it inhabits rocky outcroppings and brushy or forested habitats. Dense shrubby vegetation appears to be important to

TABLE 8. Means and 95% confidence intervals for selected external and cranial measurements for seven species of *Peromyscus* from the Trans-Pecos

Species (sample size)	Measurement (mm)						
	Total length	Tail length	Body length	Hindfoot length	Ear length	Greatest length of skull	Length of rostrum
P. boylii (16)	194.00 187.89–200.11	100.50 92.61–108.39	93.50 87.97–99.03	21.17 20.21–22.13	19.50 18.16–20.84	27.62 27.36–27.88	10.92 10.76–11.08
P. eremicus (20)	186.75 182.16–191.34	95.85 93.15–98.55	90.90 86.94–94.86	20.65 20.32–20.98	17.25 16.69–17.81	25.48 25.19–25.77	9.36 9.19–9.53
P. labecula (15)	159.20 154.91–163.49	66.67 64.03–69.31	92.53 89.70–95.36	19.93 19.05–20.81	17.67 16.67–18.58	25.29 24.41–26.17	10.01 9.80–10.22
P. laceianus (33)	191.94 188.99–194.89	100.48 98.32–102.64	91.59 89.42–93.76	21.56 21.23–21.89	17.64 17.29–17.99	27.60 27.37–27.85	10.70 10.55–10.85
P. leucopus (15)	176.40 170.98–181.82	79.80 75.60–84.00	96.60 92.53–100.67	21.40 20.91–21.89	18.00 16.91–19.09	26.94 26.50–27.38	10.57 10.16–10.98
P. nasutus (9)	209.11 203.21–215.01	110.89 108.20–113.58	98.22 94.25–102.19	22.33 21.45–23.21	21.11 19.94–22.28	28.54 28.34–28.78	11.45 11.26–11.64
P. truei (2)*	184.50	87.00	97.50	20.00	24.5	27.45	10.65

TABLE 8. (continued)

Species (sample size)	Measurement (mm)					
	Interorbital breadth	Mastoidal breadth	Depth of skull	Length of bony palate	Length of palatine foramen	Length of molar tooth row
P. boylii (16)	4.34 4.24–4.44	11.69 11.54–11.84	9.55 9.39–9.71	4.27 4.09–4.45	5.58 5.37–5.79	4.12 4.04–4.20
P. eremicus (20)	4.05 3.99–4.11	11.14 11.00–11.28	9.10 8.97–9.23	3.94 3.82–4.06	4.83 4.72–4.94	3.74 3.69–3.79
P. labecula (15)	3.92 3.83–4.01	10.80 10.70–10.90	9.25 9.11–9.39	3.68 3.59–3.77	5.58 5.39–5.77	3.66 3.56–3.74
P. laceianus (33)	4.26 4.18–4.34	11.49 11.39–11.59	9.39 9.30–9.48	4.16 4.08–4.28	5.10 5.02–5.18	3.92 3.88–3.96
P. leucopus (15)	4.13 3.77–4.39	11.29 11.09–11.49	9.68 9.56–9.80	4.04 3.69–4.39	5.50 5.26–5.74	3.88 3.72–4.04
P. nasutus (9)	4.49 4.39–4.59	12.22 12.03–12.41	10.01 9.78–10.24	4.31 4.21–4.41	5.88 5.75–6.01	4.22 4.15–4.29
P. truei (2)*	4.35	11.82	10.78	4.32	5.22	4.00

*Measurements for P. truei specimens in this table are from two subadults and are not indicative of or comparable to adult specimens.

No 95% confidence interval is presented for P. truei because of the low sample size (2).

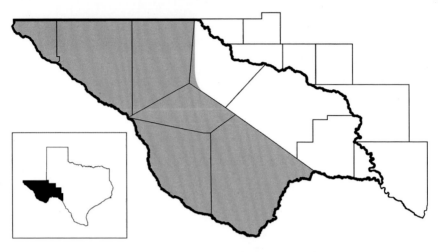

MAP 85. Distribution of the brush deermouse, *Peromyscus boylii*

this mouse, and it is uncommon in open, sparsely vegetated areas. It is especially abundant among brushy and rocky situations associated with oak and conifer woodlands at the higher elevations of the Chinati, Chisos, Davis, Franklin, and Guadalupe Mountains (Blair 1940; Baccus 1971; Cornely et al. 1981; C. Jones et al. 2011) but may occasionally be found in lowland desert habitats of BBNP, BBRSP, and GMNP (Cornely et al. 1981; Yancey 1997; Yancey et al. 2006).

The brush deermouse is active year-round, and mostly nocturnal. It feeds primarily on vegetation, and in the Trans-Pecos acorns are an especially favored food. Other plant items consumed include hackberries, juniper berries, cactus fruits, pine nuts, and Douglas fir seeds. In addition to plant material, arthropods and small amounts of mammal, bird, and reptile matter may be eaten. This mouse is a skillful climber and will scale a variety of plants from which it often gathers food.

Pregnant females have been obtained during March, April, June, August, and September (Cornely et al. 1981; Stangl et al 1994), suggesting that breeding occurs throughout much of the year. After a gestation period of about 23 days, two to five (average three) young are born. Females may produce several litters a year, with parturition occurring most frequently in spring and early summer. Hawks, owls, snakes, and mammalian predators (coyotes, foxes, mustelids, raccoons, skunks) presumably hunt this mouse. Spotted owls seem to prey especially heavily on it. *Peromyscus boylii* is a common host for agents of some important diseases including hantavirus pulmonary syndrome, Lyme disease, and the plague. The subspecies that occurs in the Trans-Pecos is *P. b. rowleyi.*

Northern Rock Deermouse
Peromyscus nasutus (J. A. Allen, 1891)

The northern rock deermouse is a relatively large, long-tailed, and long-eared *Peromyscus* (see fig. 78). Its hindfoot is large, well haired to the heel, and longer than its ears. This species is most easily confused with *P. boylii*, from which it may be distinguished as indicated in the account of that species. Average external measurements (in mm) are total length 209; tail 111; hindfoot 22; ear 21; weight is 25–40 g.

The distribution of the northern rock deermouse in the Trans-Pecos is rather spotty and is restricted to isolated mountain ranges, particularly the Guadalupe, Franklin, Davis, Chinati, and Chisos Mountains (Baccus 1971; Diersing and Hoffmeister 1974; Cornely et al. 1981) (map 86). Overall, it is uncommon compared to other species of deermice in the region. It favors mountain slopes among rocks and boulders in areas with sparse vegetation, sometimes in proximity to Texas madrone–oak and piñon-oak-juniper woodlands. Habitat suitability seems to be a function of abundance of rocks and adequate space between them (Diersing 1976), but this deermouse appears to be most abundant in rocky terrain at higher elevations (over 2,100 m) in the Guadalupe, Chinati, Davis, and Chisos Mountains (Bailey 1905; Baccus 1971; Cornely et al. 1981).

Because of its scarcity, little is known about the habits of this deermouse. There are no reports of hibernation, and it is presumed to be active mostly at night. It feeds on acorns, piñon nuts, juniper berries, grasses, mushrooms, and

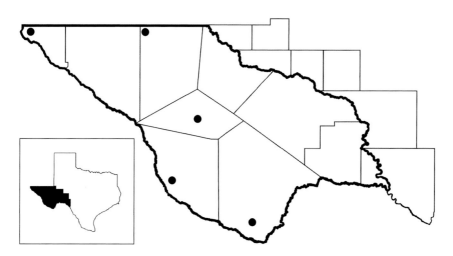

MAP 86. Distribution of the northern rock deermouse, *Peromyscus nasutus*

insects. *Peromyscus nasutus* has excellent climbing abilities and is considered semiarboreal.

Pregnant females have been recorded in June, July, and August. Gestation is about 30 days and litter size ranges from two to six. There is no information on specific predators, although presumably they include snakes, birds of prey, and mammalian carnivores.

Peromyscus nasutus was previously considered a subspecies of *P. difficilis*. However, chromosome, allozyme, and mitochondrial DNA differences have led to its elevation to specific status (Durish et al. 2004). Two subspecies of *P. nasutus* occur in the Trans-Pecos: *P. n. penicillatus* in the extreme western part of the region and *P. n. nasutus* in the remainder. The former differs from *P. n. nasutus* in having a smaller skull, longer tail, and a grayish instead of brownish-black head (Diersing and Hoffmeister 1974; Diersing 1976; Hoffmeister 1986).

Elliot's Deermouse
Peromyscus labecula Elliot, 1903

Elliot's deermouse is a small to medium-sized mouse with a pale gray, vinaceous buff, or dull ochraceous dorsal pelage and a white venter. The tail is decidedly shorter than the head and body length and sharply bicolored (brown above, white below). The ears are short and dusky (fig. 79). Average external measurements (in mm) are total length 159; tail 74; hindfoot 21; ear 17; weight is 14–25 g. *Peromyscus labecula* is distinguished from *P. leucopus* with some difficulty. In external and cranial measurements *labecula* is smaller than *leucopus* (see table 8), and the tail of *leucopus* is less sharply bicolored.

Elliot's deermouse occurs in localized areas across the Trans-Pecos (map 87), most often in desert scrub and grassland habitats that support sparse to moderate growth of vegetation. It is relatively uncommon compared to other species of deermice in the region. In BBNP it has been taken in areas of moderate grass and open vegetation on the desert plains and lower foothills of the mountains from about 565 to 1,065 m (Baccus 1971; Yancey et al. 2006). In the Davis Mountains it seems to prefer areas of relatively level, rock-free soils with a sparse cover of short grasses associated with several xeric shrubs such as yucca, bear grass, allthorn, and ephedra (Blair 1940). In both BBRSP and CMSNA, it occurs in desert scrub, desert grassland, and riparian habitats (Yancey 1997; C. Jones et al. 2011).

This deermouse is not known to hibernate and is active year-round in the Trans-Pecos. It is mostly nocturnal, foraging throughout the night primarily on grains, seeds, and insects, and to a lesser degree on fruits, roots, bark, and green

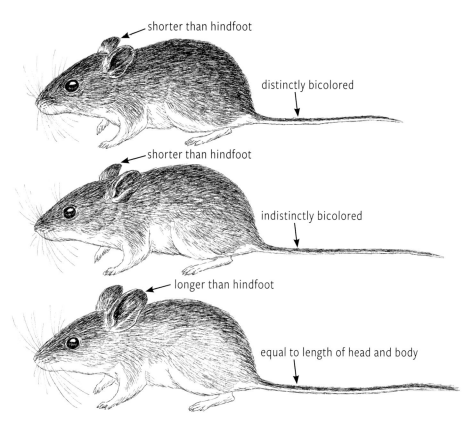

shorter than hindfoot

distinctly bicolored

shorter than hindfoot

indistinctly bicolored

longer than hindfoot

equal to length of head and body

FIG. 79. Three species of mice of the genus *Peromyscus* in which the tail is equal to or shorter than the head and body: top, *P. labecula*; center, *P. leucopus*; bottom, *P. truei*

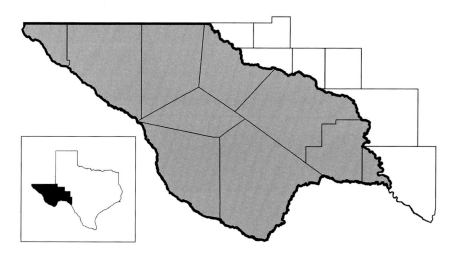

MAP 87. Distribution of Elliot's deermouse, *Peromyscus labecula*

vegetation. It is a capable climber and is known to scale vegetation in search of food (Yancey 1997).

Individuals in reproductive condition have been observed during all seasons in the Trans-Pecos (Cornely et al. 1981; Stangl et al. 1994; Yancey 1997; Yancey et al. 2006; C. Jones et al. 2011). Multiple litters of one to nine (average four) young are produced annually. Gestation ranges from 22 to 27 days. Owls, snakes, and mammalian carnivores such as long-tailed weasels, gray foxes, coyotes, and bobcats are typical predators. This species is known to have a high incidence of Sin Nombre virus infection, the causative agent of hantavirus pulmonary syndrome, and caution should be exercised when handling or coming in contact with this mouse (Glass et al. 2006).

Peromyscus labecula was formerly considered a subspecies of *P. maniculatus*. Subsequent examination of mitochondrial DNA sequence data resulted in the elevation of *P. labecula* to specific status (Bradley et al. 2019). The subspecies that occurs in the Trans-Pecos is *P. l. blandus*.

White-footed Deermouse
Peromyscus leucopus (Rafinesque, 1818)

The white-footed deermouse is a medium-sized *Peromyscus* with a dull grayish-brown dorsal pelage and a white venter (fig. 79). The tail is relatively short (shorter than head and body) and indistinctly bicolored (darker above than below), and the ears are short (see fig. 79). Average external measurements (in mm) are total length 175; tail 81; hindfoot 22; ear 17; weight is 17–26 g. This mouse is most easily confused with *P. labecula*, from which it differs as described in the account of the latter.

The white-footed deermouse ranges throughout the entire Trans-Pecos and has been documented from every county in the region (map 88). In BBNP, BBRSP, and GMNP it is partial to low-elevation riparian and desert scrub situations (Baccus 1971; Cornely et al. 1981; Yancey 1997; Yancey et al. 2006). It seems to be less abundant at higher elevations in the Chinati, Davis, Guadalupe, Apache, and Delaware Mountains, where it inhabits mixed desert scrub, grassland, riparian, juniper, and oak woodland habitats (Blair 1940; Cornely et al. 1981; Stangl et al. 1994; C. Jones et al. 2011).

The white-footed deermouse does not hibernate and has been captured during every month of the year (Yancey 1997). It is primarily nocturnal, although on occasion it may be active during the day. Its diet consists mostly of seeds, mesquite beans, berries, fruits, nuts, and insects. Carrion may be

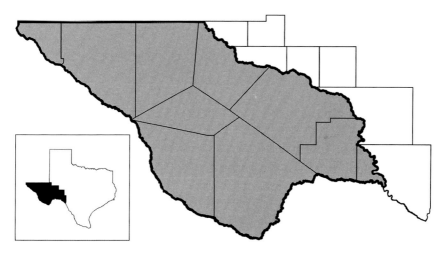

MAP 88. Distribution of the white-footed deermouse, *Peromyscus leucopus*

opportunistically consumed as well. *Peromyscus leucopus* is an adept climber, to the extent that it has been regarded as semiarboreal.

Pregnant females and males exhibiting reproductive activity have been observed during every season, indicating that breeding occurs year-round, and multiple litters are produced annually (Cornely et al. 1981; Stangl et al. 1994; Yancey 1997; Yancey et al. 2006). After a gestation period of about 22–25 days, multiple litters of one to seven (average four) pink, hairless, and blind young are born. This deermouse is a significant prey species for many predators including snakes, owls, weasels, and other mammalian carnivores. Like several other species of deermice, *P. leucopus* is a reservoir for the virus that causes hantavirus pulmonary syndrome and is also a host for a tick that transmits Lyme disease; animals should therefore be handled with care.

Two subspecies occur in the Trans-Pecos: *P. l. texanus* in the extreme eastern and southeastern portion of the region and *P. l. tornillo* throughout the remainder. *Peromyscus leucopus texanus* is measurably smaller in external and cranial measurements than *P. l. tornillo* (Osgood 1909).

Piñon Deermouse
Peromyscus truei (Shufeldt, 1855)

The piñon deermouse is a medium-sized mouse with especially large ears that are longer than its hindfeet (fig. 79). Its tail is well haired, equal in length to the head and body, and usually less than 100 mm long. The upperparts are dark

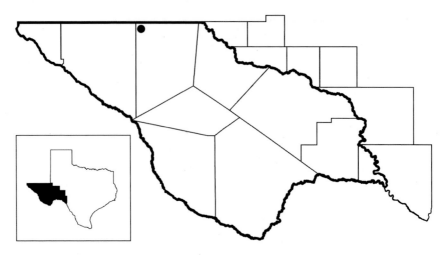

MAP 89. Distribution of the piñon deermouse, *Peromyscus truei*

buff brown, and the sides are rich buff; the underparts and feet are pure white. The skull is easily distinguished from that of other species of *Peromyscus* by its extremely large auditory bullae (reflected in the depth of skull measurement; see table 8). No external measurements of adult specimens from the Trans-Pecos are available. Average external measurements (in mm) for a series of the same subspecies from the Texas Panhandle and northwestern New Mexico are total length 190; tail 90; hindfoot 23; ear 24; weight is 24–39 g.

The piñon deermouse is peripherally known from the Trans-Pecos based on only four specimens taken at two sites in the Dog Canyon area in the northern part of GMNP (Cornely et al. 1981) (map 89). These two locations are in midelevation canyons characterized by open riparian woodlands with grassy understories. Both alligator juniper and piñon pine were prevalent at one of the sites (West Dog Canyon); alligator juniper was present, but piñon pine was absent from the other (Upper Dog Canyon).

This nocturnal deermouse is thought to feed on acorns, juniper berries, piñon seeds, and insects. Its long toes, moderately long tail, and large hindfeet suggest it is an excellent climber (Cornely et al. 1981). Breeding is thought to occur from mid-February through mid-November. The gestation period is about 26 days, and typical litter size is two to six. A single pregnant female was observed with four embryos on July 24 in GMNP (Cornely et al. 1981). Predators likely include a variety of snakes, birds of prey, and mammalian carnivores. The subspecies in the Trans-Pecos is *P. t. truei*.

Mearns's or Chihuahuan Grasshopper Mouse
Onychomys arenicola Mearns, 1896

Mearns's grasshopper mouse is a relatively long-tailed grasshopper mouse with pinkish-cinnamon or grayish-buff upperparts and pure white underparts. The tail, often white-tipped, usually measures more than half the combined length of the head and body, while in *O. leucogaster* it is commonly less than half (fig. 80). Average external measurements (in mm) are total length 147; tail 54; hindfoot 21; ear 17; average weight is 19 g.

Mearns's grasshopper mouse occurs across the entire Trans-Pecos in suitable habitat, but it is uncommon in the region compared to other cricetid rodents (map 90). It resides primarily in dry lowland desert habitats with mesquite, huisache, creosote bush, cholla, yucca, and various short grasses (Davis and Robertson 1944; Genoways et al. 1979; Yancey 1997; DeBaca 2008; C. Jones

FIG. 80. Two species of grasshopper mice, genus *Onychomys*: above, *O. arenicola* (note longer tail); below, *O. leucogaster* (note shorter tail)

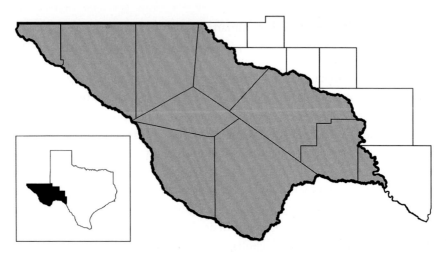

MAP 90. Distribution of Mearns's grasshopper mouse, *Onychomys arenicola*

et al. 2011) but may also be found in upland desert grasslands and juniper wood-lands (Yancey 1997).

Mearns's grasshopper mouse does not hibernate and is active at night, mostly in pursuit of animal prey. Although mostly uncommon in the region, including at BBNP, BBRSP, and CMSNA (Schmidly 1977b; Yancey 1997; Yancey et al. 2006; C. Jones et al. 2011), it was among the most abundant rodents at a site near Sierra Blanca in Hudspeth County (Wu et al. 1996). As with other grasshopper mice, its diet is somewhat unusual among North American cricetid rodents in consisting mostly of animal material. Arthropods, including grasshoppers, beetles, and scorpions, seem to be the primary items consumed, but these mice will also prey on small mammals such as pocket mice, kangaroo rats, and harvest mice (Bailey 1905; Yancey 1997). They have several unique behaviors to facilitate subduing and killing their prey. When these mice attack scorpions and whip scorpions, for example, they will first bite and immobilize the venomous or noxious tail, then kill the arthropod with repeated bites to the head. Mammalian prey are typically dispatched with a bite through the spinal cord at the base of the skull. In addition to animal prey, these mice will consume a small amount of plant material, particularly grains.

Mearns's grasshopper mice are well known for their vocalizations, which have been described as high-pitched calls, chirps, or howls. These vocalizations are emitted more frequently by males and are thought to serve as territorial advertisement and spacing mechanisms.

Gravid females from the Trans-Pecos have been documented from February, March, June, July, August, and September, which indicates that at least two litters are produced each year. Litter size ranges from one to five. The gestation period is 27–30 days. Specific information on predators is not available, but they likely include hawks, owls, snakes, and mammalian carnivores.

This taxon was previously regarded as *Onychomys torridus*, but a subsequent analysis of morphological and chromosomal characters revealed that it includes two distinct species, *O. torridus* and *O. arenicola* (Hinesley 1979). As a result, individuals from the Trans-Pecos have been reassigned to *O. arenicola*. The subspecies in the Trans-Pecos is *O. a. arenicola*.

Northern Grasshopper Mouse
Onychomys leucogaster (Wied-Neuwied, 1841)

The northern grasshopper mouse is a medium-sized, stout-bodied mouse with a brownish-gray dorsal pelage and a conspicuous white tuft at the base of its short ears. Nose, cheeks, sides, and underparts are white. The tail is thick and short (< 30 percent head and body length) and often has a white tip. *Onychomys leucogaster* differs from *O. arenicola* in being larger and in having a relatively shorter tail (fig. 80). Average external measurements (in mm) are total length 164; tail 42; hindfoot 22; ear 14; weight is 27–46 g.

The northern grasshopper mouse has a disjunct distribution in the Trans-Pecos—it is known from the far western part of the region in El Paso and Hudspeth Counties and from the far eastern Trans-Pecos along the Pecos River in Pecos and Terrell Counties. It has been recorded from just east of the Pecos River in Val Verde County and is therefore, based on the local biogeography of

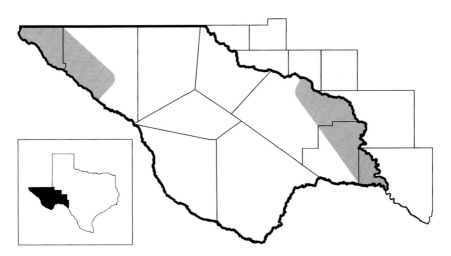

MAP 91. Distribution of the northern grasshopper mouse, *Onychomys leucogaster*

the species and the presence of suitable habitat, projected to occupy Val Verde County west of the Pecos as well (map 91). This grasshopper mouse is found primarily in sandy grassland and sandy mesquite shrubland, often in association with burrowing rodents, especially Ord's kangaroo rat. These mice are active year-round, with no evidence of hibernation. In winter they develop a thick layer of fat under the skin, and their tails become swollen with fat (Bailey 1931).

They are nocturnal and continuously wander in search of prey. Most activity occurs when the moon is below the horizon or during heavy cloud cover; activity is minimal during a full moon or sustained heavy rainfall. Like other grasshopper mice, *O. leucogaster* is highly predatory and its diet consists mostly of animal material. Common prey items include insects (especially grasshoppers, beetles, and moths), arachnids (spiders and scorpions), lizards, and small mammals (small mice and kangaroo rats). They attack and kill their prey in a manner similar to that described for *O. arenicola*. In addition to animal material, a variety of plant products (forbs, grasses, seeds) may be consumed, albeit to a much lesser degree. Like other grasshopper mice, northern grasshopper mice frequently emit loud, high-pitched vocalizations or "howls" thought to convey information regarding identity, size, sex, and location of the individual.

Breeding in northern grasshopper mice may occur throughout the year in the Trans-Pecos. After a gestation period of 32–47 days, two to six (average four) young are born. Females may produce three to six litters a year, with parturition occurring most frequently in May and June. Several animals are known to prey on northern grasshopper mice, including great horned owls, barn owls, red-tailed hawks, coyotes, and kit foxes.

Three subspecies of *O. leucogaster* are found in the Trans-Pecos: *O. l. ruidosae* in the far west (El Paso and Hudspeth Counties); *O. l. arcticeps* in the northeastern part of the region (Pecos County); and *O. l. longipes* in the southeastern sector (Terrell County). Because of the absence of specimens from the central part of the Trans-Pecos, boundaries that delineate the range of each subspecies in the region are somewhat arbitrary and ambiguous. *Onychomys leucogaster arcticeps* is paler in color than *O. l. longipes* and *O. l. ruidosae*, and *O. l. longipes* has a longer skull than *O. l. ruidosae* (Hollister 1914).

Hispid Cotton Rat
Sigmodon hispidus Say and Ord, 1825

The hispid cotton rat is the largest of the three species of cotton rats that occur in the Trans-Pecos (fig. 81). The dorsal pelage includes coarse, stiff, black guard hairs tipped in golden brown, resulting in a grizzled appearance; the underparts

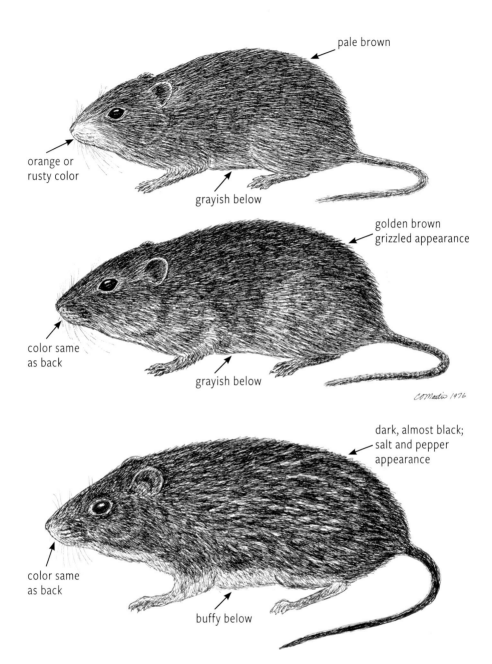

pale brown

orange or
rusty color

grayish below

golden brown
grizzled appearance

color same
as back

grayish below

C.O.Martin 1976

dark, almost black;
salt and pepper
appearance

color same
as back

buffy below

FIG. 81. Three species of cotton rats, genus *Sigmodon*: top, *S. ochrognathus*;
center, *S. hispidus*; bottom, *S. fulviventer*

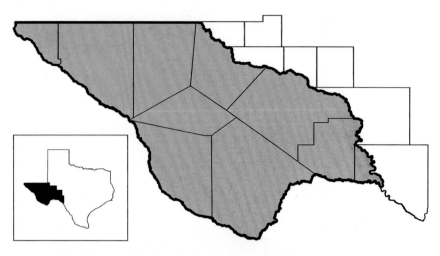

MAP 92. Distribution of the hispid cotton rat, *Sigmodon hispidus*

are grayish white. The tail is relatively short (shorter than head and body) and sparsely haired such that annulations and scales are clearly visible. External measurements (in mm) are total length 266; tail 109; hindfoot 31; ear 19; weight is 80–130 g.

The hispid cotton rat ranges across the entire Trans-Pecos and has been recorded in every county in the region (map 92). Although its distribution is widespread, the local abundance of this cotton rat is highly variable; it is common in some areas and rare in others. It occurs primarily at lower elevations, particularly in areas of moderate to dense grass cover. In BBNP, it was especially abundant in moist, grassy areas along the Rio Grande (Boeer and Schmidly 1977), as well as in other lowland areas with thick vegetation near places where water periodically accumulates (Yancey et al. 2006). It was also common in riparian areas and mesquite-cholla associations in the Davis Mountains (Blair 1940), and on some of the interior canyon floors of the Guadalupe Mountains (Genoways et al. 1979). This cotton rat was abundant in dense vegetation along washes and canyon floors with little ground cover in the Beach Mountains of Culberson County (Stangl et al. 1993). It was reported as uncommon in riparian and weedy habitats in BBRSP and CMSNA (Yancey 1997; C. Jones and Lockwood 2008; C. Jones et al. 2011), and in the Delaware Mountains, where it occurred in grassy and brushy habitats (Stangl et al. 1994).

The hispid cotton rat does not hibernate and is mostly nocturnal and crepuscular, although it may be active at any time of day. Green vegetation and seeds, especially those of grasses, are its primary foods, with some seasonal inclusion of insects. The presence of these rats is usually obvious because of the numerous trails or runways they create under grass cover or fallen vegetation.

Hispid cotton rats are prolific breeders and breed throughout the year, as indicated by reproductive data from specimens (Genoways et al. 1979; Stangl et al. 1994; Yancey 1997; Yancey et al. 2006). However, annual variations in environmental conditions may result in extreme fluctuations in numbers. During years of heavy rainfall, when suitable habitat is abundant, year-round breeding may result in prodigious population sizes. In years of low rainfall, when grassy cover is reduced, individuals do not breed as often and numbers decline significantly. Typically, multiple litters of 2 to 10 (average 5) young are produced each year after a gestation period of about 27 days. Snakes, raptors (especially barn owls and red-tailed hawks), and mammalian carnivores (particularly coyotes and bobcats) are common predators. The subspecies in the Trans-Pecos is *S. h. berlandieri*.

Yellow-nosed Cotton Rat
Sigmodon ochrognathus Bailey, 1902

The yellow-nosed cotton rat differs from the other species of cotton rats in the Trans-Pecos in its smaller size, paler dorsal coloration, and distinctly orange or rusty-colored snout (fig. 81). The tail is also hairier and somewhat bicolored, being nearly black above and grayish buff below; the underparts are grayish white. External measurements (in mm) are total length 197; tail 82; hindfoot 23; ear 16; weight is 40–80 g.

The range of the yellow-nosed cotton rat in the Trans-Pecos is restricted to the central part of the region, where it is known from Culberson, Jeff Davis, Presidio, and Brewster Counties (map 93). It occurs mostly at higher elevations in the central mountainous core of the Chinati, Chisos, Davis, and Guadalupe Mountains, Elephant Mountain, and the Sierra Vieja (Heaney et al. 1998). It favors grassy flats, rocky bunchgrass slopes, and grassland-woodland associations in these mountains but may also thrive in riparian habitats (Blair 1940; Borell and Bryant 1942; Blair and Miller 1949; Hollander et al. 1990a; C. Jones et al 2011). A single individual collected in lowland riparian habitat in BBRSP indicates that *S. ochrognathus*, on rare occasion, may inhabit nonmontane areas as well (Yancey and Jones 1996). At most places where it has been found, it is uncommon compared to the hispid cotton rat (*S. hispidus*).

The yellow-nosed cotton rat does not hibernate and is active day and night. It feeds mostly on green vegetation of grasses and other plants but will consume the fruits of prickly pear cactus when available. Like other cotton rats, it travels through grass and other vegetation via conspicuous runways.

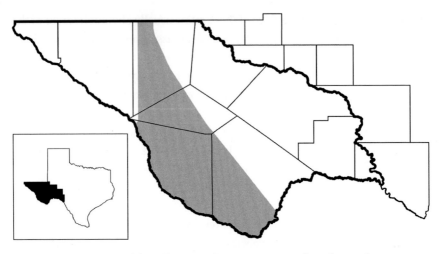

MAP 93. Distribution of the yellow-nosed cotton rat, *Sigmodon ochrognathus*

Reproduction occurs throughout the year, and litter size varies from 3 to 5 (average 3.6) (Baccus 1971). The gestation period is about 34 days, and, unlike in most other rodents, young are born rather precocial; within a few hours they look and act like a small version of an adult. A female gave birth to two young in a live trap at EMWMA, one of which had the following external measurements (in mm): total length 74; tail 27; hindfoot 11; ear 7; weight was 4.8 g (Heaney et al. 1998). Large snakes, birds of prey, and mammalian carnivores are primary predators. *Sigmodon ochrognathus* is a monotypic species, and subspecies are not recognized (Carroll et al. 2002).

Tawny-bellied Cotton Rat

Sigmodon fulviventer J. A. Allen, 1889

The tawny-bellied cotton rat is a moderately large cotton rat with a dorsal pelage of blackish underfur with relatively few overlying guard hairs. Over half the guard hairs are white, resulting in a somewhat frosted or "salt-and-pepper" appearance. The underparts are buff brown to fulvous throughout, and the tops of the feet are brown (fig. 81). The tail is relatively short (shorter than head and body), uniformly blackish, and densely covered with fine hairs to the degree that scales and annulations are not readily visible. Average external measurements (in mm) are total length 232; tail 86; hindfoot 27; ear 16; weight is 70–120 g. *Sigmodon fulviventer* may be distinguished from the other two cotton rats in the Trans-Pecos as described in the key to the cricetid rodents.

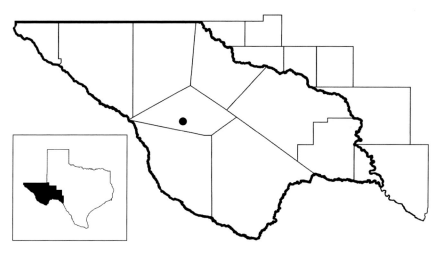

MAP 94. Distribution of the tawny-bellied cotton rat, *Sigmodon fulviventer*

The tawny-bellied cotton rat is known in Texas only from specimens taken at a single isolated locality, collected in a single year, along the southwestern flanks of the Davis Mountains near Fort Davis in Jeff Davis County (Stangl 1992a, 1992b; Stangl et al. 1994) (map 94). Specimens were obtained in March and July 1991 by Fred Stangl of Midwestern State University. Little is known of its life history. The species, in general, is known to inhabit a variety of grassy and grass-shrub habitats. Its habitat in the Trans-Pecos has been described as dense grass or grassy with scattered mesquite and catclaw acacia (Stangl 1992a).

Food habits are unknown, but green vegetation and seeds of grasses and other plants are probably its staple. Like other cotton rats, *S. fulviventer* constructs and travels along runways through thick grass (Stangl et al. 1992b), and it is believed to be a prolific breeder. Of 20 individuals taken in late March, 8 were juveniles, 2 were adult males in reproductive condition, and 10 were adult females, 4 of which were gravid and 1 of which was lactating. Three individuals captured in July included 1 juvenile, 1 male in reproductive condition, and 1 pregnant female. Litter size, based on embryo counts, ranges from 3 to 4 (average 3.8; Stangl 1992a; Stangl et al. 1994). Snakes, raptors, and mammalian carnivores are its primary predators. The subspecies that occurs in the Trans-Pecos is *S. f. dalquesti*, and it is endemic to the region (Stangl 1992b).

The status of this cotton rat in the checklist of Trans-Pecos mammals (chapter 2) is listed as enigmatic because it is not clear whether these individuals represent an established population, a locally extirpated remnant population, an accidental introduction by humans, or a recent range extension. Mammalogists at Texas Tech University have been conducting an extensive trapping survey in suitable habitat at and near the known locality in Jeff Davis County, but no additional records have been obtained to date. Given its scarcity and uncertain distribution in Texas, it has recently been listed as threatened by the TPWD.

Southern Plains Woodrat
Neotoma micropus Baird, 1855

The southern plains woodrat is a large gray rat with a relatively short tail that is distinctly bicolored, blackish above and grayish below (fig. 82). Its pelage is dense and soft, its ears are exceptionally large, and its eyes are large and bulging.

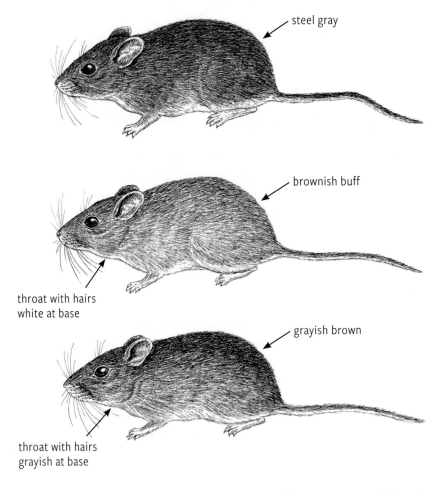

steel gray

brownish buff

throat with hairs white at base

grayish brown

throat with hairs grayish at base

FIG. 82. Three species of woodrats, genus *Neotoma*: top, *N. micropus*; center, *N. leucodon*; bottom, *N. mexicana*

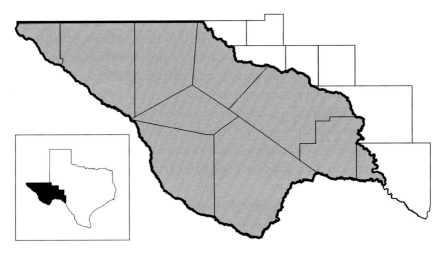

MAP 95. Distribution of the southern plains woodrat, *Neotoma micropus*

The color above is steel gray with an admixture of black hairs along the back. The underparts and feet are white. Average external measurements (in mm) are total length 299; tail 123; hindfoot 34; ear 26; average weight is 144 g. Males are larger on average than females.

The southern plains woodrat is common throughout the entire Trans-Pecos region at lower elevations (map 95). It favors desert grassland and cactus or thorny desert scrub habitats and is rarely associated with rocks or cliffs. In BBNP, it is common in brush thickets of cane and mesquite along the Rio Grande (Boer and Schmidly 1977) as well as in creosote bush–grassland habitat in the Harte Ranch area of the park (Yancey et al. 2006). In BBRSP, *N. micropus* has an affinity for desert grasslands and scrub dominated by prickly pear cactus, creosote bush, mesquite, and catclaw (Yancey 1997). The species appears to be restricted to low-elevation desert scrub habitats in both the Guadalupe and Chinati Mountains (Cornely 1979; C. Jones et al. 2011).

This woodrat is active year-round but seems to spend a greater proportion of time in its nest during winter. It is primarily nocturnal, with most activity occurring between dusk and midnight, although some foraging may occur prior to twilight in overcast conditions. This woodrat feeds mostly on plant material; primary foods include green vegetation, pads and fruits of prickly pear cactus, mesquite beans, sotol leaves, blades of lechuguilla and yucca, and acorns (Bailey 1905; Box 1959). During episodes of drought it appears to rely on cactus pulp for both food and water (Bailey 1931).

The southern plains woodrat is well known for its habit of constructing dens of piled sticks, leaves, cacti, and other kinds of debris. Houses may reach heights of 1 m, with several openings near the base that connect with well-worn trails that lead through the surrounding vegetation. Dens are typically constructed within clumps of prickly pear, cholla, or catclaw. The vernacular name "packrat" refers to the animal's habit of collecting objects, such as bones and small rodent skulls, and stowing them in its den.

Reproductive data of individuals from BBRSP and BBNP indicate that breeding in the Trans-Pecos probably occurs throughout the year (Yancey 1997; Yancey et al. 2006). Gestation is 30–39 days, and typical litter size is two to three with extremes of one and four. Primary predators include hawks and owls, roadrunners, coyotes, raccoons, gray foxes, and bobcats (Bailey 1905, 1931). Although western diamondback rattlesnakes are also significant predators, some populations of these woodrats have evolved a resistance to the snake's venom (Perez et al. 1979). The subspecies in the Trans-Pecos is *N. m. canescens*.

White-toothed Woodrat
Neotoma leucodon Merriam, 1894

The white-toothed woodrat is a large brownish-buff rat with throat hairs that are white to their bases (fig. 82). The tail is moderate in length, distinctly bicolored, and densely covered with short hairs. *Neotoma leucodon* differs from *N. micropus* in its dorsal coloration, which is steel gray in *micropus*. It is distinguished from *N. mexicana* as described in the following account. Average external measurements (in mm) are total length 328; tail 152; hindfoot 34; ear 30; weight is 136–294 g. Males are larger than females on average.

The white-toothed woodrat is common throughout the entire Trans-Pecos region in suitable habitat (map 96). It seems to prefer intermediate elevations in areas dominated by rocks or boulders, especially where prickly pear cactus is present. In BBRSP and BGWMA, it is found in desert scrub situations (Tamsitt 1954; Yancey 1997), whereas in the Chisos and Chinati Mountains it is more common in midelevation woodlands (Baccus 1971; C. Jones et al. 2011).

White-toothed woodrats are active throughout the year. They are primarily nocturnal, although there are reports of some diurnal movements. Succulent parts of prickly pear cactus are their primary foods, as these materials serve as both food and water sources. Other items include mesquite beans and pods, grasses, shrubs, and fruits. Animal material (ants, beetles, grasshoppers, birds) contributes minimally to their diet. Like most other species of woodrats, white-toothed woodrats are great builders and are well known for the elaborate middens they construct. Middens are typically assembled from sticks, cactus pads, and other materials and are often constructed within cactus patches and rock

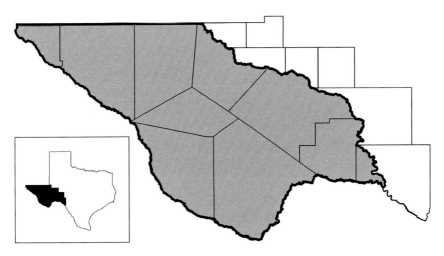

MAP 96. Distribution of the white-toothed woodrat, *Neotoma leucodon*

crevices as a means of protection. In BBRSP several middens were found in cracks or gaps in canyon walls.

Their breeding season extends throughout the year but takes place mostly from January to September. After a gestation period of 30–38 days, a litter of 1 to 3 (average 1.8) young is produced two to three times a year. Common predators include rattlesnakes, hawks, owls, coyotes, long-tailed weasels, American badgers, northern raccoons, bobcats, gray foxes, and striped, spotted, and hognosed skunks.

Neotoma leucodon was previously regarded as *N. albigula* (white-throated woodrat). Subsequent genetic analysis resulted in the splitting of *N. albigula* into two species, with populations in the Trans-Pecos now considered *N. leucodon* (C. Edwards et al. 2001). The subspecies that occurs in the Trans-Pecos is *N. l. robusta* (C. Edwards et al. 2001).

Mexican Woodrat
Neotoma mexicana Baird, 1855

The Mexican woodrat is only slightly smaller than *N. leucodon* but differs in that the hairs on its throat are grayish at their bases instead of white (fig. 82). The upperparts of *N. mexicana* are grayish brown and moderately darkened over the back by blackish hairs. The underparts are grayish white; the tail is bicolored, brownish above and white below. Average external measurements (in mm) are total length 300; tail 125; hindfoot 28; ear 26–28; weight is 140–185 g. Occasionally the grayish throat hairs are not evident, and one must resort to

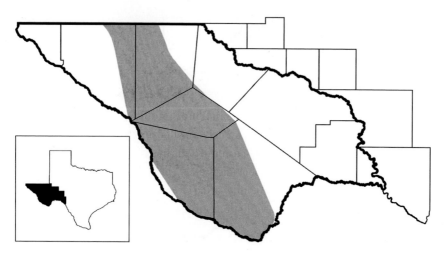

MAP 97. Distribution of the Mexican woodrat, *Neotoma mexicana*

features of the teeth in order to distinguish *mexicana* from *leucodon*. In *mexicana* the anterointernal reentrant angle of the first upper molar is deep and extends more than halfway across the crown; in *leucodon* this reentrant angle is shallow.

The Mexican woodrat is known only from Hudspeth, Culberson, Jeff Davis, Presidio, and Brewster Counties (map 97), where it occurs mostly at the highest elevations of the Guadalupe, Davis, Chinati, Rosillos, and Chisos Mountains (Blair 1940; Genoways et al. 1979; Yancey et al. 2006; C. Jones et al. 2011). It most often lives among talus slopes, rocky outcrops, rock piles, and cliffs, usually in or near juniper, pine, oak, or mixed deciduous and coniferous forest (Bailey 1905; Blair 1940; Cornely 1979; C. Jones et al. 2011). On rare occasions, it may be found in lowland desert areas as evidenced by a single individual taken from the floor of the Solitario in BBRSP (Yancey 1997).

The Mexican woodrat is active throughout the year and is mostly nocturnal, with only occasional daytime activity. Its diet is highly variable but in general consists mostly of plant material such as leaves, flowers, stems, seeds, acorns, and juniper berries. In the Guadalupe Mountains, acorns and juniper berries seem to be the principal foods (Bailey 1905, 1931). Cactus, if available, may be eaten, but they do not seem to prefer it to the same extent as white-toothed woodrats. Unlike white-toothed woodrats, they are known to cure and store large amounts of food.

Atypically for woodrats, *N. mexicana* does not usually assemble elaborate middens of sticks and other debris. Instead, it more often constructs nest chambers in sheltered rock crevices, tree cavities, or abandoned buildings. In the absence of these nesting structures, Mexican woodrats may build typical woodrat middens or move into middens built and vacated by other woodrats (Bailey 1931).

Breeding is thought to take place mostly in June and August (Genoways et al. 1979), although a lactating female trapped on February 26 in BBRSP indicates that some breeding occurs in winter (Yancey 1997). After a gestation period of 31–34 days, a litter of 2 to 5 (average 3.4) is produced, typically twice a year. Common predators are rattlesnakes, barn owls, coyotes, long-tailed weasels, bobcats, and gray foxes. The subspecies in the Trans-Pecos is *N. m. mexicana*.

Mogollon Vole
Microtus mogollonensis (Mearns, 1890)

The Mogollon vole is a small mouselike rodent with an exceptionally short tail (less than twice the length of the hindfoot) and short ears that are nearly covered by fur (fig. 83). The upperparts are a dull brownish color; the underparts, feet, and tail are brownish gray. Average external measurements (in mm) are total length 141; tail 32; hindfoot 21; weight is 29–48 g.

The Mogollon vole is known only from the higher elevations of the Guadalupe Mountains in Culberson County (map 98), where it has a decided preference for grassy meadows and open slopes of mountains in the pine-fir-oak association from about 2,375 to 2,600 m. This peripheral species is common in McKittrick Canyon in GMNP, where its runways, burrows, and winter nests are abundant under tall grass and weeds. Higher up on the open ridges,

FIG. 83. Mogollon vole, *Microtus mogollonensis*

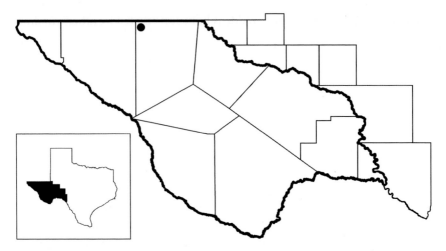

MAP 98. Distribution of the Mogollon vole, *Microtus mogollonensis*

runways have been found winding among stones and under gray oak (Bailey 1905, 1931). A fossil record of this vole, dated to 3,700 YBP, has been discovered in the Davis Mountains (Kennedy and Jones 2006).

Despite residing at high elevations where climates are relatively cool, Mogollon voles are not known to hibernate. Individuals may be active at any time, but unlike most other small mammals, they tend to be more active during the day than at night. Green vegetation, especially grass, is their principal food. Unlike most other rodents, which feed mostly on the seeds of grasses, voles readily consume the blades and stems.

Like other voles, Mogollon voles construct well-defined runways that wind through grassy fields. These small dirt pathways in tallgrass habitats often appear tunnel-like. The runways form a complex network that connects underground burrow entrances to feeding grounds. Visual evidence of runways can be useful in determining their presence and relative abundance in a particular area.

Breeding occurs primarily during the warmer months (especially April through August). Litter size may be as high as 6, but averages are only about 2.2–2.5 (Hoffmeister 1986), which is substantially lower than the 4–6 average for other North American voles. Their relatively small litter size is related to the species' reduced number of mammary glands (four as opposed to eight in other southwestern species). Multiple litters are produced annually, with about 30–40 days between litters.

Significant predators include hawks, owls, and snakes (including copper-heads and rattlesnakes), as well as several mammals such as coyotes, long-tailed weasels, badgers, bobcats, and striped, spotted, and hog-nosed skunks.

The Mogollon vole was previously regarded as a subspecies of the Mexican vole (*Microtus mexicanus*) but is now considered a distinct species. The subspecies that occurs in the Trans-Pecos is *M. m. guadalupensis*.

Common Muskrat
Ondatra zibethicus (Linnaeus, 1766)

The common muskrat is a large, stocky, semiaquatic rodent with a tail about half the length of the head and body (fig. 84). Its ears are short and barely project above the surrounding fur. Its lips close behind its incisor teeth, permitting it to gnaw underwater. The forefeet have four clawed toes and a thumb with a nail; the hindfeet possess five clawed toes that are webbed at their bases. The scaly tail is vertically flattened, a feature that serves to distinguish a muskrat from an immature beaver. The overall color is pale, dull brown. Average external measurements (in mm) are total length 516; tail 240; hindfoot 74; ear 20–21; weight of males is 923 g and of females, 839 g.

Historically the distribution of *O. zibethicus* in the Trans-Pecos is believed to have included the entire Rio Grande and Pecos River watersheds. However, reduction in the flow of parts of the Rio Grande and degradation of riparian habitats along the Pecos have led to the restriction of its range to springs and tributaries associated with the upper Rio Grande and parts of the Pecos River (map 99). For this reason, the species is regarded as rare, and there is speculation that it may have recently become extirpated from the Pecos River drainage (Falcone et al. 2019).

FIG. 84. Common muskrat, *Ondatra zibethicus*

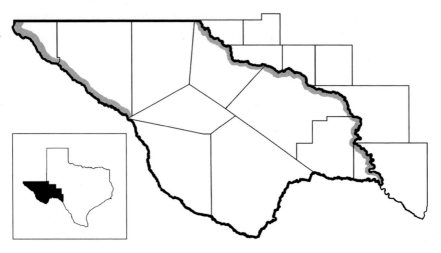

MAP 99. Distribution of the common muskrat, *Ondatra zibethicus*

In the Trans-Pecos, muskrats are active year-round and do not hibernate. Most activity occurs at night, although individuals may occasionally be active during the daytime, especially in spring and summer and on rainy days. Typical foods include the shoots, roots, rhizomes, bulbs, tubers, stems, and leaves of a wide variety of aquatic plants, especially cattails and bulrushes. In addition, this rodent will opportunistically consume animal material, including mollusks, crustaceans, fish, and young birds.

Muskrats construct two kinds of dwellings: lodges built mostly of local vegetation, and burrows dug in shoreline banks. It seems that Trans-Pecos individuals construct only burrows. Burrows prevent heat stress better than lodges and thus would be advantageous in the hot climate of the Trans-Pecos.

Breeding seems to take place from late winter through early spring. Females bear two litters per year in the region. After a gestation period of 28–30 days, an average litter of three young is born. Common predators include great horned owls, barn owls, northern harriers, raccoons, coyotes, gray foxes, badgers, long-tailed weasels, and feral dogs and cats.

Trans-Pecos specimens are small compared to other subspecies and are assigned to the subspecies *O. z. ripensis*. Falcone et al. (2019) investigated the taxonomic status of the subspecies in the region by comparing mitochondrial DNA from populations of *O. z. ripensis* to that of its neighboring subspecies to the west, *O. z. osoyoosensis*. They determined that the two are not substantially different, suggesting that perhaps they should be considered a single subspecies. Although habitat degradation continues to be a serious peril to muskrats in the Trans-Pecos, the need to manage this population as a distinct taxon may be less urgent than previously thought.

Family Erethizontidae
NEW WORLD PORCUPINES

Members of this family are large, heavyset rodents characterized by hairs modified as sharp spines or quills. Only one species, *Erethizon dorsatum*, occurs in the Trans-Pecos.

North American Porcupine
Erethizon dorsatum (Linnaeus, 1758)

The North American porcupine is a large rodent with a distinct dorsal pelage consisting of yellowish to brown guard hairs combined with prominent pale, dark-tipped, barbed quills (fig. 85). The intermixing of guard hairs and quills results in an overall yellowish-brown appearance. Quills cover all body parts except the face, abdomen, inner parts of the limbs, and underside of the tail. Because of its unique appearance, *Erethizon dorsatum* cannot be confused with any other mammal in the Trans-Pecos. Average external measurements (in mm) for males and females, respectively, are total length 808, 737; tail 235, 230; hindfoot 98, 81; weight is 3.5–18 kg.

FIG. 85. North American porcupine, *Erethizon dorsatum*

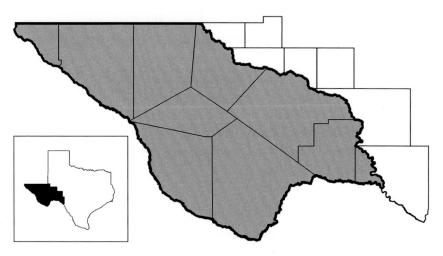

MAP 100. Distribution of the North American porcupine, *Erethizon dorsatum*

Porcupines have been recorded in variable numbers from every county in the Trans-Pecos (map 100), but in general they are uncommon in the region. They occur in a variety of habitats but are most common among rocky ridges and forested situations in mountainous areas, particularly the Davis and Guadalupe Mountains (Genoways et al. 1979; Stangl et al. 1994). They are less common in the flats, valleys, gulches, and stream beds of BBNP, BBRSP, CMSNA, and EMWMA (Yancey 1997; C. Jones et al. 2011; Yancey and Manning 2018). Occasionally they will venture into cities (Milstead and Tinkle 1958).

The North American porcupine does not hibernate and may be active day or night. It feeds primarily on herbaceous ground vegetation in the spring and summer, but in fall and winter it switches to the inner bark of piñon, oak, juniper, and willow trees. Its diet is often deficient in sodium, prompting individuals to seek out and lick or consume unusual items that contain sodium, such as plywood, paint, and dirt. Porcupines are adept climbers and often spend long periods in the same tree.

Breeding takes place in late summer and early fall and is followed by an unusually long (for rodents) gestation period of about seven months. Parturition occurs in April or May when grasses and forbs are plentiful and results in the birth of a single young (twins are rare). Predators include great horned owls, coyotes, mountain lions, and bobcats. Porcupines have been known to transmit the disease tularemia directly to humans. The subspecies in the Trans-Pecos is *E. d. couesi*.

EXTIRPATED SPECIES, SPECIES OF POSSIBLE OCCURRENCE, INTRODUCED SPECIES, AND FERAL MAMMALS

In this section we briefly explore the status of 6 native species that no longer occur in the Trans-Pecos, 10 species that could possibly be found there in the future, 10 nonnative species introduced into the region that now occur in natural habitats, and 4 kinds of mammals that exist in the feral state.

EXTIRPATED SPECIES (6)

Six species (five carnivores and one ungulate) formerly known from the Trans-Pecos were extirpated during the twentieth century, and no attempts have been made to reintroduce them back into the region.

ORDER CARNIVORA

Family Canidae
Gray Wolf
Canis lupus Linnaeus, 1758

The smallest described American subspecies of the gray wolf, the Mexican gray wolf (*C. l. baileyi*) (fig. 86), was historically found in the Trans-Pecos, where it occupied grassland habitats with the American bison, on which it relied for its primary food supply (Bogan and Melhop 1983). Today there are no resident gray

FIG. 86. Gray wolf, *Canis lupus*

wolf populations remaining in the Trans-Pecos region. The last authenticated records were of animals shot in 1970 at the Cathedral Mountain Ranch, 27 km south of Alpine in Brewster County, and at the Joe Neal Brown Ranch, located at the point where Brewster, Pecos, and Terrell Counties converge (Scudday 1972). Both of these animals were Mexican gray wolves (*C. l. baileyi*). Other historical records include six to eight wolves killed in the Davis Mountains in 1901 and 1902 (Bailey 1905), one from the Guadalupe Mountains in 1901 (Schmidly 1977b), a female killed in 1944 near Fort Davis (S. Young and Goldman 1944), and an individual killed 48 km south of Marfa in 1942 (Schmidly 1977b). Mexican gray wolves have been reintroduced into southwestern New Mexico and adjacent parts of Arizona, where they have slowly begun to establish populations, but there are no current plans to reintroduce them back into the Trans-Pecos because of ranchers' concerns about livestock depredation.

Family Felidae
Ocelot
Leopardus pardalis (Linnaeus, 1758)

The ocelot (fig. 87) occurs in the brushlands of South Texas, but only a single specimen was ever reported from the Trans-Pecos, and it represents the westernmost record of this cat in Texas. One was "killed near the Alamo de Casarae Ranch, in Brewster County, between Marfa and Terlingua, in 1903" (Bailey 1905). It is doubtful that ocelots were ever common in the Trans-Pecos, but a few may have occasionally wandered into the area from South Texas or Mexico.

FIG. 87. Ocelot, *Leopardus pardalis*

Jaguar
Panthera onca (Linnaeus, 1758)

At one time the jaguar (fig. 88) occupied nearly all of Texas and Louisiana north to the Red River, but it is now extirpated from the state. There was a report of one killed near the mouth of the Pecos River in 1889, and in 1901 another was killed east of the Pecos River south of Comstock in Val Verde County (Bailey 1905). There are additional isolated, scattered historical records of occurrence to the south and east of the Trans-Pecos (Schmidly 2002).

FIG. 88. Jaguar, *Panthera onca*

Family Mustelidae
Black-footed Ferret
Mustela nigripes (Audubon and Bachman, 1851)

The only record of a black-footed ferret (fig. 89) in the Trans-Pecos is a single specimen reported from Fort Stockton in Pecos County (Bailey 1905). However, in Bailey's account it is apparent that this animal was not collected at Fort Stockton but only maintained there. The specimen was "caught in a trap set at an old adobe house on the edge of a [prairie] dog town just north of the Pecos River at Grand Falls." Grand Falls is in Ward County, which is not part of the Trans-Pecos. It is likely, however, that black-footed ferrets did live in the region at one time because of their close association with prairie dogs, which commonly occurred there. When prairie dogs were eradicated by the poisoning campaigns of the early to mid-twentieth century, these mustelids

FIG. 89. Black-footed ferret, *Mustela nigripes*

were eliminated. The black-footed ferret is now considered endangered, and a successful breeding program has helped reestablish them in parts of their former range.

Family Ursidae
Grizzly or Brown Bear

Ursus arctos Linnaeus, 1758

The only record of a grizzly bear (fig. 90) from Texas is of a very old male recorded by Vernon Bailey from "near the head of Limpia Creek" in the Davis Mountains in 1890. This bear was hunted by dogs and later killed because it had

FIG. 90. Grizzly bear, *Ursus arctos*

killed and eaten a cow. A complete account of the hunt that produced the bear has been documented (Schmidly 2002). There is sufficient evidence to indicate that grizzly bears also once occurred in the Guadalupe Mountains (Genoways et al. 1979). In 1901, Vernon Bailey "found tracks of very large bears that were evidently of the grizzly group, though apparently no grizzlies had been killed there for some time" (Bailey 1931). Bailey also received a report from the Forest Service of grizzlies in the Guadalupes in 1909 (Genoways et al. 1979). Despite these records and reports, the grizzly bear was probably only marginal in the region and never represented by an established breeding population.

ORDER ARTIODACTYLA

Family Bovidae
American Bison

Bos bison Linnaeus, 1758

In early historic times American bison (fig. 91) undoubtedly roamed the extensive grasslands of the Trans-Pecos, including areas now in BBNP and GMNP. There are reports as early as 1854 of bison near the Salt Lakes west of the Guadalupe Mountains (Genoways et al. 1979). The only records documented by preserved materials are from 51 km east of old Fort Stockton in Pecos County and from 16 km east of Langtry in Val Verde County (Schmidly 1977b). By the mid-1880s bison were extirpated from the Trans-Pecos. It is thought that early

FIG. 91. American bison, *Bos bison*

settlers and the market and hide hunters caused their drastic decline. In 1920 a few of these animals were stocked on the Reynolds Ranch in Jeff Davis County, and as of 1963 there were 31 on that ranch (Jackson 1967). A few bison are maintained on private ranches in the Trans-Pecos today.

SPECIES OF POSSIBLE OCCURRENCE (10)

The following species have been recorded near the periphery of the Trans-Pecos and therefore may possibly be taken in the area with more field-oriented collecting or sampling. Included in this grouping are one species of shrew, five kinds of bats, and four rodents.

ORDER EULIPOTYPHLA

Family Soricidae
Least Shrew
Cryptotis parva (Say, 1823)

This diminutive shrew has been reported from the southwestern part of the Edwards Plateau in Tom Green County, and to the southeast of the Pecos River in Val Verde County (Schmidly and Bradley 2016). This shrew is frequently found in grassland habitats, and those areas near the Pecos River in the eastern Trans-Pecos should be surveyed using specialized trapping techniques (e.g., small pitfall traps) to sample for this small shrew.

ORDER CHIROPTERA

Family Phyllostomidae
Hairy-legged Vampire Bat
Diphylla ecaudata Spix, 1823

There is but one record of the species from Texas, a female found on May 24, 1967, in an abandoned railroad tunnel on the Brotherton-Calk Ranch, 7.2 km west of Comstock, Val Verde County, just to the southeast of the Trans-Pecos. This species is best treated as a record of accidental occurrence and not likely to become established in Texas, although a thorough search of the caves along the Rio Grande would be useful in confirming this.

Common Vampire Bat
Desmodus rotundus (E. Geoffroy, 1810)

The common vampire bat has been recovered from sub-Recent deposits in the Trans-Pecos (Cockerell 1930), and it could possibly recolonize the region if climates continue to warm. It has been taken in several locations in Mexico not far from the Texas border, and with more field collecting of bats along the Rio Grande it would not be surprising to discover this species in the region.

California Leaf-nosed Bat
Macrotus californicus Baird, 1858

This is another leaf-nosed bat (family Phyllostomidae) that has been recorded from sub-Recent deposits in the Trans-Pecos (Cockerell 1930), but not in modern times. The closest collection records today are from southeastern Arizona and western Chihuahua, one of the Mexican states adjacent to the Trans-Pecos. These bats are primarily cave and mine dwellers in Arizona (Hoffmeister 1986), and a careful search of caves and abandoned mines in the Trans-Pecos might document their presence in the region.

Family Vespertilionidae
Seminole Bat
Lasiurus seminolus (Rhoads, 1895)

A single specimen of the Seminole bat was collected on the western edge of the Edwards Plateau along the Devils River in Val Verde County, only a few kilometers east of the Pecos River (Brant and Dowler 2000). Recent evidence suggests this species is expanding its range westward, possibly into the Trans-Pecos (Schmidly 2002; Schmidly et al. 2022).

Southwestern Myotis
Myotis auriculus Baker and Stains, 1955

The southwestern myotis occurs in a variety of habitats, including the arid desert scrublands and woodlands of the Chihuahuan Desert. It has been documented from Chihuahua and Coahuila, Mexico, and from southwestern New Mexico and southeastern Arizona. It has not been reported from the Trans-Pecos but may be found there with continued field efforts.

ORDER RODENTIA

Family Cricetidae
Northern Pygmy Mouse
Baiomys taylori (Thomas, 1887)

This small cricetid rodent, which resembles a miniature deermouse, has expanded its range northward and westward in Texas from coastal regions during the twentieth century and now occurs in all the state's ecological regions except the Trans-Pecos. However, within the past few decades, it has been taken from just east of the Trans-Pecos in Upton, Val Verde (Schmidly and Bradley 2016), and Crockett (Krishnamoorthy et al. 2021) Counties and to the north of the region in Catron and Grant Counties in New Mexico (Geluso and Geluso 2020). This species is most often associated with grassland habitats, and, like the least shrew, should be searched for in such habitats in the Trans-Pecos.

Texas Deermouse
Peromyscus attwateri J. A. Allen, 1895

This mouse is common on the Edwards Plateau to the east of the Trans-Pecos, and it has been documented from Ward and Crockett Counties just east of the Pecos River. It may be that xeric mesquite scrubland habitats along the Pecos River drainage will impede its dispersal into the Trans-Pecos, but collectors should be aware that it could possibly be found in the region. Its morphological characters are easily confused with those of two other long-tailed deermice in the region, namely Lacey's white-ankled deermouse (*Peromyscus laceianus*) and the brush deermouse (*Peromyscus boylii*).

Family Geomyidae
Jones's Pocket Gopher
Geomys knoxjonesi Baker and Genoways, 1975

This pocket gopher is known from just northeast of the Trans-Pecos in Ward, Winkler, and Crane Counties. It prefers deep, loose sandy habitats and should be looked for along the Pecos River valley in Reeves and Pecos Counties. This species is difficult to distinguish from other species of *Geomys* without chromosomal or genetic data.

Texas Pocket Gopher
Geomys personatus True, 1889

Although this pocket gopher has not yet been documented in the Trans-Pecos, it reaches its known western range of distribution in Val Verde County and could possibly disperse west of the Pecos River in sandy soils associated with the Rio Grande.

INTRODUCED SPECIES (10)

Several nonnative mammals have been intentionally or accidentally introduced into towns, small communities, and some natural habitats throughout the Trans-Pecos during the twentieth century. Some were imported for sport hunting and subsequently escaped, dispersed, and became established. Others were introduced into other parts of Texas and later spread into the Trans-Pecos, where they have become competitors with native species. The most significant of these introduced mammals are the nutria (*Myocastor coypus*) and feral hog (*Sus scrofa*) because of their potential to wreak havoc on native species and their habitats. Included in this grouping are one carnivore, five exotic ungulates, and four rodents.

ORDER CARNIVORA

Family Canidae
Red Fox
Vulpes vulpes (Linnaeus, 1758)

The red fox was introduced to eastern and central Texas for sporting purposes in the mid-1890s. Currently this canid is found only in the eastern part of the Trans-Pecos in Pecos, Terrell, Jeff Davis, and Brewster Counties, where it seems to be well established, with breeding populations apparently present. This fox is usually found in brushy upland areas, often in association with agricultural or ranching activities.

Order Artiodactyla

Family Bovidae
Aoudad or Barbary Sheep
Ammotragus lervia (Pallas, 1777)

Aoudads are native to the highland regions within desert and subdesert zones of the Sahara in northern Africa. They were first released into the wild in Texas in 1957 and now occur in several regions of the state, where they are popular with hunters. Their preferred habitat is remote, rocky, dry, barren desert canyons and waterless areas, much like that of native bighorn sheep. They have become relatively abundant in upland habitats in Jeff Davis, Presidio, and Brewster Counties, where they often compete directly with reintroduced bighorn sheep. Aoudads are suspected of carrying diseases that are lethal to bighorns, and it should be a priority to keep them away from areas where efforts are ongoing to introduce and maintain native bighorns (Mungall and Sheffield 1994).

Blackbuck
Antilope cervicapra (Linnaeus, 1758)

The blackbuck was introduced from Pakistan or India into the Texas Hill Country for sport hunting. From there, escapees spread to other parts of the state. Its preferred habitat is grassland and open woodlands. It occurs sporadically only in Val Verde County in the southeastern part of the Trans-Pecos (Mungall and Sheffield 1994).

Scimitar-horned Oryx
Oryx dammah (Cretzschmar, 1827)

This species is native to the harsh environs of the Sahara in Africa, where it is a true desert-adapted ungulate. Like other large exotics, it was introduced for hunting purposes, having been billed as a "superexotic." In its native habitat it is critically endangered, and US herds may serve as stock for possible reintroduction to native lands (Mungall and Sheffield 1994). It occurs sparingly in the Trans-Pecos in dry, desert scrub habitats. Two individuals were recently sighted in such habitat near Terlingua in Brewster County.

Common Eland
Taurotragus oryx (Pallas, 1766)

This species is native to savannas (grasslands and open woodlands) of sub-Saharan Africa. It is uncommon and sporadic in distribution in the Trans-Pecos. One was recently photographed with a camera trap at CMSNA (Yancey and Manning 2018).

Family Cervidae
Red Deer
Cervus elaphus Linnaeus, 1758

These cervids are native to Europe and were introduced into the United States for hunting. They prefer open forests in montane areas with vegetative cover such as grass and shrubs. They are uncommon in the Trans-Pecos, with most records from the central mountainous regions (Mungall and Sheffield 1994).

ORDER RODENTIA

Family Muridae

The family Muridae, which includes Old World mice, rats, and gerbils, is the largest family of mammals, with 150 genera and 730 species worldwide. In Texas, including the Trans-Pecos, two genera and three species occur. They occupy many terrestrial habitats, usually in close association with humans.

House Mouse
Mus musculus Linnaeus, 1758

This small, scaly-tailed mouse (fig. 92) is widespread across the Trans-Pecos. It is usually found in urban areas, but it is so numerous in cane-dominated habitats along the Rio Grande that some native mice have been excluded or displaced. Elsewhere, it is occasionally encountered in remote areas around houses, outbuildings, stores, and other structures, but usually in low numbers.

FIG. 92. House mouse, *Mus musculus*

FIG. 93. Black rat, *Rattus rattus*

Norway Rat
Rattus norvegicus (Berkenhout, 1769)

The Norway rat (or brown rat), like the house mouse and black rat, is a commensal with humans and is occasionally found in rural (nonurban) areas, primarily where vegetation is tall and rank and affords adequate protection. There are not any reports of Norway rat specimens from the Trans-Pecos, but the species likely occurs in some of the population centers (Schmidly 1977b).

Black Rat
Rattus rattus (Linnaeus, 1758)

Black rats (or roof rats) (fig. 93) are largely commensals and live in close association with humans. They seldom become established in the wild as do Norway rats. They inhabit warehouses, feed stores, and poultry houses and can be common in cotton gins and associated grain warehouses. On farms and ranches, they live in barns and dilapidated outbuildings. They may live near the ground but usually frequent attics, rafters, and crossbeams of buildings. Black rats are uncommon to sporadic in distribution in the Trans-Pecos, with specimens reported only from El Paso and Brewster Counties (Schmidly 1977b). This rat is thought to be much less common than the Norway rat.

Family Echimyidae
Nutria
Myocastor coypus (Molina, 1782)

Nutrias (fig. 94) are large, semiaquatic rodents first introduced into the United States from South America as a fur resource during the 1890s. They were common in Louisiana by 1938 and subsequently spread into Texas, establishing permanent populations in aquatic habitats across the eastern two-thirds of the state (Schmidly and Bradley 2016). They have continued this westward spread into the Trans-Pecos, where they were first reported from Independence Creek, south of Sheffield in Terrell County, and from the west bank of the Pecos River at the Pandale Crossing in Val Verde County (Hollander et al. 1992). Subsequently, they were reported at Rio Grande Village in BBNP in 1993 (Raymond Skiles, personal communication).

During 2004 and 2005, more than 30 locations of nutria activity were documented along a 16 km stretch of the Rio Grande from just upstream of Rio Grande Village to Boquillas Canyon, including the Rio Grande Village

FIG. 94. Nutria, *Myocaster coypus*

beaver pond (Milholland et al. 2010). Several of these animals were captured, radio-collared, and released to learn more about their population size and movements. Population size was estimated to be between 38 and 74 nutrias, with a mean home range of about 10 ha and a maximum daily distance moved of 637 m (Milholland et al. 2010). These movement data suggest that nutrias can travel significant distances up and down the river.

The establishment of this introduced rodent in the Pecos River and Rio Grande drainages could seriously jeopardize the continued existence of the native Pecos River muskrat (*Ondatra zibethicus ripensis*) and the American beaver (*Castor canadensis*), both of which occupy much of the same habitat. For this reason, there is a real need to control this species.

FERAL MAMMALS (4)
Feral mammals are domestic mammals that have reverted to a "wild" or free-ranging state. They have become established as breeding populations in many areas of the Trans-Pecos. Many pose a potential threat to the delicate balance of the desert ecosystem by fouling water resources, destroying vegetation, and competing with native species.

ORDER ARTIODACTYLA

Family Suidae
Feral Hog
Sus scrofa Linnaeus, 1758

Feral hogs (fig. 95) were originally brought to the United States from Europe by early settlers. Over time, animals escaped and established "wild" populations. Their characteristics are varied, depending on the breed of the ancestral stock. When liberated together, European wild hogs and feral hogs interbreed readily, with the traits of European wild hogs apparently being more dominant. There is a sizable population of feral hogs in Texas (estimated at over two million), with individuals occurring in nearly every county in the state (Schmidly and Bradley 2016). These animals have become so prevalent that for all practical purposes they should be considered members of the natural fauna.

In Texas densities of feral hogs increase with increased precipitation (Adkins and Harveson 2007). Although they have occurred in western Texas since the early 1990s, because of the dry conditions throughout the Chihuahuan Desert, feral hogs were thought to be limited to riparian habitats. But recent studies have shown their abundance and distribution to be increasing in the region. Extensive studies in the Davis Mountains have revealed a viable breeding population with generalized use of habitats. Characteristics of this population include a density of 0.65 hogs/km²; large home range sizes that average

FIG. 95. Feral hog, *Sus scrofa*

about 48 km² and 34 km² for males and females, respectively; a preference for open-canopy, evergreen woodland habitat with sources of free-standing water; and summer diets dominated by herbaceous material (38.6 percent) and roots and tubers (34.3 percent) (Adkins and Harveson 2006, 2007).

Although feral hogs occur at relatively low densities in the Trans-Pecos, there is concern over their potential damage to natural resources. Feral hog damage may be direct or indirect and could include loss of soil cover, reduction of soil stability, influence on vegetation succession, predation on terrestrial fauna, interspecific competition for resources with native wildlife, and habitat disturbance (Adkins and Harveson 2007). Should they become more common, they could pose a serious threat to the native collared peccary (*Pecari tajacu*). For these reasons, populations should be controlled wherever possible (Mungall and Sheffield 1994).

ORDER PERISSODACTYLA

Family Equidae
Feral Ass or Burro
Equus asinus Linnaeus, 1758

Although not native to the United States, the feral ass has become established and fairly common over the desert regions of the Southwest. Several recent records from multiple sites in the Trans-Pecos suggest they are becoming more common in the region. Only a single individual was reported from BBRSP in 1997 (Yancey 1997), but two decades later burros were reported to be numerous and widespread in the park (Yancey and Manning 2018). In addition, Fred Stangl of Midwestern State University recently reported a small group of 8–10 individuals on property that straddles the Brewster-Presidio county line. Other groups of feral asses are known along the Rio Grande from Candelaria south, as well as east to BGWMA, including BBNP. The establishment of these animals is potentially a major problem because they can become a destructive menace. Their excrement can foul springs and creeks in a landscape where water is limited and precious. They also compete with native wildlife for limited forage, decreasing available food sources for some animals, disrupting the food chain for other species, and threatening native plants.

BBRSP is one of the major sites selected for the reintroduction of bighorn sheep, and burros are known to adversely affect desert bighorns via competition and disease transmission. In 2008 the TPWD began to remove the burros by shooting them, and a public outcry ensued. The removal program was stopped, and the TPWD is now seeking nonlethal options for removing the burros. The same problem has existed for several decades in nearby BBNP. The government is prevented from shooting the burros in the park and instead must capture and place them in adoption programs.

Feral Horse
Equus caballus Linnaeus, 1758

The mustang, or wild horse, was first brought to Texas in 1542 by early Spanish explorers. At one time they were common in the state, but wild mustangs are rare today. In BBNP, herds from Mexico will periodically cross the river and graze in the riparian areas of the park, where they trample vegetation and foul the water. This problem, however, is nothing like that seen on public lands in Arizona and Nevada, where these animals run wild in large numbers and have exceeded the carrying capacity of public lands.

Feral Mule
Equus caballus × *Equus asinus* (a hybrid not officially named)

Hybrids between a jack (male ass) and a mare (female horse) are mules, valued as working and riding animals. They are usually sterile. Mules may be found on many farms and ranches across the Trans-Pecos and usually do not pose any sort of conservation problem. However, when individuals escape confinement and populations get out of control, such as in many places in Arizona (e.g., Grand Canyon National Park), they become a nuisance and are harmful to wildlife habitat (Hoffmeister 1986). Establishment of feral mule populations requires large numbers of feral asses and feral horses for crossbreeding. As with the feral burro and feral horse, numbers should be monitored and controlled as necessary.

7

SAVING THE LAST FRONTIER

Conservation Threats and Solutions

The Trans-Pecos constitutes the last frontier of Texas and deserves to receive maximum protection for future generations of its residents as well as all Texas citizens. Its central themes are its incredible biodiversity, as we have demonstrated in the accounts of mammals in this book, plus its history and heritage, which have defined the region and its residents. As a desert, the region represents a sensitive ecosystem. Water resources are inherently rare, and desert ecosystems are unforgiving, as small disturbances or changes can have significant consequences for plant and animal life. Because of precipitation levels and topography, the landscape recovers much more slowly than it does elsewhere. Thus, resource managers have fewer tools to utilize when recovering habitat after land development. For these reasons, population growth and the associated economic activity can have a dramatic impact on desert regions.

Conservation Threats

Among the major threats to mammals, the following concerns seem to be the most relevant: rangeland conversion and water scarcity, the introduction of invasive species, environmental pollution, wildlife diseases, energy development, construction of a border wall, and climate change. Prospects for the survival of resident mammals having degrees of intolerance to one or more of these environmental disturbances can be presumed to be difficult at best. As these changes are carried to excess, it is suspected that some resident mammals will be threatened more than others.

Mammals most threatened are those that are restricted mostly to areas of water and/or moist soils and associated plant and animal life (R. H. Baker 1988). The borders of water courses in the Chihuahuan Desert have been slowly losing their characteristically green and often luxuriant natural plant growth. Intensive human occupation has often totally obliterated them through replacement with residential sites and fields of introduced agricultural crops, especially along the watersheds of the Rio Grande where it enters the Trans-Pecos in El Paso.

A variety of mammals will be impacted by such changes. The semiaquatic American beaver (*Castor canadensis*) and common muskrat (*Ondatra zibethicus*) are directly impacted by declining water resources. Riparian areas serve as the major regional habitat for the Virginia opossum (*Didelphis virginiana*), the northern raccoon (*Procyon lotor*), and, to a lesser extent, the nine-banded armadillo (*Dasypus novemcinctus*). The greenery is also a lure to small mammals, such as Elliot's deermouse (*Peromyscus labecula*), fulvous harvest mouse (*Reithrodontomys fulvescens*), and hispid cotton rat (*Sigmodon hispidus*), as well as other mammals that may also subsist in nearby desert situations.

The continued conversion of grasslands to desert scrub as a result of poor land management and climate change (see below) is putting pressure on a number of grassland species. Continued desertification of the Trans-Pecos will make living for both people and wildlife much more difficult. Grassland species that could decline under increasing desertification include the banner-tailed kangaroo rat (*Dipodomys spectabilis*), the tawny-bellied cotton rat (*Sigmodon fulviventer*), kit fox (*Vulpes macrotis*), hooded skunk (*Mephitis macroura*), mule deer (*Odocoileus hemionus*), and pronghorn (*Antilocapra americana*).

Scattered throughout the rugged topography of the Trans-Pecos are island-like, forest-covered highlands that extend upward from the desert floor to about 1,000 m. These highland habitats support many montane mammals in moist and luxuriant stands of oaks, pines, sometimes firs, and other boreal growth. In such ranges as the Chisos, Chinati, Davis, and Guadalupe Mountains, mammals are isolated by arid, inhospitable lowlands from relatives in other such highlands, creating what has been called a "sky islands and desert seas" landscape (Gehlbach 1981). Biotic resources of these islets, because of such small expanses of preferred habitat, will be highly susceptible to climate change, as they are predicted to shrink as vegetation moves upslope following hotter and drier conditions. They are already subject to human habitat manipulation caused by uncontrolled burning, mining, intense livestock grazing, and land fragmentation associated with development. Montane mammals that could be affected include Holzner's mountain cottontail (*Sylvilagus holzneri*), the Mexican woodrat (*Neotoma mexicana*), the rock deermouse (*Peromyscus nasutus*), and the yellow-nosed cotton rat (*Sigmodon ochrognathus*).

INTRODUCTION OF INVASIVE SPECIES

Another growing threat is the spread of nonnative invasive species. Without natural predators or controls, these species can flourish unchecked, degrading habitats and competing with or preying on native wildlife. Two cases involving mammals, the feral hog (*Sus scrofa*) and the nutria (*Myocastor coypus*), have the potential to be particularly harmful to the landscapes and ecological integrity of the Trans-Pecos.

The situation with the spread of feral hogs has become serious in the past 30 years. They now occur in every county of the region and are growing in numbers. They can have devastating effects on native wildlife through habitat deterioration, competition for food, transmission of disease, and even direct predation on smaller mammals, reptiles, and ground-nesting birds. They are particularly damaging in creek-side riparian habitats as well as cottonwood galleries. About the only way to control them is to remove them by trapping or hunting, and neither of those methods is foolproof because of their prolific breeding habits.

While only a recent invader to the region, the nutria can be a serious competitor with native American beaver (*Castor canadensis*) and common muskrat (*Ondatra zibethicus*) populations. Both of these species have already been steadily declining because of habitat loss, and the presence of the nutria threatens to exacerbate the problem.

ENVIRONMENTAL POLLUTION

Wildlife populations and their habitats have been and are today confronted with a bewildering array of pollutants that have been released into the environment either by intent or accident. Although the Trans-Pecos lacks the excessive levels of pollution present in the more populous regions of the state, there are problems within the region. For example, the Trans-Pecos Pipeline's Waha header compressor station in Pecos County has been identified as one of the state's worst polluters, releasing 29 million kg of pollution into the air without authorization in 2018 (Collier et al. 2018).

Among mammals, bats are highly sensitive to some insecticides, especially chlorinated hydrocarbons. Bats feeding on insects treated with these chemicals slowly accumulate pesticide residues in their fat during late summer and fall, which eventually results in their death as this fat is broken down and used during hibernation and migration (D. Clark 1981, 2001). To date only outright mortality on a local population level has been identified as a threat to bats from pesticides, but subtle—and equally devastating—effects are possible on such aspects as reproduction, acoustic behavior, and hibernation metabolism. Fortunately, pesticides in Chihuahuan Desert habitats appear to break down rapidly because of high soil temperatures, high levels of ultraviolet radiation, and alkaline soils.

Nutrient and chemical contamination of waterways is a problem, often coming from more diffuse "nonpoint" sources. Widespread metal contamination has been demonstrated in fish, reptiles, and birds along the lower Rio Grande basin, but no specific studies have been conducted on mammals upstream of the Big Bend area. Scientists working in BBNP have witnessed habitat degradation in riparian corridors because less water has been reaching this area of the border. Moreover, with less stream flow to dilute the water pollution, furthered by the lack of wastewater treatment facilities upstream in the Juarez area, the quality of the streamflows in BBNP has been worsening (Goodwin 2000).

Concern has recently been expressed about the potential polluting impact of "gas flaring" and its possible effects on human and wildlife health in oil-producing regions. Gas flaring burns associated, unwanted or excess gases and liquids released during normal or unplanned overpressuring operations in the oil and gas industry. Since 2017 gas flaring has nearly doubled in the Permian Basin, and concern has mounted about the possible harm caused by contaminants such as benzene, a known carcinogen, and methane, a greenhouse gas that contributes directly to global warming (Collier et al. 2018). There is almost no information about the impact of toxic fumes from flaring on wildlife, and this is clearly an area that needs further exploration.

Poor air quality is also becoming a problem in some parts of the Trans-Pecos. It is evident in the industrial cities of Juarez and El Paso, and the problematic air can be disseminated by atmospheric conditions to the less populated surrounding areas. Today, even the wilderness of BBNP can experience poor air quality. The visibility impairment there comes from two coal-fired electricity-generating power plants near Piedras Negras, Mexico, about 200 km southeast of the park. Neither plant is equipped with devices to control sulfur dioxide particulates, which create a white haze and play an important role in the formation of acid rain (Wauer and Fleming 2002).

The emergence of new diseases poses a particularly dire threat to wildlife. White-nose syndrome (WNS), a fungal disease that affects hibernating bats; chronic wasting disease (CWD), a contagious neurological disease affecting some of America's most iconic large game species, including mule deer (*Odocoileus hemionus*), white-tailed deer (*Odocoileus virginianus*), and elk (*Cervus canadensis*); and rabbit hemorrhagic disease (RHD), a highly infectious and lethal form of viral hepatitis that impacts lagomorphs, are three particular diseases that have recently been recorded in Texas, and all three have the potential to devastate important components of wildlife.

WNS, a fungal disease known to occur only in bats, has killed millions of hibernating bats in the eastern and midwestern United States, causing national concern. To date, WNS has not been confirmed in any bats in the Trans-Pecos region; however, the fungus responsible for the disease, *Pseudogymnoascus destructans*, has been provisionally reported from Presidio, Terrell, and Val Verde Counties (White-Nose Syndrome Response Team 2022). Moreover, a recent study using predictive modeling concluded that suitable conditions for the presence of *P. destructans* probably exist in several karst systems in the region. Even more alarming, the modeling conclusions forecast the cave system of the Guadalupe Mountains as a future WNS hotspot (Wolf et al. 2022). A major outbreak of WNS in the region would be disastrous given that it is the most species-rich area for bat diversity in the state and the country.

CWD is related to mad cow disease and Creutzfeldt-Jakob disease in humans, and it has recently been discovered in a wild mule deer (*Odocoileus hemionus*) in Hudspeth County as well as a white-tailed deer (*Odocoileus virginianus*) in Val Verde County just east of the Pecos River. An outbreak of CWD, which has no known cure, would devastate deer populations and threaten the economics of landowners who rely on hunt leasing revenue from wildlife ranching.

RHD was previously known to affect European rabbits but has recently been detected in pen-raised and natural populations of rabbits in the United States. Cases have been documented in a wild black-tailed jackrabbit (*Lepus californicus*) in Lubbock County, on the High Plains of Texas, and in a wild cottontail rabbit (*Sylvilagus* sp.) in Hudspeth County in the Trans-Pecos. The latter record is not far from the range of Holzner's mountain cottontail (*Sylvilagus holzneri robustus*), which is thought by some experts to be imperiled.

ENERGY DEVELOPMENT

As we have entered the twenty-first century, rapid expansion of the energy development sector threatens to place additional pressure on wildlife. Because the Trans-Pecos borders one of the most productive energy areas in the world, the Permian Basin, the geographic footprint of the energy sector has slowly surged its way into the region. Recent and future energy-harnessing activities and related infrastructure, such as those associated with the Delaware Basin, Alpine High, Trans-Pecos Pipeline, and a proposed wind farm in the Devils River watershed, could potentially place unprecedented pressures on wildlife resources of the region (L. Harveson 2018).

Much of the concern over the environmental impact of oil and gas development has arisen around the new technological advances in horizontal drilling and high-volume hydraulic fracturing (known as "fracking") that has proliferated in shale formations. The space and infrastructure required by these activities have the potential to transform thousands of acres in the Trans-Pecos into industrialized landscapes, and we are just beginning to learn how harmful this can be to wildlife (Lohan 2019). There is a broad range of possible effects of these activities—soil erosion; reductions in water quantity and quality; reduced flow and increased siltation in streams; habitat loss and fragmentation; changes in native vegetation; air, noise, and light pollution; and increases in heavy road traffic.

But the biggest challenge of energy development in the water-stressed Trans-Pecos is associated with the impact on water resources. Both drinkable and undrinkable sources of water are used to crack shale rock to release oil and gas, and concerns are emerging that the oil and gas boom is requiring more water than is available. There are estimates that around 38 million liters are required for each well, and much of that water becomes contaminated and returns to the surface, bringing with it heavy metals, radioactivity, toxic chemicals, and high salinity. When drilling occurs in the driest parts of the Trans-Pecos, there is little surface water available, so drillers often tap underground aquifers—some fresh, some brackish and undrinkable—for the water they need. Then there is the problem of disposing of all that wastewater, which typically gets trucked to disposal wells, where it is pumped underground far below freshwater aquifers.

Other types of energy development, although less impactful than oil and gas production, can also have landscape impacts. For example, wind farms and large solar fields also require extensive land conversion. The blades of large wind platforms are known to cause the death of significant numbers of bats and birds. So, a significant intrusion of that form of energy development is likely to impact bats that live in the region.

Unfortunately, with our current state of knowledge, it is virtually impossible to accurately predict the impact of energy development on mammals in the

Trans-Pecos. But one thing is certain—it is time for communities, landowners, energy industries, and conservation organizations to begin talking about how to make this kind of development more compatible with the conservation of the wildlife, natural resources, and heritage of the region.

BORDER PROTECTION (ILLEGAL IMMIGRATION AND A BORDER WALL)

A recent study sponsored by the National Park Service suggests that illegal immigration and the activity to stop it are having harmful effects on the environment along the US-Mexico borderlands, including the Trans-Pecos region (Wang 2019). Large numbers of people trying to cross the border, especially in remote areas, and the large number of Border Patrol agents patrolling the areas are accelerating vegetative changes that were already occurring as the result of grazing that began in the nineteenth century. A decline of vegetation cover can lead to biodiversity loss because of disturbances to wildlife habitats. In some cases, serious damage to vegetation cover can cause degradation, fragmentation, and even disappearance of habitats, thus making it more difficult for natural vegetation to come back. Trash and litter left behind by illegal activities, as well as unmaintained campfires, once accumulated in large amounts can cause serious pollution to water resources.

Another serious issue for wildlife conservation in the borderlands is the proposed extension of a border wall. Expanding the physical barriers along the border between Texas and Mexico will have substantial negative effects on wild species and ecosystems (Fowler et al. 2018). Substantial amounts of habitat would be degraded or destroyed by construction of barriers and the roads alongside them. For example, a border barrier would separate the American black bears (*Ursus americanus*) in BBNP from the population in Mexico, making the park population too small to persist (Hellgren et al. 2005). The same would be true for the reintroduced bighorn sheep (*Ovis canadensis*) populations that are now known to move back and forth to and from Mexico.

Larger mammal species would likely be the most vulnerable, but smaller mammals, reptiles, and amphibians may be blocked even if gaps are provided for animal passage (McCorkle 2011). Species cut off from the Mexican portion of their populations would have smaller effective population sizes, which in turn would further increase the probability of extirpation or extinction (Lasky et al. 2011). So much concern has been raised about the proposed wall that more than 2,500 scientists cosigned a letter describing the significant harm to wildlife posed by infrastructure on the US-Mexico border (Peters et al. 2018).

As it is currently designed and implemented, the barrier is not a continuous structure, but a series of obstructions classified as "fences" or "walls." So far in the Trans-Pecos, there are only 111 km of this fencing near El Paso–Juarez, the only large metropolitan area in the region. Mother Nature has built a "big, beautiful wall" along the Rio Grande in BBNP. That natural wall is Santa Elena Canyon, a deep canyon formed by a set of cliffs rising up to over 450 m along the Rio Grande on both sides of the international border. In addition, several other topographic features along the border provide a natural barrier and thus reduce the need for an anthropogenic structure in many parts of the Big Bend region.

CHANGING CLIMATE

On top of the aforementioned challenges, climate change has become a serious threat entering the twenty-first century, and Texas is one of the states predicted to experience the worst effects of this problem (Schmandt et al. 2011; Schmidly et al. 2022). Climate models predict average temperatures will rise at least 2.2°C by 2050 and, if global emissions continue at their present pace, by 4.4°C by the end of the century. The Trans-Pecos will experience warmer winters and summers and drier winters; annual rainfall will likely decline and droughts will become more frequent and intense; and the severity, frequency, and extent of wildfires will increase. The combination of more fires and drier conditions may expand deserts and change other parts of the landscape. It is not improbable that droughts similar to those that enabled the devastating 2019–21 wildfires in Australia and California could happen someday in the Trans-Pecos.

Changes in temperature and rainfall patterns would cause changes in the distribution of plant communities, which in turn would affect the distribution of wildlife diversity across the landscape. This could impact mammals in a number of ways because they generally utilize a variety of disjunct resources. They need places to hide, eat, drink, and breed, and in many cases these places are distinct and may change seasonally. Thus, there are many opportunities for climate change to disrupt mammalian life histories. Temperature increases and changes in precipitation can directly affect some species, depending on their physiology and tolerance of environmental change. Climate change can also alter a species' food supply or its reproductive timing, indirectly affecting its fitness. Because of their body size, morphology, and ecology, mammals would not be expected to adapt as a group to relatively rapid environmental shifts associated with climate change. Most mammals would not be able to avoid its effects, with both positive and negative impacts possible. Most mammals are also highly mobile and have relatively short life spans (generally < 20 years). Thus, if climates become unsuitable, mammalian responses could be rapid and likely detrimental.

It is important to remember that climate change represents just one of a set of stressors. When combined with other changes challenging fauna and flora, such as land development, habitat fragmentation, invasive species, chemical stressors, and direct exploitation, it becomes easy to visualize dire consequences for biodiversity. A comprehensive assessment of the ecological resources of the Trans-Pecos is needed to develop science-based management practices for wildlife and plant communities to compensate for these possible scenarios (Packard et al. 2011).

Conservation Strategies and Tactics

The following are seven immediate and long-term goals and strategies necessary for conserving the biodiversity of the Trans-Pecos, including actions focused on both private and public lands. Our hope is that these "working points" can serve as the rallying cry for a new century of change for natural history in the region.

PRACTICE RELATIONSHIP-BASED CONSERVATION

In the Trans-Pecos, where ranching and heritage are engrained in the psyche of almost every resident and where suspicion about the motives of outsiders is always on the radar, it is important to involve local stakeholders in land management decisions in order to achieve outcomes that are mutually beneficial for people and resource protection. This approach needs to be adopted by wildlife managers and other conservation professionals in the Trans-Pecos, and it needs to be applied more often and at all levels of organization.

For conservation outside protected areas to succeed, the utilization of wildlife and their habitat should result in some economic benefit to local people. Currently, we see this in the Trans-Pecos in the form of ecotourism and hunting, which have been established successfully throughout the region. This is one area where the region is way ahead of the curve, primarily because of the growing importance of wildlife resources to the economic success of the rural ranching and farming sector.

SUPPORT PRIVATE LANDOWNERS

The future of most wildlife habitat in the Trans-Pecos is in private hands, which makes it more important than ever to provide more assistance to private landowners because of their role in keeping wildlife populations sustainable. And the good news is that with the right information and motivation, private landowners have demonstrated their willingness to develop wildlife habitat and protect species. We cannot emphasize enough the importance of continuing to support landowners in their conservation efforts.

The El Carmen Land and Conservation Company (ECLCC) represents an excellent example of a private land initiative for biodiversity protection. Comprising 10,522 ha between BBNP and BGWMA, ECLCC is focused on restoring lower desert landscapes and protecting ecological corridors in a transboundary area between Mexico and the Trans-Pecos. Many projects are underway to conserve native wildlife through land restoration, water developments, and habitat enrichment. Primary partners in this endeavor include CEMEX USA, Mr. Josiah Allen, and the TPWD.

ENHANCE SUCCESSFUL PARTNERSHIPS

A number of partnerships have been established among conservation organizations, state and federal agencies, private businesses, and private landowners to address natural resource challenges, and they are having success and positive impacts. A few have been targeted to individual species of mammals that have either become extinct in the region or happen to be in steep decline. For example, the Trans-Pecos Pronghorn Restoration Project was established to bolster declining populations of pronghorn through wildlife management practices including translocation, habitat improvements, and strategic predator management. One of the most positive conservation programs in the region has been the restoration of the desert bighorn, a project made possible largely through the funding and collaboration of the Texas Bighorn Society, private landowners, and the TPWD.

A number of environmental organizations and NGOs have been focusing on important conservation issues in the Trans-Pecos. One in particular, The Nature Conservancy (TNC), has made a huge impact as a result of efforts to protect the Davis Mountains watershed from land fragmentation. In 1997 TNC established the 13,385 ha Davis Mountains Preserve. Subsequent land acquisitions combined with conservation easements on adjoining property have protected over 41,500 ha of the Davis Mountains, including the core of the mountain range as well as the iconic Sawtooth Mountain along the Davis Mountains Scenic Loop. The Davis Mountains make up one of only four of the unique sky islands—cooler, wetter landscapes surrounded by arid lowland desert—in the Trans-Pecos, making this one of the most important and successful conservation projects in the region.

MAINTAIN INTERNATIONAL COOPERATION WITH MEXICO

Because Texas shares a lengthy border with Mexico, international cooperation is essential, as wildlife do not respect politically drawn boundaries. A relatively large expanse of public lands exists along the Front Range of the Trans-Pecos, beginning with BBNP, BBRSP, and BGWMA along the border, extending northward through CMSNA, EMWMA, and the Davis Mountains Preserve, and ending with GMNP along the Texas–New Mexico border. This region constitutes a virtual continuous corridor along the Front Range from Mexico to New Mexico. The conservation potential of this vast corridor, given the large number of rare plants and animals in this region, is immense.

An effort is now underway to link the southern end of this Trans-Pecos corridor with a major conservation program in Mexico, known as the El Carmen–Maderas del Carmen Project in Coahuila, Mexico, just across the river from BBNP (McKinney 2006, 2012; McKinney and Villalobos 2014). The purpose of the project is to provide and protect a corridor on both sides of the Rio Grande that will allow wildlife to move freely within an intact ecosystem. Progress is being made, as bighorn sheep (*Ovis canadensis*), American black bear (*Ursus americanus*), mule deer (*Odocoileus hemionus*), and American beaver (*Castor canadensis*) are thriving in the corridor, and a number of other rare species, such as the white-nosed coati (*Nasua narica*), are beginning to make a comeback.

A significant problem with the development of future cooperative wildlife projects between the United States and Mexico would be the implementation of a border wall separating the two countries along the Rio Grande. A fully completed steel wall would interfere with the migration of many species of wildlife, effectively ending the current cooperative efforts that are underway (Masters 2019).

CONTINUE THE ACQUISITION OF PROTECTED AREAS

Protected areas are the cornerstone of biodiversity conservation. Where networks of protected areas are large, connected, well managed, and distributed across diverse habitats, they sustain large populations of threatened and functionally important species and ecosystems more effectively than other land uses. Protected areas also play an important role in climate-change mitigation, but they are often too small, too isolated, or both to buffer species from human influence outside their borders. Therefore, wildlife movement corridors that connect protected areas are also vital to long-term success. Corridors can be established using a variety of strategies, such as land easements and land trusts, or through the efforts of land trusts and NGOs to purchase development rights.

Some world leaders in conservation have recently called for protecting 30 percent of the planet's land and water by 2030 in order to stabilize the loss of biodiversity and help fight climate change (Smith 2020). Applying this formula to the Trans-Pecos would require an almost fivefold increase in public lands in the region, or about another 1.9 million ha. While this number may seem impossible to achieve in today's world, given the reluctance of the state to purchase private land for public purposes, continued acquisition will be important to long-term conservation success.

STRENGTHEN SCIENTIFIC RESEARCH AND SCIENCE-BASED DECISION MAKING

If we are to conserve wildlife diversity in the Trans-Pecos, it will be necessary to have an adequate information base on which to generate management decisions. Science has essential, invaluable roles in resource management: to provide basic knowledge; to yield objective and unambiguous information on what is possible; to help develop sound strategies to meet goals; and to show the costs and consequences of alternative strategies. Unfortunately, the only groups of wildlife species for which sufficient information or resources exist to effectively manage them at the present time are the commercially important species, primarily game and large charismatic species. For most nongame species, the information base and resources have been lacking to acquire the additional information needed for effective long-term conservation. We must learn more about all the species in the region and their interactions within ecosystems. This will require accelerating efforts to discover, to describe, and to conduct natural history studies to provide the kind of baseline data necessary to assess wildlife population trends and habitat conditions for the lesser-known components of our fauna.

INCREASE FUNDING FOR CONSERVATION PROJECTS AND RESEARCH

None of the above goals can be achieved without sustainable funding sources. There is "no free lunch" when it comes to providing for conservation programs. The cost to acquire land for conservation purposes is high, and many at-risk species have not been adequately studied to provide the information and knowledge to improve decision-making about their status. In this century, more funds are going to be needed to purchase more land for protected areas, parks, and wildlife management areas, as well as to fund research, education, and other initiatives. The highest levels of local and state government, the private sector, and NGOs and other conservation organizations must be brought together to make this happen on the scale that is necessary to protect biological resources.

Conclusion

Trans-Pecos habitats and landscapes have undergone thousands of years of changes generated by both Native American occupants and generations of western settlers. Although some fish and wildlife species are thriving thanks to careful, science-based management, many face increasing challenges and have begun to decline. Invasive species, emerging diseases, habitat loss and fragmentation, extreme weather, and other factors threaten many fish and wildlife populations at a scale inconceivable just a century ago. It behooves all of us to work together to ensure that the encompassing wilderness area that we call the Trans-Pecos, with its spectacular vistas and natural and cultural history, is preserved and protected for the future. It is toward this end that we have prepared this book.

APPENDIX 1

THE MEASUREMENT SYSTEM

For those who are less familiar with the metric system, the following table of metric-to-imperial conversion factors may be helpful.

TO CONVERT	MULTIPLY BY
millimeters to inches	0.039
centimeters to inches	0.394
meters to feet	3.281
meters to yards	1.094
kilometers to miles	0.621
square kilometers to square miles	0.386
hectares to acres	2.471
hectares to square miles	0.004
liters to quarts	1.057
liters to gallons	0.264
grams to ounces	0.035
kilograms to pounds	2.205
metric tons to tons	1.102
kilograms to tons	0.0011
degrees C to degrees F	1.8 + 32

APPENDIX 2

PLANTS MENTIONED IN THE TEXT

Plant names mentioned in text are listed alphabetically by vernacular name (in some cases an alternative or regional vernacular name is included in parentheses), followed by the scientific name. When specific designations could not be assigned or when multiple species of a plant genus may occur in the area, the generic name followed by "sp." is listed in lieu of a scientific name. Vernacular and scientific names were derived from the following sources: Correll and Johnston (1979); Plants of Texas Rangelands, Texas A&M AgriLIFE Extension (https://rangeplants.tamu.edu/); PLANTS Database, USDA Natural Resources Conservation Service (https://plants.usda.gov/java/); and the Lady Bird Johnson Wildflower Center (https://www.wildflower.org).

VERNACULAR NAME	SCIENTIFIC NAME
acacia	*Acacia* sp.
agave	*Agave* sp.
alfalfa	*Medicago sativa*
algerita (agarito)	*Mahonia trifoliolata*
alkali sacaton	*Sporobolus airoides*
alligator juniper	*Juniperus deppeana*
allthorn (crown of thorns, crucifixion thorn)	*Koeberlinia spinosa*
Apache plume	*Fallugia paradoxa*
ash	*Fraxinus* sp.
Ashe's juniper (Ashe juniper)	*Juniperus ashei*
bear grass	*Nolina* sp.
bindweed	*Convolvulus* sp.
bitterweed	*Hymenoxys* sp.
blackbrush acacia	*Vachellia rigidula*

VERNACULAR NAME	SCIENTIFIC NAME
black grama	*Bouteloua eriopoda*
bladderpod	*Lesquerella* sp.
blue grama	*Bouteloua gracilis*
bluegrass (Texas bluegrass)	*Poa arachnifera*
broom snakeweed	*Gutierrezia sarothrae*
broomweed (prairie)	*Amphiachyris dracunculoides*
buckthorn	*Rhamnus* sp.
buckwheat	*Eriogonum* sp.
buffalo gourd	*Cucurbita foetidissima*
buffalograss	*Bouteloua dactyloides*
bulrush	*Scirpus* sp.
burrograss	*Scleropogon brevifolius*
cane (giant cane)	*Arundo donax*
cane bluestem	*Bothriochloa barbinodis*
catclaw mimosa	*Mimosa aculeaticarpa*
cattail	*Typha* sp.
cedar	*Juniperus* sp.
century plant	*Agave* sp.
chino grama	*Bouteloua ramosa*
chinquapin oak	*Quercus muehlenbergii*
Chisos Mountains century plant	*Agave glomerulifora*
Chisos red oak	*Quercus gravesii*
cholla (tree cholla)	*Cylindropuntia imbricata*
cocklebur (rough cocklebur)	*Xanthium strumarium*
common reed	*Phragmites australis*
condalia (javelina bush)	*Condalia ericoides*
coneflower (upright prairie coneflower)	*Ratibida columnifera*
cottonwood	*Populus* sp.
creosote bush	*Larrea tridentata*
croton	*Croton* sp.
crown of thorns (crucifixion thorn)	*Koeberlinia spinosa*
curlyleaf muhly	*Muhlenbergia setifolia*
currant	*Ribes* sp.

VERNACULAR NAME	SCIENTIFIC NAME
currant, wild (algerita, agarita, agarito)	*Mahonia trifoliolata*
cutleaf daisy (Engelmann's daisy)	*Engelmannia peristenia*
dalea (prairie clover)	*Dalea* sp.
datil yucca (banana yucca)	*Yucca baccata*
deerbush	*Ceanothus* sp.
deervetch (deer pea vetch)	*Vicia* sp.
desert vine (Berlandier's wolfberry)	*Lycium berlandieri*
desert willow	*Chilopsis linearis*
dropseed	*Sporobolus* sp.
Douglas fir	*Pseudotsuga menziesii*
elm	*Ulmus* sp.
Emory oak	*Quercus emoryi*
ephedra (longleaf jointfir)	*Ephedra trifurca*
eriogonum (buckwheat)	*Eriogonum* sp.
evolvulus (dwarf morning-glory)	*Evolvulus* sp.
false willow (baccharis)	*Baccharis* sp.
feather bluestem (cane bluestem)	*Bothriochloa barbinodis*
fir (Douglas fir)	*Pseudotsuga menziesii*
fleabane	*Erigeron* sp.
fluffgrass (Texas fluffgrass)	*Tridens texanus*
giant cane	*Arundo donax*
giant dagger	*Yucca faxoniana*
giant reed	*Arundo donax*
globemallow (copper globemallow)	*Sphaeralcea angustifolia*
goathead (puncturevine)	*Tribulus terrestris*
grama grass	*Bouteloua* sp.
gray oak	*Quercus grisea*
groundsel (ragwort)	*Senecio* sp.
guayacan	*Guaiacum angustifolium*
hackberry	*Celtis* sp.
hairy grama	*Bouteloua hirsuta*
honey mesquite	*Prosopis glandulosa*
hoptree	*Ptelea* sp.

VERNACULAR NAME	SCIENTIFIC NAME
huisache (sweet acacia)	*Vachellia farnesiana*
ironwood (hophornbeam)	*Ostrya virginiana*
javelina bush	*Condalia ericoides*
juniper	*Juniperus* sp.
leadtree (littleleaf leadtree)	*Leucaena retusa*
leatherstem	*Jatropha dioica*
lechuguilla	*Agave lechuguilla*
limber pine	*Pinus fexilis*
live oak	*Quercus fusiformis*
lotebush	*Ziziphus obtusifolia*
lupine (bluebonnet)	*Lupinus* sp.
mescal (mescal bean)	*Sophora secundifora*
mesquite	*Prosopis* sp.
Mexican buckeye	*Ungnadia speciosa*
Mexican walnut (little walnut)	*Juglans microcarpa*
mimosa	*Mimosa* sp.
morning-glory	*Evolvulus* sp.
mountain cottonwood	*Populus* sp.
mountain grass (woolyspike balsamscale)	*Elionurus barbiculmis*
mountain laurel (Texas mountain laurel)	*Sophora secundifora*
mountain mahogany	*Cercocarpus montanus*
mountain maple (bigtooth maple)	*Acer grandidentatum*
mountain snowberry	*Symphoricarpos oreophilus*
muhly	*Muhlenbergia* sp.
needlegrass	*Achnatherum* sp.
nightshade (Texas nightshade)	*Solanum triquetrum*
oak	*Quercus* sp.
ocotillo	*Fouquieria splendens*
one-seeded juniper	*Juniperus monosperma*
paperflower (woolly paperflower)	*Psilostrophe tagetina*
peppergrass (pepperweed)	*Lepidium* sp.
persimmon (Texas persimmon)	*Diospyros texana*
plains bristlegrass	*Setaria vulpiseta*

VERNACULAR NAME	SCIENTIFIC NAME
pine	*Pinus* sp.
piñon pine (Mexican piñon pine)	*Pinus cembroides*
ponderosa pine	*Pinus ponderosa*
prickly pear	*Opuntia* sp.
quaking aspen	*Populus tremuloides*
red-berry juniper (Pinchot's juniper)	*Juniperus pinchotii*
red grama	*Bouteloua trifda*
Russian olive	*Elaeagnus angustifolia*
Russian thistle (prickly Russian thistle)	*Salsola* sp.
sacahuista (Texas sacahuista, sacahuiste)	*Nolina texana*
sagebrush	*Artemisia* sp.
saltbush	*Atriplex* sp.
salt cedar (tamarisk)	*Tamarix* sp.
saltgrass	*Distichlis spicata*
sand sagebrush	*Artemisia flifolia*
screwbean mesquite	*Prosopis pubescens*
scrub oak (Sonoran scrub oak)	*Quercus turbinella*
scrub sumac (littleleaf sumac)	*Rhus microphylla*
seepwillow	*Baccharis salicifolia*
shin oak (Havard shin oak)	*Quercus havardii*
sideoats grama	*Bouteloua curtipendula*
silk tree	*Ceiba* sp.
skunkbush sumac	*Rhus trilobata*
slim tridens	*Tridens muticus*
snakeweed	*Gutierrezia* sp.
soaptree yucca	*Yucca elata*
sotol	*Dasylirion* sp.
Spanish dagger (Torrey's yucca)	*Yucca torreyi*
spurge	*Euphorbia* sp.
sumac	*Rhus* sp.
sunflower	*Helianthus* sp.
sycamore	*Platanus occidentalis*
tarbush (American tarwort)	*Flourensia cernua*

VERNACULAR NAME	SCIENTIFIC NAME
tasajillo (Christmas cholla, pencil cholla)	*Cylindropuntia leptocaulis*
Texas madrone	*Arbutus xalapensis*
Texas pistache	*Pistacia* sp.
Thompson's yucca	*Yucca thompsoniana*
threeawn grass	*Aristida* sp.
tobosagrass	*Pleuraphis mutica*
tridens	*Tridens* sp.
vetch	*Vicia* sp.
walnut	*Juglans* sp.
wheat	*Triticum aestivum*
Wheeler's sotol (Trans-Pecos sotol)	*Dasylirion wheeleri*
whitebrush	*Aloysia gratissima*
wild buckwheat (Abert's buckwheat)	*Eriogonum abertianum*
wild tobacco	*Nicotiana* sp.
willow	*Salix* sp.
wolfberry	*Lycium* sp.
woolly senecio (threadleaf groundsel)	*Senecio faccidus*
yucca	*Yucca* sp.

GLOSSARY

aestivate. To pass the summer or dry season in a dormant condition. Same as estivate.

alluvial. Pertaining to soil, sand, and gravel deposited by a running watercourse where it issues from a canyon or gorge onto an open plain.

Alpine High. The petroleum play in the Delaware Basin that extends near the Davis Mountains and the city of Alpine.

annual survival rate. The estimated proportion of animals alive in year t that is still alive in year $t+1$. Usually expressed in units of deaths per 1,000 individuals per year; thus, a mortality rate of 9.5 in a population of 1,000 would mean 9.5 deaths per year, or 0.95 percent of the total population.

aquatic. Inhabiting or frequenting water.

aquifer. An underground layer of water-bearing permeable rock, rock fractures, or unconsolidated materials (gravel, sand, or silt).

arboreal. Inhabiting or frequenting trees.

bajada. An alluvial fan formed by the deposition of down-washed sediment at the base of a mountain.

benzene. A colorless volatile liquid hydrocarbon present in coal tar and petroleum and used in chemical synthesis. Its use as a solvent has been reduced because of its carcinogenic properties.

biodiversity. The diversity and frequency of organisms in a given area.

biogeographic barrier. A barrier that prevents the migration of species. Such barriers may be climatic, involving temperature and the availability of water, or physical, such as mountain ranges or rivers.

biogeographic transition zone. The boundary between biogeographic regions, representing areas of biotic overlap, which are promoted by historical and ecological changes that allow the mixing of taxa belonging to different biotic components.

biogeography. The study of the distribution of different species around the planet and the factors that influence their distribution.

bony palate. The bony roof of the mouth formed by parts of the premaxillary, maxillary, and palatine bones (fig. 96).

boreal forest. A forest that grows in regions with cold temperatures. Made up mostly of cold-tolerant species such as spruce and fir.

bottomland. Lowland along streams and rivers, usually on alluvial floodplains that are periodically flooded.

browse. Leaves, twigs, or other high-growing vegetation eaten by animals.

butte. A small, isolated hill with relatively steep sides and a flat top. Mesas are larger than buttes, and plateaus are tablelands occupying huge expanses of land.

caecal pellets. Green pellets of partially digested plant material (bolus) produced in the caecum of the digestive tract. These pellets are excreted and then reingested for complete digestion. When digestion is complete, pellets are excreted as brown fecal pellets. Also see **coprophagy.**

Chihuahuan Desert. A desert and ecoregion covering part of northern Mexico and the southwestern United States, including Trans-Pecos Texas.

Chihuahuan desert scrub. A desert habitat, receiving less than 25 cm of rainfall annually, that is common at elevations between 1,050 and 1,550 m and has recently expanded into former desert grasslands at higher elevations. Stands typically occur in flat to gently sloping desert basins extending onto alluvial plains and bajadas. Common plants include creosote bush, tarbush, ocotillo, and various species of acacia and cacti.

ciénaga. A spring that is usually a wet, marshy area at the foot of a mountain, in a canyon, or on the edge of a grassland where groundwater bubbles to the surface.

condylobasal length. A measurement of the skull taken from the front of the base of the incisor teeth to the back of the rounded condyles that border the large opening at the back of the skull (point A to point A′ in fig. 96).

coniferous forest. Vegetation composed primarily of cone-bearing, needle-leaved or scale-leaved evergreen trees (pines, spruces, firs) in areas with longer winters and moderate to high annual precipitation.

conservation easement. A legally binding restriction placed on a piece of property to protect its associated resources, most commonly through an agreement between a landowner and land trust or unit of government designed to limit certain types of uses or prevent development from taking place on the land in perpetuity while the land remains in private hands.

conservationist. A person who advocates or acts for the protection and preservation of the environment and wildlife.

coprophagy. The act of an animal eating its own caecal pellets (partially digested plant material that is reingested for complete digestion). Caecal pellets should not be confused with fecal pellets (fully digested plant material).

coronoid process. The dorsalmost projection on the posterior portion of the mandible (fig. 96).

crepuscular. Active during the twilight periods of dusk and dawn.

deciduous forest. Vegetation composed primarily of broad-leaved trees that shed all their leaves during one season.

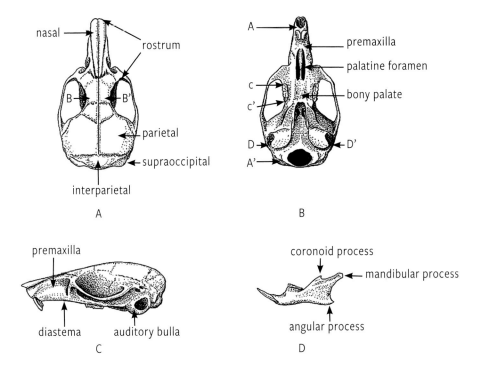

FIG. 96. Views of the skull and lower jaw of Lacey's white-ankled deermouse, *Peromyscus laceianus*, showing the bones and indicating the measurements used in identifying species

Delaware Basin. A subbasin of the Permian Basin that extends into the Trans-Pecos, where it abuts the Guadalupe Mountains and then extends south to the area around Alpine.

den. A cave, hollow log, burrow, or other cavity used by a mammal for shelter.

desertification. A particular type of land degradation in which an area becomes increasingly arid and assumes the environmental characteristics of a desert.

desert scrub. A type of habitat characterized by drought-resistant plants that grow over the ground with many bare spots between them.

desert seas. The desert and grassland habitats that surround sky island (montane) habitats and thus isolate them.

diastema. A space between adjacent teeth; for example, the space between incisors and premolars in rodents (fig. 96).

disjunct population. A population of a species that is separated geographically from the main range of the species.

dispersal. The permanent emigration of individuals from a population.

diurnal. Active by day, as opposed to by night (nocturnal).

ecoregion. An area defined by its environmental conditions, especially climate, landforms, soils, topography, geology, rainfall, and plant and animal communities.

ecosystem. A biological community of interacting organisms and their physical environment.

edaphic. Pertaining to or influenced by soil, rather than climate.

Edwards Plateau. A region of west-central Texas known as the Hill Country, bordered on the east and south by the Balcones Escarpment and on the west by the Pecos River.

Endangered Species Act (ESA). US law, passed in 1973 and subsequently amended, that regulates the capture, possession, and sale of threatened and endangered species of plants and animals.

endemic. Pertaining to a mammal that occurs only in a certain region.

endocrine disruptors. Natural and human-made chemicals that may mimic or interfere with the body's hormones, known as the endocrine system.

epizootic. A severe outbreak of a disease in which large numbers of animals die in a short time. Equivalent to an epidemic in humans.

ewe. The female of a sheep species.

exotic. A species living outside its native distributional range after being introduced there by deliberate or unintentional human activity.

extant. Taxa such as species, genera, or families with members still alive, as opposed to being extinct.

extirpation. The absence of a species from part of its former range.

feral. Pertaining to formerly domesticated animals now living in a wild state.

fluvial. Soil systems found primarily in association with rivers and streams.

forbs. Herbaceous plants other than grasses or sedges.

fossorial. Pertaining to life under the ground surface.

fragmentation. Division of a large piece of habitat into a number of smaller, isolated patches.

genetic species concept. Groups of genetically compatible interbreeding natural populations that are genetically isolated from other such groups. Genetic isolation implies that each species is genetically distinct from other species and that the integrity of the gene pools is maintained.

gestation. Length of time from fertilization until birth of a fetus or fetuses; the period of pregnancy.

global warming (or climate change). An increase in the average temperature of the earth's atmosphere, especially a sustained increase sufficient to cause climate change.

granivorous. Feeding on grains or seeds.

grazers. Mammals that feed on grass and other herbage by cropping and nibbling.

harem breeding system. A breeding system describing an animal group with one or two males, a number of females, and their offspring. The dominant male mates with the females as they become sexually active and drives off competitors until he is replaced by another male.

herb. Any seed-bearing plant that does not have a woody stem and dies down to the ground after flowering.

herbivore. An animal that consumes plant material as the primary component of its diet.

hibernation. A period, especially during winter, when an animal's body temperature approximates that of its environment, and the rate of respiration and heartbeat usually become much lower than in an active state.

Holocene. The name given to the last 11,700 years of the earth's history; the time since the end of the last major glacial epoch, or "ice age."

host. An organism harboring a parasite.

hybridization. Production of offspring from interbreeding between individuals of different species or subspecies.

insectivorous. Eating insects; preying or feeding on insects.

interorbital breadth. A measurement of the skull taken at the least diameter of the frontal bones between the orbital openings (point B to point B′ in fig. 96).

interparietal. An unpaired bone on the dorsal part of the braincase between the parietals and just anterior to the supraoccipital (fig. 96).

isolated population. A population that is separated by some sort of barrier from the main body of the species.

IUCN. Acronym for the International Union for Conservation of Nature, founded in 1964, which is the world's main authority on the conservation status of species. The *IUCN Red List of Threatened Species*, also known as the *Red List*, is the world's most comprehensive inventory of the global conservation status of biological species.

karyotype. An arrangement of chromosomes of a cell according to shape, centromere position, and number; used in the identification of species and subspecies.

keystone species. A species that has an extremely high impact on a particular ecosystem relative to its population.

length of molar tooth row. A measurement of the alveolar distance from the anterior border of the anterior molar to the posterior border of the posterior molar (point C to point C′ in fig. 96).

mad cow disease (bovine spongiform encephalopathy). A brain disorder in adult cattle caused by a prion (a type of protein) that may be spread to humans through diseased meat, resulting in a variant form of Creutzfeldt-Jakob disease, which may lead to prominent neurologic, psychiatric, and sensory abnormalities.

mast. The edible vegetative or reproductive parts (nuts, seeds, buds, or fruits) produced by woody species of plants (trees and shrubs) that wildlife consume as a food source.

mastoidal breadth. A measurement of the greatest width across the skull between the two mastoid processes (point D to point D′ in fig. 96).

mesic. Characterized by a moderate amount of moisture.

metabolic rate. The rate at which an organism metabolizes energy.

metabolic water. The water produced as a by-product of metabolism.

metatarsal gland. Gland on the outside of the lower hind extremities in some mammals, especially some artiodactyls such as deer.

methane. A colorless, odorless, flammable gas that is the main constituent of natural gas. It is among the most potent of the greenhouse gases.

Mexican tablelands. Another name for the Mexican Plateau, which is a large, arid to semiarid region occupying much of northern and central Mexico.

microhabitat. A constrained site where the environmental conditions differ enough from those in the surrounding habitat to provide suitable conditions for certain organisms.

microsatellites. Repetitive segments of DNA scattered throughout the genome in noncoding regions between or within genes. These regions are inherently genetically unstable and susceptible to mutations. Thus, they are often used as markers to detect genetic differences among populations.

migration. A movement of animals involving a journey to a definite area and a return journey to the area from which the movement arose, often in response to seasonal patterns of resource availability.

mitochondrial DNA. The DNA contained in the mitochondria of a cell; because offspring typically inherit only their mother's mitochondria, mitochondrial DNA is useful in tracing maternal lineages.

monotypic. Consisting of only one type; for example, a monotypic species has no subspecies.

montane. Pertaining to a biogeographic zone of the mountains.

nasals. The paired bones on the rostrum that roof the nasal passages (fig. 96).

nectivorous (nectarivorous). Feeding on nectar.

neonicotinoids (sometimes shortened to neonics). A class of neuroactive insecticides chemically similar to nicotine.

nocturnal. Active by night, as opposed to by day (diurnal).

nongovernmental organization (NGO). A nonprofit, citizen-based group that functions independently of government; for example, The Nature Conservancy.

nonpoint source pollution. Pollution resulting from many diffuse sources, in direct contrast to point source pollution, which results from a single source.

nuchal patch. A patch of fur on the back of the neck (nape) of rabbits and hares that differs in color from the rest of the body.

ochraceous. Reddish yellow.

omnivorous. Pertaining to those animals that eat quantities of both animal and plant foods.

palatine foramen. Either of a pair of openings through the anterior end of the hard palate between the premaxilla and maxilla (fig. 96).

palmation. A characteristic of some antlers in which the distal portions are broad, flat, and lobed.

parietals. Paired bones on the roof of the skull posterior to the frontal bones (fig. 96).

parturition. The act of giving birth to young.

Permian Basin. A shale basin about 400 km wide and 500 km long, spanning parts of West Texas and southeastern New Mexico. It is one of the oldest and most widely recognized oil and gas producing regions in the United States.

phylogeny. The evolutionary history of an organism or groups of related organisms.

Pleistocene. Relating to or denoting the first epoch of the Quaternary period, between the Pliocene and Holocene epochs.

pluvial. Characterized by a prolonged period of wet climate.

predator. An animal that lives by killing and consuming other animals.

premaxillae. The paired bones at the anterior end of the rostrum that hold the incisor teeth (fig. 96).

prey. An animal that is pursued by a predator.

protected areas. Regions or zones of land that are reserved for conserving nature and biodiversity.

ram. The male of a sheep species.

refugia. Areas with relatively unaltered climate that are inhabited by plants and animals during a period of continental climate change (such as glaciation) and that remain a center for relict forms, from which new dispersion and speciation may occur after climatic readjustment.

Rio Grande (or South Texas) Plains. The part of Texas that stretches from the Hill Country into the subtropical regions of the Lower Rio Grande Valley. Much of the area is dry and covered with grasses and thorny brush such as mesquite and prickly pear cactus.

riparian. Associated with the banks of a natural watercourse, such as a river or stream.

rostrum. The facial region of the skull anterior to the orbits (fig. 96).

rut. The mating season of ruminant mammals, such as deer, sheep, and pronghorn.

savanna. A grassland containing scattered trees and drought-resistant undergrowth.

scat. The excrement of a wild animal.

science-based decision-making (and management practices). A process for making decisions about a program, practice, or policy that is grounded in the best available research evidence and informed by experimental evidence from the field and relevant contextual evidence.

sedge. A rushlike or grasslike plant that grows in wet ground.

seep. A moist or wet place where water, usually groundwater, reaches the surface from an underground aquifer.

semiaquatic. Pertaining to organisms that are partially, but not fully, adapted to life in water; for example, common muskrats and American beavers are semi-aquatic mammals in the Trans-Pecos.

shrub. A woody plant that is smaller than a tree and has several main stems arising at or near the ground.

sky islands. Isolated mountains tall enough to develop communities greatly dissimilar from those in the surrounding area and well separated from mountains elsewhere. Commonly high-elevation forests surrounded by lowland deserts.

species. Groups of actually (or potentially) interbreeding natural populations that are reproductively isolated from other such groups. Reproductive isolation implies that interbreeding between individuals of two species is prohibited by intrinsic factors.

steppe. Vast tract of land that is generally level and without forests.

Stockton Plateau. The portion of the Hill Country that extends west of the Pecos River; the physiographic boundary separating the arid Trans-Pecos from the hilly Edwards Plateau.

strychnine. A highly toxic, colorless, bitter, crystalline alkaloid used as a pesticide, particularly for killing small vertebrates such as birds and rodents.

subspecies. A geographically defined aggregate of local populations that differs taxonomically from other such subdivisions of the species.

substrate. The surface of material on or from which an organism lives, grows, or obtains its nourishment.

succulent. A plant having thick, fleshy parts adapted to retain water.

supraoccipital. The dorsal element of the occipital bone located above the foramen magnum and the occipital condyles (fig. 96).

sylvatic plague. A bacterial disease transmitted by fleas that afflicts many mammalian species, especially rodents (e.g., prairie dogs), and can be transmitted to humans.

talus. The pile of rocks that accumulates at the base of a cliff, chute, or slope. A talus slope is formed by an accumulation of rock debris.

Tamaulipan thornscrub habitat. A region of southern Texas and northeastern Mexico with unique plant and animal communities containing tree- and brush-covered dunes, wind tidal flats, and dense native brushland.

taxonomy. The science of classifying organisms.

tending-bond mating system. A mating system characterized by polygyny, wherein a single male forms a tending bond with a single female, courting, guarding, and ultimately mating with her during estrus, before moving on to another female.

terrestrial. Referring to animals that live on land as opposed to living in water.

tinajas. Surface pockets (depressions) formed in bedrock that occur below waterfalls; they are carved out by spring flow or seepage, or created by the scouring of sand and gravel in intermittent streams (arroyos).

torpid. Without most of the power of exertion; dormant.

Trans-Pecos Pipeline. A recently built pipeline for transporting natural gas to Mexico that extends 230 km from the Waha Hub storage facility near Fort Stockton to the Rio Grande at Presidio-Ojinaga, passing through Alpine and the greater Big Bend region along the way.

transpiration. The evaporative loss of water through the stems, leaves, and flowers of plants.

trespass livestock. Livestock that graze on, or are driven across, land without permission of the landowner.

ungulate. A mammal having hooves, not claws.

urban sprawl. The unrestricted growth in many urban and suburban areas of housing, commercial development, and roads over large expanses of land, with little concern for urban planning and green areas.

vegetative canopy. The total mass of plant life that occupies a given area, such as a canopy of trees covering a stream bed.

watershed. An area that contains a common set of streams and rivers that drains into a single larger body of water, such as a large river, a lake, or an ocean.

wetland. Area inundated or saturated by surface water or groundwater at a frequency and duration sufficient to support a prevalence of vegetation typically adapted for life in saturated soil conditions.

wildlife (big game) ranching. The intentional raising of wildlife, especially ungulates, for any purpose, including hunting. Domestic livestock may be raised simultaneously with game on a big game ranch.

wild mustang. A free-roaming feral horse of the western United States, descended from once-domesticated horses originally brought to the Americas by the Spanish.

wild species. All organisms that grow or live wild in an area without being introduced by humans.

woody plants. Trees and shrubs whose shoots are durable (with wood as their structural tissue) and survive over a period of years. They can be further classified into deciduous and evergreen plants.

xeric. Characterized by a dry, desertlike climate.

xerophytic. Pertaining to a plant structurally adapted for life and growth with a limited water supply (in xeric habitats).

zoonosis. A disease that can be transmitted to humans from animals.

LITERATURE CITED

Ackerson, B. K., and L. A. Harveson. 2006. Characteristics of a ringtail (*Bassariscus astutus*) population in Trans Pecos, Texas. *Texas Journal of Science* 58:169–84.

Adkins, R. N., and L. A. Harveson. 2006. Summer diets of feral hogs in the Davis Mountains, Texas. *Southwestern Naturalist* 51:578–80.

———. 2007. Demographic and spatial characteristics of feral hogs in the Chihuahuan Desert, Texas. *Human-Wildlife Conflicts* 1:152–60.

Allen, J. A. 1891. On a collection of mammals from southern Texas and northeastern Mexico. *Bulletin of the American Museum of Natural History* 3:219–28.

———. 1893. On a collection of mammals from the San Pedro Martir region of lower California, with notes on other species, particularly of the genus *Sitomys*. *Bulletin of the American Museum of Natural History* 5:181–202.

Ammerman, L. K. 2005. Noteworthy records of the eastern pipistrelle, *Perimyotis subflavus*, and silver-haired bat, *Lasionycteris noctivagans*, (Chiroptera: Vespertilionidae) from the Chisos Mountains, Texas. *Texas Journal of Science* 57:202–7.

Ammerman, L. K., C. L. Hice, and D. J. Schmidly. 2012. *Bats of Texas*. College Station: Texas A&M University Press.

Ammerman, L. K., R. M. Rodríguez, J. L. Higginbotham, and A. K. Matthews. 2002. Recent records of bats from the lower canyons of the Rio Grande River of west Texas. *Texas Journal of Science* 54:369–74.

Andersen, B. R., K. Geluso, H. W. Otto, and L. Bishop-Boros. 2017. Westward expansion of the evening bat (*Nycticeius humeralis*) in the United States, with notes on the first record from New Mexico. *Western North American Naturalist* 77:223–29.

Anderson, A. W. 1949. Early summer foods and movements of the mule deer (*Odocoileus hemionus*) in the Sierra Vieja range of southwest Texas. *Texas Journal of Science* 1:45–49.

Axtell, R. W. 1961. An additional record for the bat *Tadarida molossa* from Trans-Pecos Texas. *Southwestern Naturalist* 6:52–53.

Baccus, J. 1971. The influence of a return of native grasslands upon the ecology and distribution of small rodents in Big Bend National Park. PhD diss., North Texas State University.

———. 1978. Notes on the distribution of some mammals from Coahuila, Mexico. *Southwestern Naturalist* 23:706–8.

Bailey, V. 1905. Biological survey of Texas. *North American Fauna* 25:1–222.

———. 1931. Mammals of New Mexico. *North American Fauna* 53:1–412.

Baird, A. B., J. K. Braun, M. A. Mares, J. C. Morales, J. C. Patton, C. Q. Tran, and J. W. Bickham. 2015. Molecular systematic revision of tree bats (Lasiurini): Doubling the native mammals of the Hawaiian Islands. *Journal of Mammalogy* 96:1255–74.

Baker, R. H. 1951. Two new moles (genus *Scalopus*) from Mexico and Texas. *University of Kansas Publications, Museum of Natural History* 5:17–24.

———. 1956. Mammals of Coahuila, Mexico. *University of Kansas Publications, Museum of Natural History* 9:125–335.

———. 1988. Future prospects for the depletion of mammalian populations in the Chihuahuan Desert region. In *Third Symposium on Resources of the Chihuahuan Desert Region, United States and Mexico,* edited by A. M. Powell, R. R. Hollander, J. C. Barlow, W. B McGillivray, and D. J. Schmidly, 71–79. Alpine, TX: Chihuahuan Desert Research Institute.

Baker, R. J., J. C. Patton, H. H. Genoways, and J. W. Bickham. 1988. Genic studies of *Lasiurus* (Chiroptera: Vespertilionidae). *Occasional Papers, Museum of Texas Tech University* 117:1–15.

Balin, L. 2009. Mexican long-tongued bat (*Choeronycteris mexicana*) in El Paso, Texas. *Southwestern Naturalist* 54:225–26.

Ballinger, S. W., L. H. Blankenship, J. W. Bickham, and S. A. Carr. 1992. Allozyme and mitochondrial analysis of a hybrid zone between white-tailed deer and mule deer (*Odocoileus*) in west Texas. *Biochemical Genetics* 30:1–11.

Barela, I. A., and J. K. Frey. 2016. Habitat and forage selection by the American beaver (*Castor canadensis*) on a regulated river in the Chihuahuan Desert. *Southwestern Naturalist* 61:286–93.

Baumgardner, G. D., and D. J. Schmidly. 1981. Systematics of the southern races of two kangaroo rats (*Dipodomys compactus* and *D. ordii*). *Occasional Papers, Museum of Texas Tech University* 73:1–27.

Beauchamp-Martin, S. L., F. B. Stangl Jr., D. J. Schmidly, R. D. Stevens, and R. D. Bradley. 2019. Systematic review of Botta's pocket gopher (*Thomomys bottae*) from Texas and southeastern New Mexico, with description of a new taxon. In *From Field to Laboratory: A Memorial Volume in Honor of Robert J. Baker,* edited by R. D. Bradley, H. H. Genoways, D. J. Schmidly, and L. C. Bradley, 515–39. Special Publications, Museum of Texas Tech University 71.

Bissonette, J. A. 1978. The influence of extremes of temperature on activity patterns of peccaries. *Southwestern Naturalist* 23:339–46.

———. 1982. *Ecology and Social Behavior of the Collared Peccary in Big Bend National Park.* Scientific Monograph Series Number 16. Washington, DC: US Department of the Interior, National Park Service.

Black, H. L. 1972. Differential exploitation of moths by the bats *Eptesicus fuscus* and *Lasiurus cinereus. Journal of Mammalogy* 53:598–601.

Blair, W. F. 1940. A contribution to the ecology and faunal relationships of the mammals of the Davis Mountain region, southwestern Texas. *Miscellaneous Publications, University of Michigan Museum of Zoology* 46:1–39.

Blair, W. F., and C. E. Miller Jr. 1949. The mammals of the Sierra Vieja region, southwestern Texas, with remarks on the biogeographic position of the region. *Texas Journal of Science* 1:67–92.

Boeer, W. J., and D. J. Schmidly. 1977. Terrestrial mammals of the riparian corridor in Big Bend National Park. In *Importance, Preservation and Management of Riparian Habitat: A Symposium,* coordinated by R. Roy Johnson and Dale A. Jones, 212–17. USDA Forest Service General Technical Report RM-43. Fort Collins, CO.

Bogan, M. A. 1974. Identification of *Myotis californicus* and *Myotis leibii* in southwestern North America. *Proceedings of the Biological Society of Washington* 87:49–56.

Bogan, M. A., and P. Melhop. 1983. Systematic relationships of gray wolves (*Canis lupus*) in southwestern North America. *Occasional Papers, Museum of Southwestern Biology, University of New Mexico* 1:1–21.

Bogdanowicz, W., S. Kasper, and R. D. Owen. 1998. Phylogeny of plecotine bats: Reevaluation of morphological and chromosomal data. *Journal of Mammalogy* 79:78–90.

Borell, A., and M. D. Bryant. 1942. Mammals of the Big Bend area of Texas. *University of California Publications in Zoology* 48:1–62.

Box, T. W. 1959. Density of plains wood rat dens on four plant communities in south Texas. *Ecology* 40:715–16.

Bradley, R. D., F. C. Bryant, L. C. Bradley, M. L. Haynie, and R. J. Baker. 2003. Implications of hybridization between white-tailed deer and mule deer. *Southwestern Naturalist* 48:654–60.

Bradley, R. D., D. S. Carroll, M. L. Clarey, C. W. Edwards, I. Tiemann-Boege, M. J. Hamilton, R. A. Van Den Bussche, and C. Jones. 1999. Comments on some small mammals from the Big Bend and Trans-Pecos regions of Texas. *Occasional Papers, Museum of Texas Tech University* 193:1–6.

Bradley, R. D., J. Q. Francis, R. N. Platt II, T. J. Soniat, D. Alvarez, and L. L. Lindsey. 2019. *Mitochondrial DNA Sequence Data Indicate Evidence for Multiple Species within* Peromyscus maniculatus, 1–59. Special Publications, Museum of Texas Tech University 70.

Bradley, R. D., A. Pham, E. K. Roberts, T. J. Soniat, C. M. Poehlein, M. N. Mills, M. Ballard, R. Pitts, L. L. Lindsey, M. K. Halsey, D. A. Ray, R. D. Stevens, D. J. Schmidly, and E. A. Wright. 2023. *Genetic identification of pocket gophers (Genera Cratogeomys, Geomys, and Thomomys) in Texas and surrounding areas.* Special Publications, Museum of Texas Tech University, in press.

Bradley, R. D., D. J. Schmidly, B. R. Amman, R. N. Platt II, K. M. Neumann, H. M. Huynh, R. Muñiz-Martínez, C. López-González, and N. Ordóñez-Garza. 2015. Molecular and morphologic data reveal multiple species in *Peromyscus pectoralis. Journal of Mammalogy* 96:446–59.

Brant, J. G., and R. C. Dowler. 2000. Noteworthy record of the Seminole bat, *Lasiurus seminolus* (Chiroptera: Vespertilionidae), in Val Verde County, Texas. *Texas Journal of Science* 52:353–55.

Brant, J. G., J. L. Higginbotham, and C. Jones. 2002. Noteworthy records of the silver-haired bat, *Lasionycteris noctivagans* (Chiroptera: Vespertilionidae), in Presidio County, Texas. *Southwestern Naturalist* 47:633–35.

Brant, J. G., and T. E. Lee Jr. 2006. Morphological analysis of *Perognathus flavus* and *P. merriami* (Rodentia: Heteromyidae). *Southwestern Naturalist* 51:79–86.

Brewer, C. E. 2002. History, status, ecology, and management of desert bighorn sheep in Texas. In *Proceedings of the Trans-Pecos Wildlife Conference*, edited by L. A. Harveson, P. M. Harveson, and C. Richardson, 41–45. Alpine, TX: Sul Ross State University.

Brewer, C. E., and L. A. Harveson. 2007. Diets of bighorn sheep in the Chihuahuan Desert, Texas. *Southwestern Naturalist* 52:97–103.

Browning, E. S. 2014. Seasonal and habitat-based prey diversity of bobcats, *Lynx rufus*, in Big Bend National Park, Texas. PhD diss., University of Texas at Arlington.

Brune, G. 1975. *Major and Historical Springs of Texas.* Texas Water Development Board Report 189. Austin: Texas Water Development Board.

Brune, G. M. 2002. *Springs of Texas.* College Station: Texas A&M University Press.

Buechner, H. K. 1950. Life history, ecology, and range use of the pronghorn antelope in Trans-Pecos Texas. *American Midland Naturalist* 43:257–354.

Burke, R. A., J. K. Frey, A. Ganguli, and K. E. Stoner. 2019. Species distribution modeling supports "nectar corridor" hypothesis for migratory nectivorous bats and conservation of tropical dry forest. *Diversity and Distributions* 25:1399–415. https://doi.org/10.1111/ddi.12950.

Camargo, I., and S. T. Álvarez-Castañeda. 2019. Analyses of predation behavior of the desert shrew *Notiosorex crawfordi. Mammalia* 83:276–80.

Canon, S. K. 1993. Fawn survival and bed-site characteristics of Trans-Pecos pronghorn. PhD diss., Texas Tech University.

Canon, S. K., and F. C. Bryant. 1997a. Bed-site characteristics of pronghorn fawns. *Journal of Wildlife Management* 61:1134–41.

———. 1997b. Home ranges of pronghorn in the Trans-Pecos region of Texas. *Texas Journal of Agriculture and Natural Resources* 10:87–92.

———. 2006. Home range and survival of pronghorn in the Trans-Pecos region of Texas. In *Pronghorn Symposium 2006,* edited by K. A. Cearley and S. Nelle, 23–40. College Station: Texas Cooperative Extension.

Carpenter, C. 1961. Climb high for elk. *Texas Game and Fish* 19:8–9.

Carpenter, J. A. 1993. Movements, habitat use, and population dynamics of elk in Guadalupe Mountains National Park, Texas. PhD diss., Texas A&M University.

Carr, S. M., S. W. Ballinger, J. N. Derr, L. H. Blankenship, and J. W. Bickham. 1986. Mitochondrial DNA analysis of hybridization between sympatric white-tailed deer and mule deer in west Texas. *Proceedings of the National Academy of Sciences* 83:9576–80.

Carroll, D. S., L. L. Peppers, C. Jones, and R. D. Bradley. 2002. *Sigmodon ochrognathus* is a monotypic species: Evidence from DNA sequences. *Southwestern Naturalist* 47:494–97.

Carson, B. 1945. *Final Report—Bighorn Sheep Status.* Austin: Texas Game, Fish and Oyster Commission.

Carter, W. T., M. W. Beck, H. M. Smith, H. W. Hawker, E. H. Templin, and T. C. Reitch. 1928. *Soil Survey (Reconnaissance) of the Trans-Pecos Area, Texas.* USDA Bureau of Chemistry and Soils, Soil Survey Report 35. Washington, DC.

Cathey, J. C., J. W. Bickham, and J. C. Patton. 1998. Introgressive hybridization and nonconcordant evolutionary history of maternal and paternal lineages in North American deer. *Evolution* 52:1224–29.

Ceballos, G., ed. 2014. *Mammals of Mexico.* Baltimore: Johns Hopkins University Press.

Chapman, B. R., and E. G. Bolen. 2018. *The Natural History of Texas.* College Station: Texas A&M University Press.

Clark, D. R., Jr. 1981. *Bats and Environmental Contaminants: A Review.* US Fish and Wildlife Service Special Scientific Report, Wildlife 235. Washington, DC.

———. 2001. DDT and the decline of free-tailed bats (*Tadarida brasiliensis*) at Carlsbad Cavern, New Mexico. *Archives of Environmental Contamination and Toxicology* 40:537–43.

Clark, W. K. 1953. The baculum in the taxonomy of *Peromyscus boylei* and *P. pectoralis. Journal of Mammalogy* 34:189–92.

Cleveland, A. G. 1970. The current geographic distribution of the armadillo in the United States. *Texas Journal of Science* 22:90–92.

Cockerell, T. D. A. 1930. An apparently extinct *Euglandina* from Texas. *Proceedings of the Colorado Museum of Natural History* 9:52–53.

Collier, K., J. S. Hopkins, and R. Leven. 2018. As oil and gas exports surge, West Texas becomes the world's "extraction colony." *Texas Tribune*, October 11, 2018.

Constantine, D. G. 1961. Locality records and notes on western bats. *Journal of Mammalogy* 42:404–5.

Cook, R. L. 1994. *A Historical Review of Reports, Field Notes, and Correspondence: The Desert Bighorn Sheep in Texas.* Special report to the Desert Bighorn Sheep Advisory Committee. Contribution to Federal Aid Project Number W-127-R and W-123-D. Austin: Texas Parks and Wildlife Department.

Cooper, J. D., P. M. Waser, D. Gopurenko, E. C. Hellgren, T. M. Gabor, and J. A. DeWoody. 2010. Measuring sex-biased dispersal in social mammals: Comparisons of nuclear and mitochondrial genes in collared peccaries. *Journal of Mammalogy* 91:1413–24.

Cooper, J. D., P. M. Waser, E. C. Hellgren, T. M. Gabor, and J. A. DeWoody. 2011. Is sexual monomorphism a predictor of polygynandry? Evidence from a social mammal, the collared peccary. *Behavioral Ecology and Sociobiology* 65:775–85.

Cornely, J. E. 1979. Ecological distribution of woodrats (genus *Neotoma*) in Guadalupe Mountains National Park, Texas. In *Biological Investigations in the Guadalupe Mountains National Park, Texas*, edited by H. H. Genoways and R. J. Baker, 373–94. National Park Service Proceedings and Transactions Series Number 4. Washington, DC.

Cornely, J. E., D. J. Schmidly, H. H. Genoways, and R. J. Baker. 1981. Mice of the genus *Peromyscus* in Guadalupe Mountains National Park, Texas. *Occasional Papers, Museum of Texas Tech University* 74:1–35.

Correll, D. S., and M. C. Johnston. 1979. *Manual of the Vascular Plants of Texas.* Dallas: University of Texas at Dallas.

Coykendall, A. 1990. 1988 elk translocations in the Davis Mountains, Jeff Davis County, and the Wylie Mountains, Culberson County, Texas. Master's thesis, Sul Ross State University.

Coyner, B. S., T. E. Lee Jr., D. S. Rogers, and R. A. Van Den Bussche. 2010. Taxonomic status and species limits of *Perognathus* (Rodentia: Heteromyidae) in the southern Great Plains. *Southwestern Naturalist* 55:1–10.

Culver, M., W. E. Johnson, J. Pecon-Slattery, and S. J. O'Brien. 2000. Genomic ancestry of the American puma (*Puma concolor*). *Journal of Heredity* 91:186–97.

Dalquest, W. W., and F. B. Stangl Jr. 1986. Post-Pleistocene mammals of the Apache Mountains, Culberson County, Texas, with comments on zoogeography of the Trans-Pecos Front Range. *Occasional Papers, Museum of Texas Tech University* 104:1–35.

Davis, W. B. 1940a. Critical notes on Texas beavers. *Journal of Mammalogy* 21:84–86.

———. 1940b. Mammals of the Guadalupe Mountains of western Texas. *Occasional Papers of the Museum of Zoology, Louisiana State University* 7:69–84.

———. 1960. *The Mammals of Texas.* Game and Fish Commission Bulletin No. 41. Austin: Texas Game and Fish Commission.

———. 1961. Vanished: A commentary on the extinct and threatened mammals of Texas. *Texas Game and Fish*, December 1961, 15–22.

———. 1966. *The Mammals of Texas*. Rev. ed. Texas Parks and Wildlife Department Bulletin No. 41. Austin: Texas Parks and Wildlife Department.

———. 1974. *The Mammals of Texas*. Rev. ed. Texas Parks and Wildlife Department Bulletin No. 41. Austin: Texas Parks and Wildlife Department.

Davis, W. B., and J. L. Robertson. 1944. The mammals of Culberson County, Texas. *Journal of Mammalogy* 25:254–73.

Davis, W. B., and W. P. Taylor. 1939. The bighorn sheep of Texas. *Journal of Mammalogy* 20:440–55.

DeBaca, R. S. 2008. Distribution of mammals in the Davis Mountains, Texas and surrounding areas. PhD diss., Texas Tech University.

DeBaca, R. S., and C. Jones. 2002. The ghost-faced bat, *Mormoops megalophylla*, (Chiroptera: Mormoopidae) from the Davis Mountains, Texas. *Texas Journal of Science* 54:89–91.

Decker, S. K., D. M. Krejsa, L. L. Lindsey, R. P. Amoateng, and L. K. Ammerman. 2020. Updated distribution of three species of yellow bats (*Dasypterus*) in Texas based on specimen records. *Western Wildlife* 7:2–8.

Dennison, C. C., P. M. Harveson, and L. A. Harveson. 2016. Assessing habitat relationships of mountain lions and their prey in the Davis Mountains, Texas. *Southwestern Naturalist* 61:18–27.

Denyes, H. A. 1956. Natural terrestrial communities of Brewster County, Texas, with special reference to the distribution of the mammals. *American Midland Naturalist* 55:289–320.

Diersing, V. E. 1976. An analysis of *Peromyscus difficilis* from the Mexican-United States boundary area. *Proceedings of the Biological Society of Washington* 89:451–66.

Diersing, V. E., and D. F. Hoffmeister. 1974. The rock mouse, *Peromyscus difficilis*, in western Texas. *Southwestern Naturalist* 19:213.

Diersing, V. E., and D. E. Wilson. 2021. Systematics of the mountain-inhabiting cottontails (*Sylvilagus*) from southwestern United States and northern Mexico (Mammalia: Lagomorpha: Leporidae). *Proceedings of the Biological Society of Washington* 134:42–79.

Dolan, B. F. 2006. Water development and desert bighorn sheep: Implications for conservation. *Wildlife Society Bulletin* 34:642–46.

Dooley, T. J. 1974. Bats of El Paso County, Texas, with notes on habitat, behavior, and ectoparasites. Master's thesis, University of Texas at El Paso.

Dowler, R. C., R. C. Dawkins, and T. C. Maxwell. 1999. Range extensions for the evening bat (*Nycticeius humeralis*) in west Texas. *Texas Journal of Science* 51:193–95.

Downie, A. 1978. *Terrell County, Texas: Its Past, Its People*. San Angelo, TX: Anchor.

Dragoo, J. W., R. L. Honeycutt, and D. J. Schmidly. 2003. Taxonomic status of white-backed hog-nosed skunks, genus *Conepatus* (Carnivora: Mephitidae). *Journal of Mammalogy* 84:159–76.

Drew, M. L., K. M. Rudolf, A. C. S. Ward, and G. C. Weiser. 2014. Health status and microbial (Pasteurellaceae) flora of free-ranging bighorn sheep following contact with domestic ruminants. *Wildlife Society Bulletin* 38:332–40.

Duncan, N. P., S. S. Kahl, S. S. Gray, C. J. Salice, and R. D. Stevens. 2016. Pronghorn habitat suitability in the Texas Panhandle. *Journal of Wildlife Management* 80:1471–78.

Dunn, C. D., M. R. Mauldin, M. E. Wagley, J. E. Wilkinson, C. D. Phillips, and R. D. Bradley. 2017. Genetic diversity and the possible origin of contemporary elk (*Cervus canadensis*) populations in the Trans-Pecos region of Texas. *Occasional Papers, Museum of Texas Tech University* 350:1–15.

Durish, N. D., K. E. Halcomb, C. W. Kilpatrick, and R. D. Bradley. 2004. Molecular systematics of the *Peromyscus truei* species group. *Journal of Mammalogy* 85:1160–69.

Easterla, D. A. 1968. First record of the pocketed free-tailed bat from Texas. *Journal of Mammalogy* 49:515–16.

———. 1970a. First record of the pocketed free-tailed bat for Coahuila, Mexico, and additional Texas records. *Texas Journal of Science* 22:92–93.

———. 1970b. First records of the spotted bat in Texas and notes on its natural history. *American Midland Naturalist* 83:306–8.

———. 1971. Notes on young and adults of the spotted bat, *Euderma maculatum*. *Journal of Mammalogy* 52:475–76.

———. 1973. Ecology of 18 species of Chiroptera at Big Bend National Park, Texas. *Northwest Missouri State University Studies* 34:1–165.

Easterla, D. A., and J. Baccus. 1973. A collection of bats from the Fronteriza Mountains, Coahuila, Mexico. *Southwestern Naturalist* 17:424–27.

Easterla, D. A., and J. O. Whitaker. 1972. Food habits of some bats from Big Bend National Park, Texas. *Journal of Mammalogy* 53:887–90.

Ederhoff, L. T. 1971. The mammals of El Paso County, Texas, with notes on vegetation. Master's thesis, University of Texas at El Paso.

Edwards, C. W., C. F. Fulhorst, and R. D. Bradley. 2001. Molecular phylogenetics of the *Neotoma albigula* species group: Further evidence of a paraphyletic assemblage. *Journal of Mammalogy* 82:267–79.

Edwards, R. L. 1946. Some notes on the life history of the Mexican ground squirrel in Texas. *Journal of Mammalogy* 27:105–21.

Ellison, J. E., and W. F. Harwell. 1969. Mobility and home range of collared peccary in southern Texas. *Journal of Wildlife Management* 33:425–27.

Escalante, T., J. J. Morrone, and G. Rodríguez-Tapia. 2013. Biogeographic regions of North American mammals based on endemism. *Biological Journal of the Linnean Society* 110:485–99.

Falcone, J. H., P. M. Harveson, M. R. Mauldin, and R. D. Bradley. 2019. Taxonomic and conservation status of the Pecos River muskrat. *Occasional Papers, Museum of Texas Tech University* 359:1–16.

Findley, J. S. 1987. *The Natural History of New Mexican Mammals*. New Mexico Natural History Series. Albuquerque: University of New Mexico Press.

Findley, J. S., A. H. Harris, D. E. Wilson, and C. Jones. 1975. *Mammals of New Mexico*. Albuquerque: University of New Mexico Press.

Findley, J. S., and C. Jones. 1965. Comments on spotted bats. *Journal of Mammalogy* 46:679–80.

———. 1967. Taxonomic relationships of bats of the species *Myotis fortidens, M. lucifugus*, and *M. occultus*. *Journal of Mammalogy* 48:429–44.

Fowler, N., T. Keitt, O. Schmidt, M. Terry, and K. Troutt. 2018. Border wall: Bad for diversity. *Frontiers in Ecology and the Environment* 16:137–38.

Freeman, P. W. 1981. A multivariate study of the family Molossidae (Mammalia: Chiroptera): Morphology, ecology, evolution. *Fieldiana: Zoology* 7:1–173.

French, J. T. 2015. Pronghorn diet, nutrition, and habitat assessment in Trans-Pecos, Texas. Master's thesis, Sul Ross State University.

French, J. T., R. O'Shaughnessy, L. A. Harveson, B. J. Warnock, T. O. Garrison, and S. S. Gray. 2019. Selection of ecological sites by pronghorn in Trans-Pecos region of Texas. *Southwestern Naturalist* 63:96–101.

Frey, J. K., J. Iglesias, and K. Herman. 2013. Eastern fox squirrel (*Sciurus niger*): New threat to pecan orchards in far west Texas. *Western North American Naturalist* 73:382–85.

Fulbright, T. E., W. F. Robbins, E. C. Hellgren, R. W. DeYoung, and I. D. Humbreys. 2001. Lack of diet partitioning by sex in reintroduced desert bighorn sheep. *Journal of Arid Environments* 48:49–57.

Galetti, M., H. Camargo, T. Siqueira, A. Keuroghlian, C. I. Donatti, M. L. S. P. Jorge, F. Pedrosa, C. Z. Kanda, and M. C. Ribeiro. 2015. Diet overlap and foraging activity between feral pigs and native peccaries in the Pantanal. *PLoS ONE* 10(11): e0141459. https://doi.org/10.1371/journal.pone.0141459.

Gardner, A. L. 1973. *The Systematics of the Genus Didelphis (Marsupialia: Didelphidae) in North and Middle America*, 1–81. Special Publications, Museum of Texas Tech University 4.

Garrison, T. O. 2015. Post-release survival and movements of translocated pronghorn in Trans-Pecos, Texas. Master's thesis, Sul Ross State University.

Gedir, J. V., J. W. Cain III, G. Harris, and T. T. Turnbull. 2015. Effects of climate change on long-term population growth of pronghorn in an arid environment. *Ecosphere* 6:1–20.

Gehlbach, F. R. 1981. Mountain islands and desert seas: A natural history of the US Mexican borderlands. College Station: Texas A&M University Press.

Geluso, K., T. R. Mollhagen, J. M. Tigner, and M. A. Bogan. 2005. Westward expansion of the eastern pipistrelle (*Pipistrellus subflavus*) in the United States, including new records from New Mexico, South Dakota, and Texas. *Western North American Naturalist* 65:405–9.

Geluso, K. N. 1978. Urine concentrating ability and renal function of insectivorous bats. *Journal of Mammalogy* 59:312–23.

Geluso, K. N., and K. Geluso. 2020. New distributional records and observations of natural history for the yellow-nosed cotton rat (*Sigmodon ochrognathus*) in southwestern New Mexico. *Occasional Papers, Museum of Texas Tech University* 362:1–21.

Genoways, H. H., and R. J. Baker. 1988. *Lasiurus blossevillii* (Chiroptera: Vespertilionidae) in Texas. *Texas Journal of Science* 40:111–13.

Genoways, H. H., R. J. Baker, and J. E. Cornely. 1979. Mammals of the Guadalupe Mountains National Park, Texas. In *Biological Investigations in the Guadalupe Mountains National Park, Texas*, edited by H. H. Genoways and R. J. Baker, 271–332. National Park Service Proceedings and Transactions Series Number 4. Washington, DC.

Gilad, O., J. E. Janecka, F. Armstrong, M. E. Tewes, and R. L. Honeycutt. 2011. Cougars in the Guadalupe Mountains National Park, Texas: Estimates of occurrence and distribution using analysis of DNA. *Southwestern Naturalist* 56:297–304.

Gill, R. B., C. Gill, R. Peel, and J. Vasquez. 2016. Are elk native to Texas? Historical and archaeological evidence for the natural occurrence of elk in Texas. *Journal of Big Bend Studies* 28:205–70.

Glass, G. E., T. M. Shields, R. R. Parmenter, D. Goade, J. N. Mills, J. Cheek, J. Cook, and T. L. Yates. 2006. Predicted hantavirus risk in 2006 for the southwestern US. *Occasional Papers, Museum of Texas Tech University* 255:1–14.

Goetze, J. R., R. W. Manning, and F. D. Yancey II. 2018. Non-geographic variation in *Chaetodipus eremicus* and *Chaetodipus nelsoni* from the Chinati Mountains State Natural Area, Presidio County, Texas, with comparison to populations of the same species from Brewster County, Texas. *Occasional Papers, Museum of Texas Tech University* 356:1–9.

Goldman, E. A. 1950. Raccoons of North and Middle America. *North American Fauna* 60:1–153.

Goodwin, S. L. 2000. Conservation connections in a fragmented desert environment: The US-Mexico border. *Natural Resources Journal* 40:989–1016.

Grace, K. T. 1983. A preliminary ecological study of elk in the Glass Mountains, Brewster County, Texas. Master's thesis, Sul Ross State University.

Green, M. C., L. A. Harveson, and L. E. Loomis. 2001. Habitat selection by collared peccaries in Trans-Pecos Texas. *Southwestern Naturalist* 46:246–51.

Gruver, K. S., and J. W. Guthrie. 1996. Parasites and selected diseases of collared peccaries (*Tayassu tajacu*) in the Trans-Pecos region of Texas. *Journal of Wildlife Diseases* 32:560–62.

Guevara, A. 2009. Assessing landowner attitudes toward elk and elk management in Trans-Pecos, Texas. Master's thesis, Sul Ross State University.

Guevara, A., B. R. Witt, and L. A. Harveson. 2008. Elk research in Texas: Past, present, and future. In *Proceedings of the Trans-Pecos Wildlife Conference*, edited by B. Warnock, P. H. Dickerson, and L. A. Harveson, 30–33. Alpine, TX: Sul Ross State University.

Hafner, D. J., M. S. Hafner, G. L. Hasty, T. A. Spradling, and J. W. Demastes. 2008. Evolutionary relationships of pocket gophers (*Cratogeomys castanops* species group) of the Mexican Altiplano. *Journal of Mammalogy* 89:190–208.

Hafner, J. C., and M. S. Hafner. 1983. Evolutionary relationships of heteromyid rodents. *Great Basin Naturalist Memoirs* 7:3–29.

Hailey, T. L. 1975. Report from Texas. In *Wild Sheep in Modern North America: Proceedings of the Workshop on the Management Biology of North American Wild Sheep*, edited by J. B. Trefethen, 71–72. Missoula: University of Montana.

Hailey, T. L., D. DeArment, and P. Evans. 1964. Pronghorn decline. *Texas Game and Fish* 22:22–23.

Hall, E. R. 1981. *The Mammals of North America*. 2nd ed. New York: John Wiley and Sons.

Hall, E. R., and K. R. Kelson. 1951. Comments on the taxonomy and geographic distribution of some North American rabbits. *University of Kansas Publications, Museum of Natural History* 5:49–58.

———. 1959. *Mammals of North America*. New York: Ronald Press.

Halsey, M. K., J. D. Stuhler, M. A. Madden, E. E. Bohlender, S. C. Brothers, A. N. Kildow, S. C. de la Piedra, et al. 2018. New distributional records of mammals in Texas: Orders Chiroptera, Carnivora, and Rodentia. *Occasional Papers, Museum of Texas Tech University* 354:1–6.

Hanttula, M. K., and E. W. Valdez. 2021. First record and diet of the tri-colored bat (*Perimyotis subflavus*) from Guadalupe Mountains National Park and Culberson County, Texas. *Western North American Naturalist* 81:131–34.

Harveson, L. A. 2018. Conserving the last frontier: The case for the Trans-Pecos. Respect Big Bend Coalition. https://respectbigbend.org/articles/conserving-the-last-frontier-the-case-for-the-trans-pecos?rq=Conserving%20the%20last%20frontier%3A%20the%20case%20for%20the%20Trans-Pecos.

Harveson, L. A., B. Route, F. Armstrong, N. J. Silvy, and M. E. Tewes. 1999. Trends in populations of mountain lions in Carlsbad Caverns and Guadalupe Mountains National Parks. *Southwestern Naturalist* 44:490–94.

Harveson, L. A., M. E. Tewes, N. J. Silvy, and J. Rutledge. 1997. Mountain lion research in Texas: Past, present, and future. In *Proceedings of the Fifth Mountain Lion Workshop*, edited by W. D. Padley, 40–43. San Diego: California Department of Fish and Game.

Harveson, P. M., C. Dennison, B. Geary, D. Milani, D. Rumbelow, and L. A. Harveson. 2016. *Mountain Lion Ecology and Predator-Prey Dynamics in the Davis Mountains*. Report of the Borderlands Research Institute. Alpine, TX: Sul Ross State University.

Harveson, P. M., L. A. Harveson, C. Dennison, D. Rumbelow, S. Stevens, B. Geary, and D. Milani. 2014. *Evaluating Potential for Mountain Lion–Human Conflict in Big Bend National Park*. Report of the Borderlands Research Institute. Alpine, TX: Sul Ross State University.

Harveson, P. M., L. A. Harveson, L. Hernandez-Santin, M. E. Tewes, N. J. Silvy, and M. T. Pittman. 2012. Characteristics of two mountain lion *Puma concolor* populations in Texas, USA. *Wildlife Biology* 18:58–66.

Harvey, M. J., J. S. Altenbach, and T. R. Best. 2011. *Bats of the United States and Canada*. Baltimore: Johns Hopkins University Press.

Heaney, M. R., E. J. Cook, and R. W. Manning. 1998. Noteworthy record of the yellow-nosed cotton rat (*Sigmodon ochrognathus*) from Trans-Pecos Texas. *Texas Journal of Science* 50:347–49.

Heffelfinger, J. R. 2000. Status of the name *Odocoileus hemionus crooki* (Mammalia: Cervidae). *Proceedings of the Biological Society of Washington* 113:319–33.

———. 2008. Desert mule deer ecology and hybridization with whitetails. In *Proceedings of the Trans-Pecos Wildlife Conference*, edited by B. J. Warnock, P. H. Dickerson, and L. A. Harveson, 42–52. Alpine, TX: Sul Ross State University.

Heinen, R. A., and S. K. Canon. 1997. Food habits of mountain lions in the Trans-Pecos region of Texas. In *Thirteenth Great Plains Wildlife Damage Control Workshop Proceedings*, edited by C. D. Lee and S. E. Hygnstrom, 101–95. Manhattan: Kansas State University Agricultural Experiment Station and Cooperative Extension Service.

Helgen, K. M. 2005. Family Castoridae. In *Mammal Species of the World: A Taxonomic and Geographic Reference*, 3rd ed., edited by D. E. Wilson and D. M. Reeder, 842–43. Baltimore: Johns Hopkins University Press.

Helgen, K. M., F. R. Cole, L. E. Helgen, and D. E. Wilson. 2009. Generic revision in the Holarctic ground squirrel genus *Spermophilus*. *Journal of Mammalogy* 90:270–305.

Helgen, K. M., and D. E. Wilson. 2005. A systematic and zoogeographic overview of the raccoons of Mexico and Central America. In *Contribuciones mastozoológicas en homenaje a Bernardo Villa*, edited by V. Sanchez-Cordero and R. A. Medellin, 20:221–36. Mexico City: Instituto de Biología and Instituto de Ecología, UNAM; CONABIO.

Hellgren, E. C. 1993. Status, distribution, and summer food habits of black bears in Big Bend National Park. *Southwestern Naturalist* 38:77–80.

Hellgren, E. C., D. P. Onorato, and J. R. Skiles Jr. 2005. Dynamics of a black bear population within a desert metapopulation. *Biological Conservation* 122:131–40.

Henke, S. E., and F. C. Bryant. 1999. Effects of coyote removal on the faunal community in western Texas. *Journal of Wildlife Management* 63:1066–81.

Hermann, J. A. 1950. The mammals of the Stockton Plateau of northeastern Terrell County, Texas. *Texas Journal of Science* 2:368–93.

Hernandez-Santin, L., P. M. Harveson, and L. A. Harveson. 2012. Suitable habitats for cougars (*Puma concolor*) in Texas and northern Mexico. *Southwestern Naturalist* 57:314–18.

Higginbotham, J. L., and L. K. Ammerman. 2002. *Chiropteran Community Structure and Seasonal Dynamics in Big Bend National Park*, 1–44. Special Publications, Museum of Texas Tech University 44.

Higginbotham, J. L., L. K. Ammerman, and M. T. Dixon. 1999. First record of *Lasiurus xanthinus* (Chiroptera: Vespertilionidae) in Texas. *Southwestern Naturalist* 44:343–47.

Higginbotham, J. L., R. S. DeBaca, J. G. Brant, and C. Jones. 2002. Noteworthy records of bats from the Trans-Pecos region of Texas. *Texas Journal of Science* 54:277–82.

Hinesley, L. L. 1979. Systematics and distribution of two chromosome forms in the southern grasshopper mouse, genus *Onychomys*. *Journal of Mammalogy* 60:117–28.

Hoffman, J. K. 2015. Post-release survival and movements of translocated pronghorn in Trans-Pecos, Texas. Master's thesis, Sul Ross State University.

Hoffmeister, D. F. 1986. *Mammals of Arizona*. Tucson: University of Arizona Press.

Hoffmeister, D. F., and W. W. Goodpaster. 1962. Life history of the desert shrew *Notiosorex crawfordi*. *Southwestern Naturalist* 7:236–52.

Holbrook, J. D., R. W. DeYoung, J. E. Janecka, M. E. Tewes, R. L. Honeycutt, and J. H. Young. 2012. Genetic diversity, population structure, and movements of mountain lions (*Puma concolor*) in Texas. *Journal of Mammalogy* 93:989–1000.

Hollander, R. R. 1990. *Biosystematics of the Yellow-Faced Pocket Gopher,* Cratogeomys castanops *(Rodentia: Geomyidae) in the United States*, 1–62. Special Publications, Museum of Texas Tech University 33.

Hollander, R. R., B. N. Hicks, and J. F. Scudday. 1990a. Distributional records of the yellow-nosed cotton rat, *Sigmodon ochrognathus* Bailey, in Texas. *Texas Journal of Science* 42:101–2.

Hollander, R. R., and K. M. Hogan. 1992. Occurrence of the opossum, *Didelphis virginiana* Kerr, in the Trans-Pecos of Texas. *Texas Journal of Science* 44:127–28.

Hollander, R. R., C. Jones, J. K. Jones Jr., and R. W. Manning. 1990b. Preliminary analysis of the effects of the Pecos River on the geographic distribution of small mammals in western Texas. *Journal of Big Bend Studies* 2:97–107.

Hollander, R. R., R. N. Robertson, and R. J. Kinucan. 1992. First records of the nutria, *Myocastor coypus*, in the Trans-Pecos region of Texas. *Texas Journal of Science* 44:119.

Hollister, N. 1914. A systematic account of the grasshopper mice. *Proceedings of the United States National Museum* 47:427–89.

Holtcamp, W. 2008. In search of America's lion. *Texas Parks and Wildlife Magazine*, April 2008.

Hoofer, S. R., and R. A. Van Den Bussche. 2003. Molecular phylogenetics of the chiropteran family Vespertilionidae. *Acta Chiropterologica* 5 (Supplement):1–63.

Hoofer, S. R., R. A. Van Den Bussche, and I. Horacek. 2006. Generic status of the American pipistrelles (Vespertilionidae) with description of a new genus. *Journal of Mammalogy* 87:981–92.

Howell, A. H. 1938. Revision of the North American ground squirrels, with a classification of the North American Sciuridae. *North American Fauna* 56:1–256.

IUCN (International Union for Conservation of Nature). 2022. *The IUCN Red List of Threatened Species*. https://www.iucnredlist.org.

Jackson, A. W. 1967. Present numbers of bison in Texas. *Journal of Mammalogy* 48:145–46.

Jaquish, V. G., and L. K. Ammerman. 2021. Agave flower visitation by pallid bats, *Antrozous pallidus*, in the Chihuahuan Desert. *Journal of Mammalogy* 102:1101–9.

Jefferson, K. P., S. L. A. Garcia, D. M. Krejsa, J. C. Perkins, S. Stevens, R. S. Matlack, and R. C. Dowler. 2022. Noteworthy records, range extensions, and conservation status of skunk species in Texas. *Occasional Papers, Museum of Texas Tech University* 384:1–13.

Johnson, K. 1981. Social organization in a colony of rock squirrels (*Spermophilus variegatus*, Sciuridae). *Southwestern Naturalist* 26:237–42.

Jones, C., and R. D. Bradley. 1999. Notes on red bats, *Lasiurus* (Chiroptera: Vespertilionidae), of the Davis Mountains and vicinity, Texas. *Texas Journal of Science* 51:341–44.

Jones, C., L. Hedges, and K. Bryan. 1999. The western yellow bat, *Lasiurus xanthinus* (Chiroptera: Vespertilionidae), from the Davis Mountains, Texas. *Texas Journal of Science* 51:267–69.

Jones, C., and M. W. Lockwood. 2008. Additions to the mammalian fauna of Big Bend Ranch State Park, Texas. *Occasional Papers, Museum of Texas Tech University* 282:1–3.

Jones, C., M. W. Lockwood, T. R. Mollhagen, F. D. Yancey II, and M. A. Bogan. 2011. Mammals of the Chinati Mountains State Natural Area, Texas. *Occasional Papers, Museum of Texas Tech University* 300:1–29.

Jones, C., and D. A. Parish. 2001. Effects of the Pecos River on the geographic distributions of mammals in western Texas. *Occasional Papers, Museum of Texas Tech University* 204:1–11.

Jones, J. K., Jr., D. M. Armstrong, and J. R. Choate. 1985. *Guide to Mammals of the Plains States*. Lincoln: University of Nebraska Press.

Jones, J. K., Jr., and C. Jones. 1992. Revised checklist of Recent land mammals of Texas, with annotations. *Texas Journal of Science* 44:53–74.

Jones, J. K., Jr., C. Jones, and D. J. Schmidly. 1988. Annotated checklist of Recent land mammals of Texas. *Occasional Papers, Museum of Texas Tech University* 119:1–26.

Jones, J. K., Jr., and M. R. Lee. 1962. Three species of mammals from western Texas. *Southwestern Naturalist* 7:77–78.

Jones, J. K., Jr., and R. W. Manning. 1991. Comments on distribution of two species of pocket mice (genus *Chaetodipus*) along the Pecos River, Texas. *Texas Journal of Science* 43:99–101.

Jones, J. K., Jr., R. W. Manning, F. D. Yancey II, and C. Jones. 1993. Records of five species of small mammals from western Texas. *Texas Journal of Science* 45:104–5.

Judd, F. W. 1967. Notes on some mammals from Big Bend National Park. *Southwestern Naturalist* 12:192–94.

Kasper, S. 2014. *Rhinocheilus lecontei tessellatus* (Texas long-nosed snake): Predation. *Herpetological Review* 45:344.

Kasper, S., and F. D. Yancey II. 2018a. *Sceloporus merriami* (canyon lizard): Predation by pallid bat. *Herpetological Review* 49:746–47.

———. 2018b. Year-round bridge colony of Mexican free-tailed bats (*Tadarida brasiliensis mexicana*) in Trans-Pecos Texas. *Texas Journal of Science* 70:57–69.

———. 2020. Bighorn sheep (*Ovis canadensis*) ram associating with a herd of aoudad (*Ammotragus lervia*) at Big Bend Ranch State Park, Texas. *Western Wildlife* 7:38–41.

Keleher, R. C. 2010. Genetic variation of pronghorn populations in Texas. Master's thesis, Sul Ross State University.

Kennedy, S., and C. Jones. 2006. Two new records of mammals from the Davis Mountains, Jeff Davis County, Texas. *Occasional Papers, Museum of Texas Tech University* 252:1–4.

Kirk, C. A. 2019. Age distribution and survival of coyotes and gray foxes in western Texas. Master's thesis, West Texas A&M University.

Kirkpatrick, C. M. 1956. Coprophagy in the cottontail. *Journal of Mammalogy* 37:300.

Kitchener, A. C., C. Breitenmoser-Würsten, E. Eizirik, A. Gentry, L. Werdelin, A. Wilting, N. Yamaguchi, et al. 2017. *A Revised Taxonomy of the Felidae.* Final report of the Cat Classification Task Force of the IUCN/SSC Cat Specialist Group. *Cat News*, Special Issue 11.

Krausman, P. R. 1978. Forage relationships between two deer species in Big Bend National Park, Texas. *Journal of Wildlife Management* 42:101–7.

Krausman, P. R., and E. D. Ables. 1981. *Ecology of the Carmen Mountains White-Tailed Deer.* National Park Service Scientific Monograph Series 15. Washington, DC.

Krausman, P. R., J. T. Avey, and J. C. Tull. 1999. Associations between desert mule deer and collared peccaries. *Southwestern Naturalist* 44:543–44.

Krausman, P. R., D. J. Schmidly, and E. D. Ables. 1978. Comments on the taxonomic status, distribution, and habitat of the Carmen Mountains white-tailed deer (*Odocoileus virginianus carminis*) in Trans-Pecos Texas. *Southwestern Naturalist* 23:577–90.

Krejsa, D. M., S. K. Decker, and L. K. Ammerman. 2020. Noteworthy records of 14 bat species in Texas including the first record of *Leptonycteris yerbabuenae* and the second record of *Myotis occultus*. *Occasional Papers, Museum of Texas Tech University* 368:1–10.

Krishnamoorthy, M. A., C. J. Garcia, E. Guest, M. K. Halsey, J. A. Parlos, T. J. Soniat, J. Stuhler, et al. 2021. County records for 20 mammal species across Texas from the orders Cingulata, Lagomorpha, Soricomorpha, Chiroptera, Carnivora, and Rodentia. *Occasional Papers, Museum of Texas Tech University* 372:1–11.

Krysl, L. J. 1979. Food habits of mule deer and elk, and their impact on vegetation in Guadalupe Mountains National Park. Master's thesis, Texas Tech University.

Krysl, L. J., and F. C. Bryant. 2001. Food habits and dietary overlap of elk and mule deer in Guadalupe Mountains National Park, Texas. *Texas Journal of Agriculture and Natural Resources* 14:84–90.

Kuban, J. F., and G. G. Schwartz. 1985. Nectar as a diet item of the ringtailed cat. *Southwestern Naturalist* 30:311–12.

Lasky, J. R., W. Jetz, and T. H. Keitt. 2011. Conservation biogeography of the US-Mexico border: A transcontinental risk assessment of barriers to animal dispersal. *Diversity and Distributions* 17:673–87.

Lawrence, R. K., S. Demarais, R. A. Relyea, S. P. Haskell, W. B. Ballard, and T. L. Clark. 2004. Desert mule deer survival in southwest Texas. *Journal of Wildlife Management* 68:561–69.

Layton, D. R. 1973. An ecological study of the rock squirrel (*Spermophilus variegatus*) in Brewster County, Texas. Master's thesis, Sul Ross State University.

Lee, D. N., R. S. Pfau, and L. K. Ammerman. 2010. Taxonomic status of the Davis Mountains cottontail, *Sylvilagus robustus*, revealed by amplified fragment length polymorphism. *Journal of Mammalogy* 91:1473–83.

Lee, T. E., Jr., J. W. Bickham, and M. D. Scott. 1994. Mitochondrial DNA and allozyme analysis of North American pronghorn populations. *Journal of Wildlife Management* 58:307–18.

Lee, T. E., Jr., J. N. Derr, J. W. Bickham, and T. L. Clark. 1989. Genetic variation in pronghorn from west Texas. *Journal of Wildlife Management* 53:890–96.

Lee, T. E., Jr., and M. D. Engstrom. 1991. Genetic variation in the silky pocket mouse (*Perognathus flavus*) in Texas and New Mexico. *Journal of Mammalogy* 72:273–85.

Lee, T. E., Jr., B. R. Riddle, and P. L. Lee. 1996. Speciation in the desert pocket mouse (*Chaetodipus penicillatus* Woodhouse). *Journal of Mammalogy* 77:58–68.

Leopold, B. D. 1984. Ecology of the desert mule deer in Big Bend National Park, Texas. PhD diss., University of Arizona.

Leopold, B. D., and P. R. Krausman. 1986. Diets of 3 predators in Big Bend National Park, Texas. *Journal of Wildlife Management* 50:290–95.

———. 1987. Diets of two desert mule deer herds in Big Bend National Park, Texas. *Southwestern Naturalist* 32:449–55.

———. 1991. Factors influencing desert mule deer distribution and productivity in southwestern Texas. *Southwestern Naturalist* 36:67–74.

Levenson, H., R. S. Hoffmann, C. F. Nadler, L. Deutsch, and S. D. Freeman. 1985. Systematics of the Holarctic chipmunks (*Tamias*). *Journal of Mammalogy* 66:219–42.

Light, J. E., A. S. Keane, and J. W. Evans. 2021. Updating the distribution of American black bears (*Ursus americanus*) in Texas using community science, state agencies, and natural history collections. *Western North American Naturalist* 81:396–406.

Lightfoot, S. 2010. The pronghorn prognosis. *Texas Parks and Wildlife Magazine*, September 2010.

Locke, S. L., C. E. Brewer, and L. A. Harveson. 2005. Identifying landscapes for desert bighorn sheep translocations in Texas. *Texas Journal of Science* 57:25–34.

Lohan, T. 2019. We're just starting to learn how fracking harms wildlife. EcoWatch, October 5, 2019. https://www.ecowatch.com/how-fracking-harms-wildlife-2640821015.html.

MacMillen, R. E. 1965. Aestivation in the cactus mouse, *Peromyscus eremicus*. *Comparative Biochemistry and Physiology* 16:227–48.

Madsen, W. C., Jr. 1997. Food habits of the gray fox (*Urocyon cinereoargenteus*) and red fox (*Vulpes vulpes*) in west Texas. Master's thesis, Sul Ross State University.

Manning, R. W., C. Jones, and F. D. Yancey II. 2008. Annotated checklist of Recent land mammals of Texas, 2008. *Occasional Papers, Museum of Texas Tech University* 278:1–18.

Manning, R. W., F. D. Yancey II, and C. Jones. 1996. Nongeographic variation and natural history of two sympatric species of pocket mice, *Chaetodipus nelsoni* and *C. eremicus*, from Brewster County, Texas. In *Contributions in Mammalogy: A Memorial Volume Honoring Dr. J. Knox Jones, Jr.*, edited by H. H. Genoways and R. J. Baker, 191–95. Lubbock: Museum of Texas Tech University.

———. 2006. Morphometric variation in two populations of the cactus mouse (*Peromyscus eremicus*) from Trans-Pecos Texas. *Occasional Papers, Museum of Texas Tech University* 262:1–5.

Masters, B. 2019. The river and the wall: A journey down the Rio Grande. *Explorers Journal*, Fall 2019, 14–33.

McAlpine, S. 1990. Continued decline of elk populations within Guadalupe Mountains National Park, Texas. *Southwestern Naturalist* 35:362–63.

McClinton, S. F., P. L. McClinton, and J. V. Richerson. 1992. Food habits of black bears in Big Bend National Park. *Southwestern Naturalist* 37:433–35.

McCorkle, R. 2011. Wildlife and the wall: What is the impact of the border fence on Texas animals? *Texas Parks and Wildlife Magazine*, August 2011.

McDonald, B., B. Geiger, and S. Vrla. 2020. Ultraviolet vision in Ord's kangaroo rat (*Dipodomys ordii*). *Journal of Mammalogy* 101:1257–66.

McDonough, M. M., A. W. Ferguson, R. C. Dowler, M. E. Gompper, and J. E. Maldonado. 2022. Phylogenomic systematics of the spotted skunks (Carnivora, Mephitidae, *Spilogale*): additional species diversity and Pleistocene climate change as a major driver of diversification. *Molecular Phylogenetics and Evolution* 167:107266.

McKinnerney, M. 1978. Carrion communities in the northern Chihuahuan Desert. *Southwestern Naturalist* 23:563–76.

McKinney, B. R. 2006. Room to roam. *Texas Parks and Wildlife Magazine*, November 2006.

———. 2012. *In the Shadow of the Carmens: Afield with a Naturalist in the Northern Mexican Mountains*. Lubbock: Texas Tech University Press.

McKinney, B. R., and M. T. Pittman. 2000. *Habitat, Diet, Home Range, and Seasonal Movement of Resident and Relocated Black Bears in West Texas*. Project WER57-STATE. Austin: Texas Parks and Wildlife Department.

McKinney, B. R., and J. D. Villalobos. 2014. Overview of El Carmen Project, Maderas del Carmen, Coahuila, México. In *Proceedings of the Sixth Symposium on the Natural Resources of the Chihuahuan Desert Region, October 14–17, 2004*, edited by C. A. Hoyt and J. Karges, 37–45. Fort Davis, TX: Chihuahuan Desert Research Institute.

McLaughlin, M. A. 1979. Density, distribution, and status of the kit fox in Trans-Pecos, Texas. PhD diss., Texas A&M University.

Mead, R. A. 1968. Reproduction in western forms of the spotted skunk (genus *Spilogale*). *Journal of Mammalogy* 49:373–90.

Milholland, M. T., J. P. Schumate, T. R. Simpson, and R. W. Manning. 2010. Nutria (*Myocastor coypus*) in Big Bend National Park: A non-native species in desert wetlands. *Texas Journal of Science* 62:205–22.

Miller, G. S., Jr., and R. Kellogg. 1955. List of North American Recent mammals. *Bulletin of the United States National Museum* 205:1–954.

Milstead, W. W., and D. W. Tinkle. 1958. Notes on the porcupine (*Erethizon dorsatum*) in Texas. *Southwestern Naturalist* 3:236–37.

Mitchell, F. S., D. P. Onorato, E. C. Hellgren, J. R. Skiles Jr., and L. A. Harveson. 2005. Winter ecology of American black bears in a desert montane island. *Wildlife Society Bulletin* 33:164–71.

Mollhagen, T. R. 1973. Distributional and taxonomic notes on some west Texas bats. *Southwestern Naturalist* 17:427–30.

Moody, J. D., and C. D. Simpson. 1979. Population status, habitat and movement of elk in Guadalupe Mountains National Park, Texas. In *Proceedings of the 31st Annual Conference of the Southeastern Association of Fish and Wildlife Agencies, October 9 – October 12, 1977*, 151–58. San Antonio: Southeastern Association of Fish and Wildlife Agencies.

Mullican, T. R., and J. T. Baccus. 1990. Horizontal and vertical movements of the white-ankled mouse (*Peromyscus pectoralis*) in central Texas. *Journal of Mammalogy* 71:378–81.

Mungall, E. C., and W. J. Sheffield. 1994. *Exotics on the Range: The Texas Example*. College Station: Texas A&M University Press.

Nadler, C. F., R. S. Hoffmann, J. H. Honacki, and D. Pozin. 1977. Chromosomal evolution in chipmunks, with special emphasis on A and B karyotypes of the subgenus *Neotamias*. *American Midland Naturalist* 98:343–53.

Nalls, A. V., L. K. Ammerman, and R. C. Dowler. 2012. Genetic and morphologic variation in the Davis Mountains cottontail (*Sylvilagus robustus*). *Southwestern Naturalist* 57:1–7.

Neiswenter, S. A., and R. C. Dowler. 2007. Habitat use of western spotted skunks and striped skunks in Texas. *Journal of Wildlife Management* 71:583–86.

Neiswenter, S. A., D. J. Hafner, J. E. Light, G. D. Cepeda, K. C. Kinzer, L. F. Alexander, and B. R. Riddle. 2019. Phylogeography and taxonomic revision of Nelson's pocket mouse (*Chaetodipus nelsoni*). *Journal of Mammalogy* 100:1847–64.

Nelle, S. 2006. Food habits of pronghorn antelope in the Trans-Pecos. In *Pronghorn Symposium 2006*, edited by K. A. Cearley and S. Nelle, 13–22. College Station: Texas Cooperative Extension.

Nelson, E. W. 1909. The rabbits of North America. *North American Fauna* 29:1–314.

———. 1925. *Status of the Pronghorned Antelope, 1922–1924*. US Department of Agriculture Bulletin 1346. Washington, DC.

Oaks, E. C., P. J. Young, G. L. Kirkland Jr., and D. F. Schmidt. 1987. *Spermophilus variegatus*. *Mammalian Species* 272:1–8.

O'Brien, J. M., C. S. O'Brien, C. McCarthy, and T. E. Carpenter. 2014. Incorporating foray behavior into models estimating contact risk between bighorn sheep and areas occupied by domestic sheep. *Wildlife Society Bulletin* 38:321–31.

O'Connell, M. A. 1979. Coexistence of two species of kangaroo rats (genus *Dipodomys*) in the Guadalupe Mountains National Park, Texas. In *Biological Investigations in the Guadalupe Mountains National Park, Texas*, edited by H. H. Genoways and R. J. Baker, 349–71. National Park Service Proceedings and Transactions Series Number 4. Washington, DC.

Ohlendorf, H. M. 1972. Observations on a colony of *Eumops perotis* (Molossidae). *Southwestern Naturalist* 17:297–300.

Onorato, D. P., E. C. Hellgren, F. S. Mitchell, and J. R. Skiles Jr. 2003. Home range and habitat use of American black bears on a desert montane island in Texas. *Ursus* 14:120–29.

Onorato, D. P., E. C. Hellgren, R. A. Van Den Bussche, and D. L. Doan-Crider. 2004. Phylogeographic patterns within a metapopulation of black bears (*Ursus americanus*) in the American southwest. *Journal of Mammalogy* 85:140–47.

Osgood, W. H. 1909. Revision of the mice of the American genus *Peromyscus*. *North American Fauna* 28:1–285.

Packard, J. M. 1991. *Behavior of High Risk Mountain Lions in Big Bend National Park, Texas*. Final report to USDI, Cooperative Agreement no. 702990004. Santa Fe, NM: National Park Service, Southwest Regional Office.

Packard, J. M., W. Gordon, and J. Clarkson. 2011. Biodiversity. In *The Impact of Global Warming on Texas*, 2nd ed., edited by J. Schmandt, G. R. North, and J. Clarkson, 124–56. Austin: University of Texas Press.

Patterson, B. D., H. E. R. Chavez, J. F. Vilela, A. E. R. Soares, and F. Grewe. 2021. On the nomenclature of the American clade of weasels (Carnivora: Mustelidae). *Journal of Animal Diversity* 3:1–8. URL: http://jad.lu.ac.ir/article-1-132-en.html

Patton, R. F. 1974. Ecological and behavioral relationships of the skunks of Trans Pecos Texas. PhD diss., Texas A&M University.

Perez, J. C., S. Pichyangkul, and V. E. Garcia. 1979. The resistance of three species of warm-blooded animals to western diamondback rattlesnake (*Crotalus atrox*) venom. *Toxicon* 17:601–7.

Peters, R., W. J. Ripple, C. Wolf, M. Moskwik, G. Carreón-Arroyo, G. Ceballos, A. Córdova, et al. 2018. Nature divided, scientists united: US-Mexico border wall threatens biodiversity and binational conservation. *Bioscience* 68:740–43.

Piaggio, A. J., E. W. Valdez, M. A. Bogan, and G. S. Spicer. 2002. Systematics of *Myotis occultus* (Chiroptera: Vespertilionidae) inferred from sequences of two mitochondrial genes. *Journal of Mammalogy* 83:386–95.

Platt, S. G., P. R. Manning, and T. R. Rainwater. 2014. Consumption of desert gourds by collared peccary suggests the fruit is not an ecological anachronism. *Southwestern Naturalist* 59:141–44.

Pohler, P. S., L. A. Harveson, and P. M. Harveson. 2014. Demographic characteristics of elk in the Glass Mountains, Texas. *Wildlife Society Bulletin* 38:466–72.

Porter, R. D. 1962. Movements, populations, and habitat preferences of three species of pocket mice (*Perognathus*) in the Big Bend region of Texas. PhD diss., Texas A&M University.

———. 2011. *Movements, Populations, and Habitat Preferences of Three Species of Pocket Mice (Perognathinae) in the Big Bend Region of Texas*. Edited by C. A. Porter, 1–107. Special Publications, Museum of Texas Tech University 58.

Punzo, F. 2003a. Natural history and ecology of the desert shrew, *Notiosorex crawfordi*, from the northern Chihuahuan Desert, with notes on captive breeding. *Mammalia* 67:541–50.

———. 2003b. Observations on the diet composition of the gray shrew *Notiosorex crawfordi* (Insectivora), including interactions with large arthropods. *Texas Journal of Science* 55:75–86.

Punzo, F., and R. Lopez. 2003. Observations on nest site selection and litter size in the gray shrew (*Notiosorex crawfordi*) from Presidio County, Texas. *Texas Journal of Science* 55:169–74.

Reich, H. J., IV. 2015. Rio Grande beaver (*Castor canadensis mexicanus*) survey in Big Bend National Park. Master's thesis, Texas State University.

Reichman, O. J., and R. J. Baker. 1972. Distribution and movements of two species of pocket gophers (Geomyidae) in an area of sympatry in the Davis Mountains, Texas. *Journal of Mammalogy* 53:21–33.

Richardson, C. 2003. *Trans-Pecos Vegetation: A Historical Perspective*. Trans-Pecos Wildlife Management Series Leaflet No. 7. Austin: Texas Parks and Wildlife Department.

———. 2006. Pronghorn habitat requirements. In *Pronghorn Symposium 2006*, edited by K. A. Cearley and S. Nelle, 5–12. College Station: Texas Cooperative Extension.

Roberts, K. J., F. D. Yancey II, and C. Jones. 1997. Predation by great-horned owls on Brazilian free-tailed bats in north Texas. *Texas Journal of Science* 49:215–18.

Ruedas, L. A. 1998. Systematics of *Sylvilagus* Gray, 1867 (Lagomorpha: Leporidae) from southwestern North America. *Journal of Mammalogy* 79:1355–78.

Russ, W. B. 1997. The status of the mountain lion in Texas. In *Proceedings of the Fifth Mountain Lion Workshop*, edited by W. D. Padly, 69–73. San Diego: California Department of Fish and Game.

Russell, R. J. 1968. Revision of the pocket gophers of the genus *Pappogeomys*. *University of Kansas Publications, Museum of Natural History* 16:581–776.

Schmandt, J., G. R. North, and J. Clarkson, eds. 2011. *The Impact of Global Warming on Texas*. 2nd ed. Austin: University of Texas Press.

Schmidly, D. J. 1977a. Factors governing the distribution of mammals in the Chihuahuan Desert region. In *Transactions of the Symposium on the Biological Resources of the Chihuahuan Desert Region, United States and Mexico*, edited by R. H. Wauer and D. H. Riskind, 163–92. National Park Service Transactions and Proceedings Series Number 3. Washington, DC.

———. 1977b. *The Mammals of Trans-Pecos Texas Including Big Bend National Park and Guadalupe Mountains National Park*. College Station: Texas A&M University Press.

———. 1991. *The Bats of Texas*. College Station: Texas A&M University Press.

———. 2002. *Texas Natural History: A Century of Change*. Lubbock: Texas Tech University Press.

———. 2018. *Vernon Bailey: Writings of a Field Naturalist on the Frontier*. College Station: Texas A&M University Press.

Schmidly, D. J., and R. D. Bradley. 2016. *The Mammals of Texas*. 7th ed. Austin: University of Texas Press.

Schmidly, D. J., R. D. Bradley, and L. C. Bradley. 2022. *Texas Natural History in the 21st Century*. Lubbock: Texas Tech University Press.

Schmidly, D. J., R. D. Bradley, L. C. Bradley, F. D. Yancey II, and J. Bateman. 2023. Catalog 1: Type specimens, type localities, synonymies, and authors/collectors of Recent mammals described exclusively from Texas. In *Taxonomic Catalogs for the Recent Terrestrial Vertebrates (Species and Subspecies) Described from Texas*, edited by D. J. Schmidly, R. D. Bradley, L. C. Bradley, and F. D. Yancey II, 23–126. Special Publications, Museum of Texas Tech University 77.

Schmidly, D. J., R. B. Ditton, W. J. Boeer, and A. R. Graefe. 1979. Interrelationships among visitor usage, human impact, and the biotic resources of the riparian ecosystem in Big Bend National Park. In *Proceedings of the First Conference on Scientific Research in the National Parks*, edited by R. M. Linn, 261–68. National Park Service Proceedings and Transactions Series Number 5. Washington, DC.

Schmidly, D. J., and C. Jones. 2001. 20th century changes in mammals and mammalian habitats along the Rio Grande/Rio Bravo from Fort Quitman to Amistad. In *Proceedings of the Rio Grande/Rio Bravo Binational Symposium: Fort Quitman to Amistad Reservoir, June 14, 2000, Ciudad Juárez, Mexico*, 177–204. Washington, DC: US Department of the Interior, Secretariat of Environment and Natural Resources, International Boundary and Water Commission.

Schmidly, D. J., J. Karges, and R. Dean. 2016a. Distribution records and reported sightings of the white-nosed coati (*Nasua narica*) in Texas, with comments on the species' population and conservation status. In *Contributions in Natural History: A Memorial Volume in Honor of Clyde Jones*, edited by R. W. Manning, J. R. Goetze, and F. D. Yancey II, 127–45. Special Publications, Museum of Texas Tech University 65.

Schmidly, D. J., W. E. Tydeman, and A. L. Gardner, eds. 2016b. *United States Biological Survey: A Compendium of its History, Personalities, Impacts, and Conflicts*, 1–123. Special Publications, Museum of Texas Tech University 64.

Schmidt, D. F. 1999. Rock squirrel | *Spermophilus variegatus.* In *The Smithsonian Book of North American Mammals,* edited by D. E. Wilson and S. Ruff, 438–40. Washington, DC: Smithsonian Institution Press.

Schmidt-Nielson, K. 1964. *Desert Animals: Physiological Problems of Heat and Water.* New York: Oxford University Press.

Scudday, J. F. 1972. Two recent records of gray wolves in west Texas. *Journal of Mammalogy* 53:598.

Setzer, H. W. 1949. Subspeciation in the kangaroo rat, *Dipodomys ordii. University of Kansas Publications, Museum of Natural History* 1:473–573.

Sides, A. R., D. C. Simpson, L. A. Harveson, and C. E. Brewer. 2006. Relationship of Trans-Pecos pronghorn to precipitation trends and 30 years of land cover change. In *Pronghorn Symposium 2006,* edited by K. A. Cearley and S. Nelle, 69–75. College Station: Texas Cooperative Extension.

Simpson, D. C., L. A. Harveson, C. E. Brewer, R. E. Walser, and A. R. Sides. 2007. Influence of precipitation on pronghorn demography in Texas. *Journal of Wildlife Management* 71:906–10.

Smith, J. 2020. 30 by 30: Is it possible to protect 30% of the planet's land and water by 2030? We're already working to make it happen. *Nature Conservancy Magazine,* Spring 2020, 46–51.

Sparks, D. W., K. J. Roberts, and C. Jones. 2000. Vertebrate predators on bats in North America north of Mexico. In *Reflections of a Naturalist: Papers Honoring Professor Eugene D. Fleharty,* edited by J. R. Choate, 229–41. Fort Hays Studies, Special Issue 1. Hays, KS: Fort Hays State University.

Stangl, F. B., Jr. 1992a. First record of *Sigmodon fulviventer* in Texas: Natural history and cyto-genetic observations. *Southwestern Naturalist* 37:213–14.

———. 1992b. A new subspecies of the tawny-bellied cotton rat, *Sigmodon fulviventer,* from Trans-Pecos Texas. *Occasional Papers, Museum of Texas Tech University* 145:1–4.

Stangl, F. B., Jr., W. W. Dalquest, and R. R. Hollander. 1994. *Evolution of a Desert Mammalian Fauna: A 10,000-Year History of Mammals in Culberson and Jeff Davis Counties, Trans-Pecos Texas.* Wichita Falls, TX: Midwestern State University Press.

Stangl, F. B., Jr., W. W. Dalquest, and S. Kuhn. 1993. Mammals from the Beach Mountains of Culberson County, Trans-Pecos Texas. *Texas Journal of Science* 45:87–96.

Stangl, F. B., Jr., and J. R. Goetze. 1991. Comments on pelage and molt of the spotted ground squirrel, *Spermophilus spilosoma* (Rodentia: Sciuridae). *Texas Journal of Science* 43:305–8.

Stangl, F. B, Jr., S. Henry-Langston, N. Lamar, and S. Kasper. 2014. Sexual dimorphism in the ringtail (*Bassariscus astutus*) from Texas. *Occasional Papers, Museum of Texas Tech University* 328:1–10.

Stevens, R. D., C. J. Garcia, E. E. Guest, A. Hargrove, M. A. Krishnamoorthy, C. F. Rickert, E. M. Sanchez, et al. 2021. Seasonal use of bridges as day-roosts by bats in the Trans-Pecos of Texas. *Therya* 12:207–12.

Stubblefield, S. S., R. J. Warren, and B. R. Murphy. 1986. Hybridization of free-ranging white-tailed and mule deer in Texas. *Journal of Wildlife Management* 50:688–90.

Sullins, M. R. 2002. Factors affecting pronghorn antelope populations in Trans-Pecos, Texas. In *Proceedings of the Trans-Pecos Wildlife Conference,* edited by L. A. Harveson, P. M. Harveson, and C. Richardson, 29–36. Alpine, TX: Sul Ross State University.

Sumner, M. L. 2002. Factors affecting mule deer numbers in west Texas. In *Proceedings of the Trans-Pecos Wildlife Conference*, edited by L. A. Harveson, P. M. Harveson, and C. Richardson, 37–40. Alpine, TX: Sul Ross State University.

Sumner, M. L., and L. A. Harveson. 2008. Evaluating home range and habitat use of desert mule deer in the Apache Mountains of the Chihuahuan Desert, Texas. In *Proceedings of the Trans-Pecos Wildlife Conference*, edited by B. J. Warnock, P. H. Dickerson, and L. A. Harveson, 55. Alpine, TX: Sul Ross State University.

Tabor, F. W. 1940. Range of the coati in the United States. *Journal of Mammalogy* 21:11–14.

Tabor, S. P., and R. E. Thomas. 1986. The occurrence of plague (*Yersinia pestis*) in a bobcat from the Trans-Pecos area of Texas. *Southwestern Naturalist* 31:135–36.

Tamsitt, J. R. 1954. The mammals of two areas in the Big Bend region of Trans-Pecos Texas. *Texas Journal of Science* 6:33–61.

Taylor, W. P. 1948. Jack rabbits experience a population "high" in the Trans-Pecos region, Texas. *Journal of Mammalogy* 29:186–87.

Thompson, R., P. M. Harveson, L. A. Harveson, D. Milani, and K. Dennison. 2012. *Ecology of Mountain Lions in the Davis Mountains: Assessing Impacts on Prey Populations*. Report of the Borderlands Research Institute. Alpine, TX: Sul Ross State University.

Thorington, R. W., Jr., and R. Hoffmann. 2005. Family Sciuridae. In *Mammal Species of the World: A Taxonomic and Geographic Reference*, 3rd ed., edited by D. E. Wilson and D. M. Reeder, 754–818. Baltimore: Johns Hopkins University Press.

Thorington, R. W., Jr., J. L. Koprowski, M. A. Steele, and J. F. Whatton. 2012. *Squirrels of the World*. Baltimore: Johns Hopkins University Press.

Tipps, T. M., B. Mayes, and L. K. Ammerman. 2011. New county records for six species of bats (Vespertilionidae and Molossidae) in Texas. *Texas Journal of Science* 63:141–52.

Truett, J. C., D. P. Gober, A. E. Ernst, R. List, H. Whitlaw, C. L. Hayes, G. Schmitt, and W. E. Van Pelt. 2014. Prairie dogs in the Chihuahuan Desert: History, ecology, conservation. In *Proceedings of the Sixth Symposium on the Natural Resources of the Chihuahuan Desert Region*, edited by C. A. Hoyt and J. Karges, 211–40. Fort Davis, TX: Chihuahuan Desert Research Institute.

Tucker, R. D., and G. W. Garner. 1983. Habitat selection and vegetational characteristics of antelope fawn bed sites in west Texas. *Journal of Range Management* 36:110–13.

Tumlison, R., and M. E. Douglas. 1992. Parsimony analysis and the phylogeny of the plecotine bats (Chiroptera: Vespertilionidae). *Journal of Mammalogy* 73:276–85.

Van Pelt, A. F. 1977. A mountain lion kill in southwest Texas. *Southwestern Naturalist* 22:271.

Vestal, A. L. 2005. Genetic variation in the Davis Mountains cottontail (*Sylvilagus robustus*). Master's thesis, Angelo State University.

Walker, C. W., L. A. Harveson, M. T. Pittman, M. E. Tewes, and R. L. Honeycutt. 2000. Microsatellite variation in two populations of mountain lions (*Puma concolor*) in Texas. *Southwestern Naturalist* 45:196–203.

Wang, H. 2019. Change of vegetation cover in the US-Mexico border region: Illegal activities or climatic variability? *Environmental Research Letters* 14:1–12.

Warnock, B. J., and L. Loomis. 2002. Was the Trans-Pecos a grassland? Past, present, and potential. In *Proceedings of the Trans-Pecos Wildlife Conference*, edited by L. A. Harveson, P. M. Harveson, an C. Richardson, 94–97. Alpine, TX: Sul Ross State University.

Wauer, R. H., and C. M. Fleming. 2002. *Naturalist's Big Bend: An Introduction to the Trees and Shrubs, Wildflowers, Cacti, Mammals, Birds, Reptiles and Amphibians, Fish, and Insects*. College Station: Texas A&M University Press.

Weins, J. J. 2011. The niche, biogeography and species interaction. *Philosophical Transactions of the Royal Society B* 366:2336–50.

White-Nose Syndrome Response Team. 2022. US Fish and Wildlife Service. www.whitenose syndrome.org.

Wilkins, K. T., and D. J. Schmidly. 1979. Identification and distribution of three species of pocket mice (genus *Perognathus*) in Trans-Pecos Texas. *Southwestern Naturalist* 24:17–32.

Williams, D. F. 1978a. Karyological affinities of the species groups of silky pocket mice (Rodentia, Heteromyidae). *Journal of Mammalogy* 59:599–612.

———. 1978b. Systematics and ecogeographic variation of the Apache pocket mouse (Rodentia: Heteromyidae). *Bulletin of the Carnegie Museum of Natural History* 10:1–57.

Williams, M. 2020. Mule deer's antlers unlike any other. Outdoors, *Dallas Morning News*, February 1, 2020. https://dallasnews.com.

Wilson, D. E. 1973. The systematic status of *Perognathus merriami* Allen. *Proceedings of the Biological Society of Washington* 86:175–92.

Wilson, D. E., and D. M. Reeder, eds. 2005. *Mammal Species of the World: A Taxonomic and Geographic Reference*. 3rd ed. Baltimore: Johns Hopkins University Press.

Wilson, D. E., and S. Ruff, eds. 1999. *The Smithsonian Book of North American Mammals*. Washington, DC: Smithsonian Institution Press.

Wolf, L. K., M. B. Meierhofer, M. L. Morrison, D. M. Cairns, and T. E. Lacher Jr. 2022. Modeling the suitability of Texas karst regions for infection by *Pseudogymnoascus destructans* in bats. *Journal of Mammalogy* 103:503–11.

Woodward, L. A. 2020. Pronghorns: A story of survival. *Texas Wildlife, Magazine of the Texas Wildlife Association*, July 2020. https://www.texas-wildlife.org/resources/publications /pronghorns-a-story-of-survival.

Wu, Y., K. Rylander, and D. Wester. 1996. Rodent habitat associations in a Chihuahuan Desert grassland community in Trans-Pecos Texas. *Texas Journal of Science* 48:68–74.

Yancey, F. D., II. 1996. Diversity, distribution, and natural history of the mammals of Big Bend Ranch State Park, Texas. PhD diss., Texas Tech University.

———. 1997. *The Mammals of Big Bend Ranch State Park, Texas*, 1–210. Special Publications, Museum of Texas Tech University 39.

———. 2016. Ecological distribution and foraging activity of the ghost-faced bat (*Mormoops megalophylla*) in Big Bend Ranch State Park, Texas. In *Contributions in Natural History: A Memorial Volume in Honor of Clyde Jones*, edited by R. W. Manning, J. R. Goetze, and F. D. Yancey II, 147–55. Special Publications, Museum of Texas Tech University 65.

Yancey, F. D., II, and C. Jones. 1996. Notes on three species of small mammals from the Big Bend region of Texas. *Texas Journal of Science* 48:247–50.

———. 1999. Alopecia in the white-ankled mouse, *Peromyscus pectoralis* (Mammalia: Rodentia), in Texas. *Texas Journal of Science* 51:271–72.

———. 2000. Distribution and ecologic relationships of pocket mice (*Chaetodipus*) in the Big Bend region of Texas. *Occasional Papers, Museum of Texas Tech University* 195:1–14.

———. 2006. Changes in distributions of bats in Texas. *Occasional Papers, Museum of Texas Tech University* 258:1–5.

Yancey, F. D., II, C. Jones, and J. R. Goetze. 1995a. Notes on harvest mice (*Reithrodontomys*) of the Big Bend region of Texas. *Texas Journal of Science* 47:263–68.

Yancey, F. D., II, C. Jones, and R. W. Manning. 1995b. The eastern pipistrelle, *Pipistrellus subflavus* (Chiroptera: Vespertilionidae), from the Big Bend region of Texas. *Texas Journal of Science* 47:229–31.

Yancey, F. D., II, J. K. Jones Jr., and R. W. Manning. 1993. Individual and secondary sexual variation in the Mexican ground squirrel, *Spermophilus mexicanus*. *Texas Journal of Science* 45:63–68.

Yancey, F. D., II, and M. W. Lockwood. 2017. Rare account of the Virginia opossum (*Didelphis virginiana*) from Trans-Pecos Texas. *Texas Journal of Science* 69:190–94.

———. 2018. First record of the American black bear (*Ursus americanus*) from the Chinati Mountains of western Texas. *Southwestern Naturalist* 63:133–36.

Yancey, F. D., II, and R. W. Manning. 2018. *Update on the Mammals of Big Bend Ranch State Park and Chinati Mountains State Natural Area, with Additions to the Verified Mammalian Faunas and a Current Checklist for Each Site*. Final report, Natural Resources Program. Austin: Texas Parks and Wildlife Department.

Yancey, F. D., II, R. W. Manning, J. R. Goetze, L. L. Lindsey, R. D. Bradley, and C. Jones. 2017. The hooded skunk (*Mephitis macroura*) from the Davis Mountains of west Texas: Natural history, morphology, molecular characteristics, and conservation status. *Texas Journal of Science* 69:87–95.

Yancey, F. D., II, R. W. Manning, and C. Jones. 2006. Mammals of the Harte Ranch area of Big Bend National Park, Brewster County, Texas. *Occasional Papers, Museum of Texas Tech University* 253:1–15.

Yancey, F. D., II, R. W. Manning, S. Kasper, M. W. Lockwood, J. R. Goetze, and N. E. Havlik. 2019. Update on the mammals of Chinati Mountains State Natural Area, Texas. In *From Field to Laboratory: A Memorial Volume in Honor of Robert J. Baker*, edited by R. D. Bradley, H. H. Genoways, D. J. Schmidly, and L. C. Bradley, 697–712. Special Publications, Museum of Texas Tech University 71.

Yancey, F. D., II, K. J. Roberts, and C. Jones. 1996. Prairie falcon predation on Brazilian free-tailed bats. *Prairie Naturalist* 28:146.

Yates, T. L., and D. J. Schmidly. 1977. Systematics of *Scalopus aquaticus* (Linnaeus) in Texas and adjacent states. *Occasional Papers, Museum of Texas Tech University* 45:1–36.

Young, J. H., M. E. Tewes, A. M. Haines, G. Guzman, and S. J. DeMaso. 2010. Survival and mortality of cougars in the Trans-Pecos region. *Southwestern Naturalist* 55:411–18.

Young, S. P., and E. A. Goldman. 1944. *The Wolves of North America*. Washington, DC: American Wildlife Institute.

Zervanos, S. M., and N. F. Hadley. 1973. Adaptational biology and energy relationships of the collared peccary (*Tayassu tajacu*). *Ecology* 54:759–74.

INDEX

wind farms, 303
Wisconsin glaciation, effects, 5–6
WNS (white-nose syndrome), 302
wolves, 25, 136, 281–82
woodlands: changes summarized, 5–6, 49–53, 299; mammal distribution patterns, 28–29, 30, 33–34; vegetation patterns, 11, 16–18
woodrats (*Neotoma* spp.), 29, 270–75, color plate 31

Xerospermophilus spilosoma (spotted ground squirrel), 46–47, 202, 203–204

yellow bat, western (*Dasypterus xanthinus*), 116, 119–20
yellow-faced pocket gopher (*Cratogeomys castanops*), 36, 212, 213, 216–19
yellow-nosed cotton rat (*Sigmodon ochrognathus*), 28–29, 265, 267–68, color plate 30
Yersinia pestis (plague), 145, 194
Yuma myotis (*Myotis yumanensis*), 101, 103, 104–105